Lecture Notes in Computer Science 2390

Edited by G. Goos, J. Hartmanis, and J. van Leeuwen

W0107668

Springer

Berlin
Heidelberg
New York
Barcelona
Hong Kong
London
Milan
Paris
Tokyo

Dorothea Blostein   Young-Bin Kwon (Eds.)

Graphics Recognition

Algorithms and Applications

4th International Workshop, GREC 2001
Kingston, Ontario, Canada, September 7-8, 2001
Selected Papers

 Springer

Series Editors

Gerhard Goos, Karlsruhe University, Germany
Juris Hartmanis, Cornell University, NY, USA
Jan van Leeuwen, Utrecht University, The Netherlands

Volume Editors

Dorothea Blostein
Queen's University, Computing and Information Science
Kingston, Ontario, Canada K7L 3N6
E-mail: blostein@cs.queensu.ca

Young-Bin Kwon
Chung-Ang University, Department of Computer Engineering
Seoul, 156-756, South Korea
E-mail: ybkwon@visionnet.cse.cau.ac.kr

Cataloging-in-Publication Data applied for

Die Deutsche Bibliothek - CIP-Einheitsaufnahme

Graphics recognition : algorithms and applications ; 4th international
workshop ; selected papers / GREC 2001, Kingston, Ontario, Canada,
September 7 - 8, 2001. Dorothea Blostein ; Young-Bin Kwon (ed.). - Berlin ;
Heidelberg ; New York ; Barcelona ; Hong Kong ; London ; Milan ; Paris ;
Tokyo : Springer, 2002
    (Lecture notes in computer science ; Vol. 2390)
    ISBN 3-540-44066-6

CR Subject Classification (1998): I.5, I.4, I.3.5, I.2.8, G.2.2, F.2.2

ISSN 0302-9743
ISBN 3-540-44066-6 Springer-Verlag Berlin Heidelberg New York

Springer-Verlag Berlin Heidelberg New York,
a member of BertelsmannSpringer Science+Business Media GmbH

http://www.springer.de

© Springer-Verlag Berlin Heidelberg 2002
Printed in Germany

Typesetting: Camera-ready by author, data conversion by DA-TeX Gerd Blumenstein
Printed on acid-free paper   SPIN: 10873497   06/3142   5 4 3 2 1 0

Preface

This book presents refereed and revised papers presented at GREC 2001, the 4th IAPR International Workshop on Graphics Recognition, which took place in Kingston, Ontario, Canada in September 2001. Graphics recognition is a branch of document image analysis that focuses on the recognition of two-dimensional notations such as engineering drawings, maps, mathematical notation, music notation, tables, and chemical structure diagrams. Due to the growing demand for both off-line and on-line document recognition systems, the field of graphics recognition has an exciting and promising future.

The GREC workshops provide an opportunity for researchers at all levels of experience to share insights into graphics recognition methods. The workshops enjoy strong participation from researchers in both industry and academia. They are sponsored by IAPR TC-10, the Technical Committee on Graphics Recognition within the International Association for Pattern Recognition. Edited volumes from the previous three workshops in this series are available as Lecture Notes in Computer Science, Vols. 1072, 1389, and 1941.

After the GREC 2001 workshop, authors were invited to submit enhanced versions of their papers for review. Every paper was evaluated by three reviewers. We are grateful to both authors and reviewers for their careful work during this review process. Many of the papers that appear in this volume were thoroughly revised and improved, in response to reviewers' suggestions.

This book is organized into eight sections, reflecting the session topics in the GREC 2001 workshop: technical drawings and forms; validation, user interfaces; symbol segmentation and recognition; perceptual organization; map recognition; graphics recognition technology; vectorization and early processing; and math notation, charts, music notation. As is traditional in the GREC workshops, each session included ample time for panel discussions. A summary of these panel discussions, prepared by Karl Tombre and Atul Chhabra, appears at the end of this book.

Graphics recognition contests are held at GREC workshops to foster the development of high-performance algorithms, and the development of methods for performance evaluation. GREC 2001 featured an arc segmentation contest, organized by Liu Wenyin and Dov Dori. During the workshop, contestants ran their programs on given test images, using evaluation criteria developed by the contest organizers. A special section of this volume is devoted to the arc segmentation contest. The three papers in this section describe the contest, the images, the evaluation criteria, and the algorithms developed by the contestants.

We gratefully acknowledge the support provided by the sponsors of GREC 2001: the IAPR (International Association for Pattern Recognition), Queen's University, Canada (Department of Computing and Information Science, Office of Research Services, and Faculty of Arts and Science), CSERIC (Computer Science and Engineering Research Information Center, Korea), the Xerox Foundation (Xerox Palo Alto Research Center, California), and CITO (Communications and Information Technology Ontario).

The next Graphics Recognition Workshop (www.cvc.uab.es/grec2003/) will be held in Barcelona, Spain, in July 2003.

April 2002   Dorothea Blostein
                                                                Young-Bin Kwon

Table of Contents

Perceptual Organization

Map Recognition

Graphics Recognition Technology

Vectorization and Early Processing

Math Notation, Charts, Music Notation

Arc Segmentation Contest

3D Reconstruction of Paper Based Assembly Drawings: State of the Art and Approach

El-Fathi El-Mejbri, Hans Grabowski, Harald Kunze,
Ralf-Stefan Lossack, and Arno Michelis

Institute for Applied Computer Science in Mechanical Engineering
Adenauerring 20, D-76131 Karlsruhe, University of Karlsruhe, Germany
{mejbri,gr,kunze,lossack,michelis}@rpk.uni-karlsruhe.de
http://www-rpk.mach.uni-karlsruhe.de/

Abstract. Engineering solutions are generally documented in assembly and part drawings and bill of materials. A great benefit, qualitatively and commercially, can be achieved if these paper based storages can be transformed into digital information archives. The process of this transformation is called reconstruction. The reconstruction process of paper based assembly drawings consists of four steps: digitization; vectorization/interpretation; 3D reconstruction of the parts and the 3D reconstruction of the assembly.

This paper evaluates existing commercial systems worldwide for interpretation of paper based mechanical engineering drawings. For a complete reconstruction process a 3D reconstruction is needed. This functionality is already supported by some CAD systems to a certain extent, but it still remains a major topic of research work. One CAD system which converts 2D CAD models into 3D CAD models is presented. Finally, after the reconstruction of the parts the whole assembly can be reconstructed. Until now, no system for the automatic reconstruction of assemblies is available. In our paper we present a general approach for automatic reconstruction of 3D assembly model data by interpretation of mechanical engineering 2D assembly drawings, their part drawings, and the bill of materials.

1 Introduction

The Swiss company SULZER, a manufacturer of water turbines, has paper based engineering drawings, which are still in use for the maintenance of existing plants, and are older than 120 years [1]. This shows the need to convert paper based drawings into accessible and processable 3D CAD data.

The transformation of paper based engineering drawings into 2D/3D CAD models was a research topic during the last two decades. In this respect, a variety of research work is being undertaken in the field of engineering drawing reconstruction with the objective of finding specific solutions to process companies' information inventories. In general, the existing approaches for engineering

D. Blostein and Y.-B. Kwon (Eds.): GREC 2002, LNCS 2390, pp. 1–12, 2002.

drawing reconstruction divide the overall conversion process into three procedural stages: digitization, vectorization/interpretation, and 2D/3D reconstruction of single parts [10,11]. In contrast, the reconstruction of mechanical engineering assembly drawings is quite difficult. There are still no concepts for a complete reconstruction of paper based assembly drawings. In the field of interpretation of engineering assembly drawings just simple methods exist with the drawback of losing a substantial amount of information [13,15].

2   Reconstruction Process of Assemblies

The reconstruction process of conventionally generated assembly drawings consists of four steps: digitization; vectorization/interpretation; 2D/3D reconstruction of the parts and the 3D reconstruction of the assembly drawing (Fig. 1).

In the first step all documents of the assembly (parts, assembly, bill of material) are digitized. The result of this step are pixel images of the technical documents. The respective pixel data of the drawings is vectorized. The interpretation of the vectorized drawings provides 2D CAD models (second step). The interpretation process can be divided into the recognition of geometric elements (lines, circles, etc.), technological data (assembly information, tolerances, etc.) and meta information (position number, date, quantity of parts, etc.). After the interpretation of the CAD models the recognized data (see above) is separated, so that only the geometry data remains. Based on the 2D data a 3D CAD model reconstruction is started for all parts (third step).

After the digitization, the bill of materials is interpreted by OCR (**O**ptical **C**haracter **R**ecognition) systems. Then the physical files (pixel images, 2D and 3D CAD part models, 2D CAD assembly model) and the meta information of

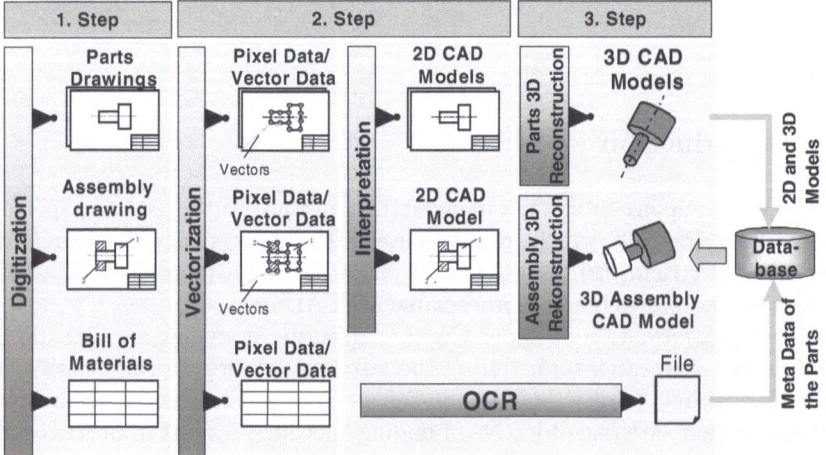

Fig. 1. Reconstruction process of assemblies

the bill of materials are stored in a database. The database includes all relevant information needed for a successful automatic reconstruction of the assembly. In the fourth step the positions of the parts are elaborated and the 3D CAD model of the assembly is built.

Since the last two decades the transformation of paper based engineering part drawings into 2D/3D CAD models is research topic. Worldwide several concepts have been developed for the different steps of the reconstruction process. Based on these concepts some systems are implemented and commercialized. The best supported steps are digitization and vectorization. For a better interpretation of engineering drawings special algorithms are needed, some of them are already integrated in commercial systems. But systems which automatically convert paper based drawings into CAD models are still not available because of the complexity of the problem.

In the following sections we evaluate commercial systems for each step of the reconstruction process. At the end we present our own approach for the reconstruction of paper based assembly drawings.

2.1   Digitization

Digitization systems are directly coupled with scanners. Today scanners can reach a resolution of more than 800 dpi (**d**ots **p**er **i**nch). But in our experience a resolution of 400 dpi is sufficient for DIN-A0 mechanical engineering drawings. The quality of the digital pixel data depends on the drawing quality and the resolution. The use of higher resolutions causes higher amount of data. A further important prerequisite is the use of digital filter processes, which could eliminate the background of the drawing and lead to better pixel data quality. Filter tools are mostly integrated in the scanning software and yield very good results.

2.2   Vectorization and Interpretation

The vectorization is the transformation of the original raster data into vector sequences. In general commercial systems provide three types of vectorization: through line middle, outlines or the combination of both.

The extraction process of the geometrical structure[1] from a digitized document is called "document analysis". The illustration process of the geometrical structure into the logical structure[2] is called "document understanding". The whole process comprising the document analysis and the document understanding is called "document interpretation".

The aim of the vectorization and interpretation steps is to generate a 2D CAD model in a standard format, e.g. STEP, IGES, DXF. In the last two decades

---

[1] The geometrical structure is a result of the segmentation of the document content into basic elements.

[2] The logical structure is a result of the segmentation of the document content into basic elements which are comprehensible by human beings.

several research works were published on the domain of interpretation of engineering drawings. Mainly these approaches are published in GREC'95, 97 and 99 [7,8,9] and ICDAR'99 [5]. But the majority of these works is focussed on single subtasks of the interpretation process and their methods. Only few works deal with the whole interpretation process (e.g. CELESSTIN [15], ANON [14], AI-MUDAMS [12]). The example drawings utilized by researchers are simple and idealized. This makes the difference to complex drawings in the industry. Furthermore existing approaches assume that drawing elements are already well extracted in vector data. Another lack is the removal of wrong interpretation during the interpretation process. Until now the interpretation of texts is not helpful for the reconstruction process.

Commercial Vectorization and Interpretation Systems: In this section we present some commercial vectorization and interpretation systems and show their functionality as well as their restrictions (Fig. 2). Except **VPStudio** all of these systems are available only as a demo version at our institute. The systems are tested on the following points:

- Recognition of different line thicknesses and types for visible and invisible geometry, symmetry lines, dimensioning (dimensioning lines, dimensioning references and dimensioning text);
- Recognition of hatch areas as faces;
- Recognition of text and symbols;
- Recognition of orthogonality of lines;
- Detection of the accuracy of vector lengths and angles;
- Separation of lines which lie close to each other.

Draftsman Plus 32 - Arbor Image Corp. - http://www.aay.com: Pixel data can be processed, e.g. removal of particles. Pixel data can be converted into raster data (generating of vectors through line middle and outlines, definition of vectorization parameters, recognition of lines, circles and arcs,...). Text recognition is also possible. Circles can be recognized, but the whole drawing contains a lot of little vectors. These vectors are traced manually. This vectorization system is not suitable for the processing of engineering drawings.

E.G.S. Quick Vector - Enhanced Graphics Solutions - http://www.egsolutions.com: This system can convert pixel data in simple edited vector data. There are no possibilities for manual interactions during vectorization. Algorithms for recognition of line thicknesses, line types and text are not implemented. The tracing of thin lines works good even when the lines are intersecting. But overall this system is not suitable to generate 2D CAD data from pixel data.

GTXRaster CAD *PLUS* - GTX Corporation - http://www.gtx.com: This system was not available, thus the specifications are based on the software producer. With **GTXRaster CAD *PLUS*** pixel data, raster data and text recognition can be processed. In addition this system is able to convert automatically pixel data into vector data. This system supports the well known file formats of pixel data, e.g. TIFF (**T**agged **I**mage **F**ile **F**ormat), and the vector data can be exported in DXF (**D**ata e**X**change **F**ormat).

ProVec 3.5 - Abakos Digital Images - http://www.abakos.com.au: With **ProVec** pixel and raster data can be processed. In addition the converting of pixel in vector data are described in the following points:

- Text recognition with the possibility of manual interactions;
- Recognition of line thickness, line type, arcs and circles;
- Separation of layers;
- Generating of vectors through line middle and outlines;
- Vectorization parameters can be set by an assistant;
- Generating of vectors with certain angles proportions.

The generation of CAD data is not possible, some geometry help elements, e.g. hatch areas, can not be recognized. **ProVec** is only suitable for DXF data from AutoCAD R13.

R2V - Able Software Co. - http://www.ablesw.com: This software system is used in cartographic and geographic domain. Pixel data, automatic vectorization and text recognition can be performed. The system can not recognize line thicknesses and line types, thus this system is not suitable for vectorization of mechanical engineering drawings.

RXSpotlight Pro - Rasterex - http://www.raterex.com: RXSpotlight Pro is a system, which is able to process pixel data and to convert pixel data into vector data. The processing of pixel data includes: removal of particles and smoothing. The recognition of lines includes the recognition of line types, line thicknesses, arcs, circles, hatch areas and symbols using a symbol library. Text recognition is also possible. The used data formats are DXF, DWG (AutoCAD **DraWinG**), TIFF, PCX (Micrografix Raster Format) and BMP (Microsoft **B**itmap **F**ormat). Diameter symbols, symmetry lines and dimensioning can not be correctly recognized.

Scan2CAD - Softcover International Limited - http://www.softcover.com: The main use of this system is text recognition. But it includes simple tools for pixel data processing, and automatic vectorization for many domains (mechanical engineering, architecture, electric schemas). The processing of lines includes generation of lines through line middle and outlines, generation orthogonal lines, entering of parameters for creation of circles and arcs. Therefore there is no distinction of line types and line thicknesses.

ScanVector - CyberSonic Technologies - http://ourworld.compuserve.com/homepages/cybersonic/: The characteristics of this system are the editing of pixel data and an automatic vectorization. The pixel image can be oriented and filtered. Lines, arcs, circles, polylines, texts and symbols can be recognized. An intelligent tool for handling of intersection points is available. Hatch areas can not be recognized.

Trac Trix 2000 - Trix System AB - http://www.trixsystems.com: Both Pixel and raster images can be edited. Lines, arcs, circles, polylines and NURBS can be generated. Pixel images can be rotated, oriented, reflected and cleaned.

Vectory (I/Vector) -Ideal Scanners & System - http://www.ideal.com: The pixel functions are: rotating, orienting, cleaning, reflecting and pixel data processing. Line thicknesses and line types, arcs, circles, dimensioning arrows, hatch areas, symbols and texts can be recognized. Furthermore the handling of the system is very simple.

VPStudio - Softelec GmbH - http://www.softelec.com: VPStudio is a powerful system for pixel data processing, vector data processing and automatic vectorization. The functions which can process the pixel image are: removal of particles, reflecting, rotating, orienting and equalizing. Pixel elements, e.g. lines and arcs, can be copied, translated and scaled. Vector elements or generally CAD elements can be generated and processed. The automatic vectorization includes the generation of lines through line middle and outlines or the combination of both. Lines, polylines, arcs, circles, ellipses, different line types and line thicknesses, texts and hatch areas can be recognized. Furthermore symbols can be searched and substituted. The processing of pixel and vector data is interactive. The system supports several pixel formats (TIFF, RLE, PCX, BMP, GIF, JPEG) and vector formats (DXF, DWG, IGES, DGN).

After the digitization of the drawing, thin lines are generated. Then the type of the drawing, e.g. mechanical engineering drawing, is set. The vectorization (postprocessing) is done by the classification of the vectors using parameters suggested by the system.

The problems are mainly caused by the recognition of texts, e.g. the diameter symbols can not be recognized. The dimensioning arrows and chamfers can not be recognized.

The separation of the recognized line thicknesses onto different layers works very well. Furthermore dimensioning can be generated interactive, which can be stored as CAD elements in an own layer. The order to the generation of the dimensioning is important (dimensioning arrow must be generated first and then the dimensioning reference), otherwise the bounding points are set wrong. A connection between the generated dimensioning and the geometry is not given, i.e. the dimensioning is stored in the CAD file as independent element.

Evaluation: The difference between these systems is the quality of generated vector data and their classification in CAD elements. The separation of edges,

		Draftsman Plus 32	E.G.S. Quick Vector	GTXRaster CAD PLUS	ProVec 3.5	R2V	RXSpotlight Pro	Scan2CAD	ScanVector	Trac Trix 2000	Vectory (I/Vector)	VPStudio
Raster Data Processing	Removal of particles	x		x	x	x	x	x	x	x	x	x
	Reflecting of the pixel image			x	x	x	x			x	x	x
	Rotating of the pixel image			x	x	x	x			x	x	x
	Aligning of the pixel image			x		x	x		x	x	x	x
	Equalizing of the pixel image			x	x	x					x	x
	Copying of pixel elements										x	x
	Translating of pixel elements										x	x
	Scaling of pixel elements										x	x
Vector Data Processing	Generating of vector elements	x					x	x	x		x	x
	Processing of vector elements										x	x
	Removal of vector elements						x	x		x		x
Raster/ Vector Converting	Text recognition	x		x	x	x	x	x	x	x	x	x
	Vectorization through line middle	x			x		x	x				x
	Vectorization through outlines	x			x		x	x				x
	Vect. through line middle and outlines						x					x
	Recognition of lines	x	x	x	x		x	x	x	x	x	x
	Recognition of dashed lines			x								x
	Recognition of dotted lines											
	Recognition of polylines		x	x						x	x	x
	Recognition of arcs			x	x	x		x	x	x	x	x
	Recognition of circles	x		x	x	x		x	x	x	x	x
	Recognition of ellipses											
	Recogntion of line types	x			x		x			x	x	x
	Recognition of line thicknesses			x			x			x	x	x
	Recognition of hatch areas				x		x				x	x
	Tracing				x		x					x
	Search and substitution of symbols						x		x	x		x
Legend	x: Functionality is supported by this system											

Fig. 2. Commercial vectorization and interpretation systems

dimensioning and help lines is not perfect. The text recognition is not satisfying, because of the different font types, font sizes and font directions in the vector data. Tolerances, e.g. form tolerances and positional tolerances, can not be recognized. The best vectorization systems which generate CAD models are RXSPOTLIGHT, VECTORY and VPSTUDIO. But they have different potencies and failings. VPSTUDIO can generate CAD models therefore needs manual interactions (postprocessing), which is time-consuming. A remodelling of the part in a 3D CAD system is more profitable.

2.3   2D/3D Reconstruction of Parts

For many years a lot of work has been done in the domain of 3D geometry reconstruction from 2D projection illustrations (e.g. [2,3,6,10]). The conversion of 2D views into 3D model can be performed today with commercial CAD systems[3].

In this paper we have a closer look on the CAD system Helix99, especially the modules "Helix Drafting" and "Helix Modelling". The "Helix Modelling" module contains a tool called "AutoSOLID" which converts 2D CAD models into 3D

---

[3] e.g. Helix99 (http://www.helixsystem.com) and HiCAD2 (http://www.isdcad.de)

CAD models. The CAD System Helix results from a joint venture between IBM and Kawasaki Heavy Industries Ltd. AutoSOLID is based on IBM' OrthoSolid reconstruction system. Helix99 contains a vectorization system which is based on GTXRaster CAD *PLUS*. In the following we give a brief introduction in OrthoSolid system. Finally, we evaluate Helix99.

The converting process consists of four steps:

1. *Generating of a wire frame*: First 3D points are calculated from triples of the 2D points. Then the 3D points are connected to edges. Afterwards the vectors are established and the 3D wire frame is built.
2. *Generating of a surface model*: In the second step a surface model is built by looping the edges to singles surfaces. The types of surfaces are planes, cylinders, cones, spheres and torus. Furthermore their equations can be calculated by investigating the couple of edges, which should not lie on the same curve. Edges are either straight edges or cylindric edges. Finally loops are searched in the calculated equations of the determined surfaces. When a loop is found, a surface is defined and can be integrated in the surface model.
3. *Generating of a cell model*: In the third step the cell model is generated from the surface model and the data structure is built. For this the cells are generated by searching regions surrounded by surfaces.
4. *Generating of a solid model*: The converting by OrthoSolid is based on ATMS approach (**A**ssumption-based **T**ruth **M**aintenance **S**ystem). The embodiment is done, when the projected views of selected topology elements are coincided with the orthographic projections. The solid model is built as a set of cells by some destined justifications.

Evaluation: The automatic reconstruction within Helix works good on certain conditions. These conditions are very restrictive. Three views with all their invisible edges must be available. Furthermore, the 2D views must be accurate like generated by CAD systems. In spite of 2D CAD models there are further problems, e.g. concerning intersection of two removed solids. Generally complex parts can not be reconstructed by Helix99.

3   Subtasks of the Assembly Reconstruction

In our approach the 3D reconstruction of assembly drawings is divided into three subtasks. At first to recognize 2D position of the parts in the assembly, then to hide the identified parts in the 2D CAD model and finally to add the identified 3D part CAD model to the 3D assembly model (Fig. 3).

3.1   2D Position Recognition of the Parts

The basic idea of position recognition is to identify the initial positions of the parts in the assembly. The position numbers and their related pointers are the

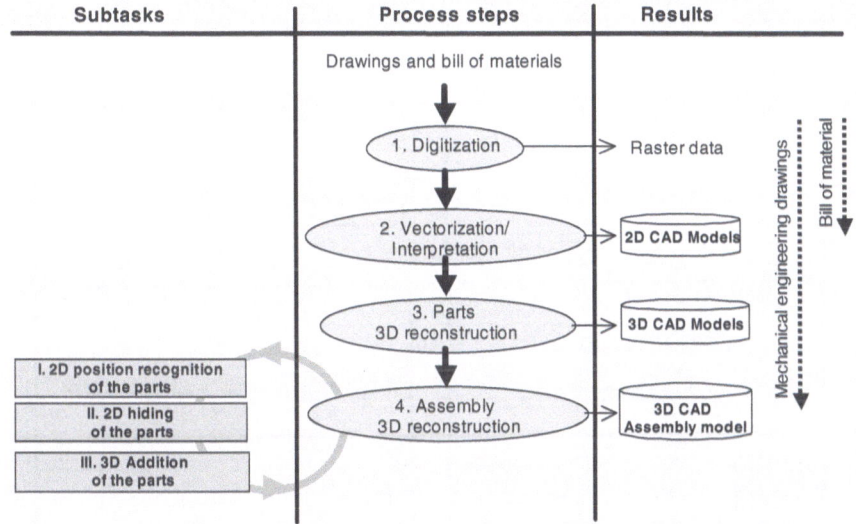

Fig. 3. The subtasks of the reconstruction of assemblies

most important features for the detection of the subarea where the part is (approximately) located in an assembly.

That means that for a fast detection the recognition must be started within this subarea. When the area is identified, further features can be used for a final recognition. For the application of further feature recognition techniques all six views of a part are needed. The different views of a part can be automatically derived from the 3D model.

One method for finding a part in an assembly drawing is to select a vertex in the part and two edges starting at this vertex. The two edges are characterized by their lengths. Recognition is done by searching the assembly drawing for these two lines. One restriction of this method is that the two edges have to be entirely visible in the assembly drawing. A strategy for the initial vertex and edge selection is to select rectangular edges (Fig. 4 a). Another strategy is to take a vertex and two edges which do not often occur in the part drawing (Fig. 4 b). Another method for finding a part in an assembly drawing is to search for a cross-section. This is done by searching fields of parallel edges which represent the hatch areas (Fig. 4 c). Another method is to search for long edges as they appear in machine cases or shafts. Long lines do not appear very often so it is very likely to find the part position (Fig. 4 d).

The final recognition of the part position is done by comparing the vector data of the respective view of the part with the vector data of the assembly drawing. The comparison of the vector data provides a probability value for each position. This value can be understood as an index of the percentage a part is recognized. Uncompleted recognition can be caused by invisible edges

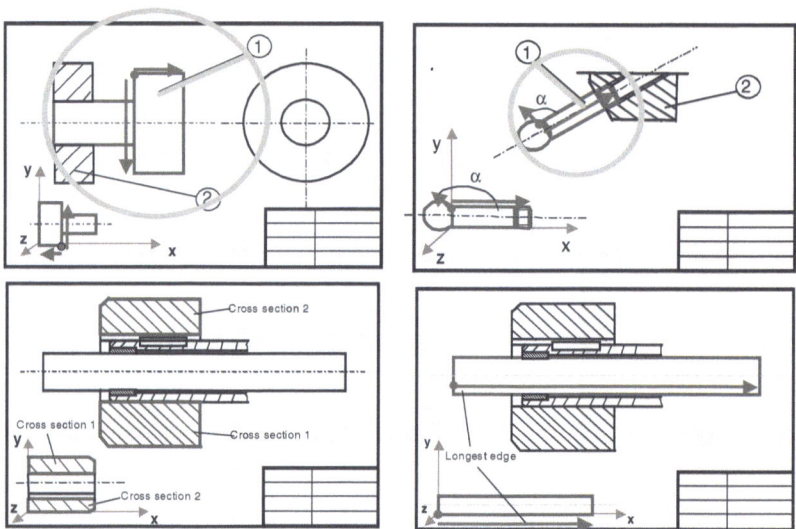

Fig. 4. 2D position recognition

and the similarity of some parts. We can affirm (Fig. 4) that using 2D position recognition of parts the following transformations are automatically generated: translation in x, y direction and rotation about the z axis. If the part view suites the assembly drawing view the x and y rotation is automatically determined. Otherwise, a further view of the assembly drawing (i.e. top view) has to be used for the detection of the x and y rotation.

3.2   2D Hiding of the Parts

The aim of the 2D hiding is to verify whether the parts are identified correctly. The verification takes place when a part is recognized. After successful recognition the parts have to be hidden from the assembly drawing.

3.3   3D Addition of the Parts

The starting point for the 3D addition are the recognized orientations of the different parts. The aim of the 3D addition is to build the 3D model of the assembly. The idea is to select the 3D model of a part from the database as a reference part[4]. The further parts can be assembled successively, based on the calculated coordinates (see above). During the 3D addition process views and cross-sections can be derived immediately which enables a comparison to the original assembly drawing.

---

[4] The selection is random, but it is recommended to choose the biggest parts, e.g. machine case, because they have the most contact surfaces with other parts.

4   Summary

In this contribution the reconstruction process of paper based engineering assembly drawings is presented. The single process steps (digitization, vectorization/interpretation, 3D reconstruction of the parts and the 3D reconstruction of the assembly) are described. Some commercial vectorization systems for the subtasks of the vectorization and interpretation are presented and evaluated. Some CAD systems for converting 2D views into 3D CAD models are discussed and evaluated.

Furthermore an approach for the automatic reconstruction of conventional generated mechanical engineering assembly drawings is discussed. The prerequisites for the reconstruction are the assembly drawing, the part drawings, the 3D models of the parts and the bill of materials. The main task is the recognition of the spatial position and orientation of the parts. A strategy is presented which identifies the 2D positions of the parts in the assembly drawing, to verify these positions, and finally generates the 3D model of the assembly by adding the 3D models of the parts.

Acknowledgment

This project is sponsored by the Deutsche Forschungsgemeinschaft DFG (German Research Foundation) under the project number GR. 582/36-1.

References

1. N. N.: Micro CADAM bei der Sulzer Hydro AG, Wasserturbinen in allen Grössen. Anwenderbericht. CAD/CAM-Report, Nr. 12, Dressler Verlag. Heidelberg. (1996) 1

2. Chen, Z. and Perng D. B.: Automatic Reconstruction of 3D Solid Objects from 2D Orthographic Views. Pattern Recognition Vol. 21, Nr. 5. (1988) 439–449 7

3. Chen, Z., Perng, D. B., Chen, C. J. and Wu, C. S.: Fast Recognition of 3D Mechanical Parts from 2D Orthographic Rules with Views. International Journal of Computer Integrated Manufacturing Vol. 5, Nr. 1. (1992) 2–9 7

4. Hans Grabowski, Chenguang Liu and Arno Michelis: Stepwise Segmentation and Interpretation of Section Representations in Vectorized Drawings. ICDAR'99 Bangalore, India. (1999)

5. 5th International Conference on Document Analysis and Recognition (ICDAR '99), ICDAR'99 Bangalore, India. (1999) 4

6. Kuo, M. H.: A Systematic Approach Towards Rconstructing 3D Curved Models from Multiple 2D Views. Graphics Recognition, Algorithms and Systems. K. Chahbra and K. Tombre. Springer Verlag. Nancy, France. (1997) 265–279 7

7. Graphics Recognition: Methods and Applications. GREC'95. R.Kasturi and K. Tombre (Eds.). (1995) 4

8. Graphics Recognition, Algorithms and Systems. K. Chahbra and K. Tombre (Eds.). Springer Verlag. Nancy, France. (1997) 4

9. Graphics Recognition, Recent Advances. Third International Workshop, GREC'99, Jaipur, India, September 1999, K. Chahbra and D. Dov (Eds.). Springer Verlag. Nancy, France. (1997) 4

10. Liu, Baoyuan: Entwicklung einer Methode zur automatischen 2D/3D-Produktmodellwandlung. Universität Fridericiana Karlsruhe (TH). Shaker Verlag, Aachen. (1994) 2, 7

11. Liu, Chenguang: Ein Beitrag zur Interpretation technischer Zeichnungen mittels Graphentheorie und Statistik. Universität Fridericiana Karlsruhe (TH). Shaker Verlag, Aachen. (2000) 2

12. Lu, W.; Ohsawa, Y. and Sakauchi, M.: A Database Capture System for Mechanical Drawings Using an Efficient Multi-dimensional Data Structure. 9th ICPR. (1988) 266–269 4

13. Maria A. Capellades and Octavia I. Camps: Functional parts detection in engineering drawings, Looking for the screws. Graphics Recognition, Methods and applications. R. Kasturi and K. Tombre. Springer Verlag. Heidelberg (Germany). (1995) 307–317 2

14. Joseoh, U., Pridmore, T. P. and Dunn M. E.: Towards the Automatic Interpretation of Mechanical Engineering Drawings. SDIA 92. (1992) 25–56 4

15. Vaxivière P. and Tombre K.: Knowledge Organization and Interpretation Process in Engineering Drawing Interpretation. Document Analysis Systems. A. L. Spitz and A. Dengel. World Scientific. (1995) 307–317 2, 4

Interpretation of Low-Level CAD Data for Knowledge Extraction in Early Design Stages

Hans Grabowski, Harald Kunze, Ralf Lossack, and Arno Michelis

Institute for Applied Computer Science in Mechanical Engineering (RPK)
University of Karlsruhe, Karlsruhe, Germany
{gr,kunze,lossack,michelis}@rpk.mach.uni-karlsruhe.de

Abstract. In this paper we introduce a method for knowledge extraction in early stages of design by the interpretation of low-level CAD data, i.e. the geometry of mechanical products. Therefore, only the 3D CAD model of the product, including its components, is needed. The information about physical principles and functions is extracted automatically. First a more detailed introduction to the design process in mechanical engineering is presented. This enables the reader to understand the information which is accumulated during product design and which must be recognized during the interpretation process, i.e. the redesign. Then the approach of low-level CAD data interpretation is presented in detail. This includes the recognition of physical principles and product functions and the automatic construction of effective geometries and function structures. At the end the approach is verified by presenting an example of a product geometry, which is interpreted stepwise.

1 Introduction and Background

Product development as a whole becomes an important factor for industrialized nations. It becomes more and more difficult to meet customers' demand and to compete on international markets with high quality and good value products, which have to be produced faster and faster to cut down time to market.

During its life time, a product passes several product-life-cycle stages (*Fig. 1*). Different stages of the product-life-cycle are supported by isolated CAx-systems. Today it is important to *integrate* the *different views among the product-life-cycle stages*. Such an integration requires both a *logically integrated product model* as well as *modeling methods* to work with the integrated product model.

In general the development of products begins with the product planning stage, where the idea of a product function is born. Then, in the product design stage, the future product is modeled at different abstraction levels to end up with an embodiment, which has to be manufactured. At the end of the product-life-cycle the product has to be recycled (*Fig. 1*).

The design process in the product design stage has been analyzed by several German design methodologists [4,5,6]. The result of this analysis is a methodology subdivided into four design stages in order to support a more discursive than

D. Blostein and Y.-B. Kwon (Eds.): GREC 2002, LNCS 2390, pp. 13–24, 2002.
© Springer-Verlag Berlin Heidelberg 2002

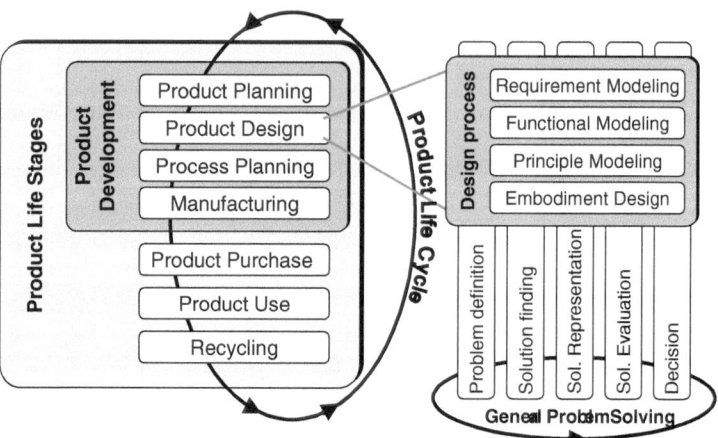

Fig. 1. Product Life Stages

intuitive form of designing: *requirement modeling, functional modeling, principle modeling* and *embodiment design.*

Within the different design stages a general problem solving cycle with several design activities is performed. These design activities are: *problem definition, solution finding, solution representation, solution evaluation* and *decision.* As can be seen in *Fig. 1* the design process is iterative.

The problem is that in reality the designer gets the requirements and starts with the CAD system doing the embodiment design right away. The designer does not explicitly develop the stages of functional modeling and principle modeling, thus no knowledge about product functions and physical principles is available. But this explicit knowledge is needed, e.g. to better understand a product, to discover dependencies between product components, and to find good starting points for product optimization.

In the next section a more detailed description of the methodological design process will be given. This enables the reader to understand the information which is accumulated during product design and which must be recognized during the interpretation process, i.e. the redesign.

2   Design Process

Each design process starts with a clarification of the design task. As a first approach a *requirement list* is given by a customer or the marketing branch. Mostly the requirement list is incomplete, either requirements are missing, or they imply other requirements. At the *requirement modeling stage* the given requirements are decomposed and completed. The output of the requirement modeling stage is a consistent and complete *requirement model.*

Based on the requirement model, a designer continues the design process with *functional modeling.* A functional model describes at a formal level in which way a technical object fulfills functional requirements. PAHL and BEITZ [5] defined a function as follows:

A function describes the *general* and *desired context* between the input and output of a system, with the goal to fulfill a task.

In a first step the *overall function* can be derived from the requirement model. For very complex products the overall function is decomposed in *sub-functions* until a level is reached where a further decomposition is not possible or useful. By going this way a *function structure* will be constructed.

ROTH [6] distinguishes between *general functions* and *special functions*. General functions have the three basic quantities: *energy, material* and *information*. He found the five basic verbs: *store, channel, change, transform* and *join* to describe each design at a formal level by a general function structure. With the three basic quantities and the five verbs it is possible to combine 30 general functions.

General functions guarantee that no domain is excluded from the design space. But engineers do not like to think in general functions, as they think more in physical quantities like forces and torques. Therefore, *special functions* are needed. A special function has a physical quantity as in-/output and a function verb. The function verb describes the transition from the physical quantity input to the physical quantity output. The combination of special functions to special function structures is the input for principle modeling.

The principle modeling is divided in two steps [6]:

1. searching and assigning a *physical effect* to a function and
2. assigning and adapting an *effective geometry* of a physical effect to the geometrical environment.

A *physical effect* is described by a *physical law* and its *mathematical equation*. The combination of physical effects results in a *physical effective structure*. An *effective geometry* consists of *effective points, effective lines, effective surfaces* and *effective spaces* and shows the schematic construction of a part. A combination of effective geometries builds the *effective geometry structure*.

In order to facilitate the building of *physical effective structures* and *effective geometry structures* the concept of the *morphological box* had been established [7]. In the first column of the morphological box single *functions* of a *function structure* are located (*Fig. 2*). One or more *physical effects* with their *effective geometries* are assigned to each function. They build the rows. The designer connects the cells of the morphological box (the physical effects) to each other, with the supposition of their technical compatibility. These combinations of physical effects and effective geometries result in several design solutions, represented by effective geometry structures and physical effective structures with their mathematical equations (*Fig. 2*).

Based on a physical effective structure and its function flow, it is possible to analyze and simulate the behavior of an engineering design. It is called *static*

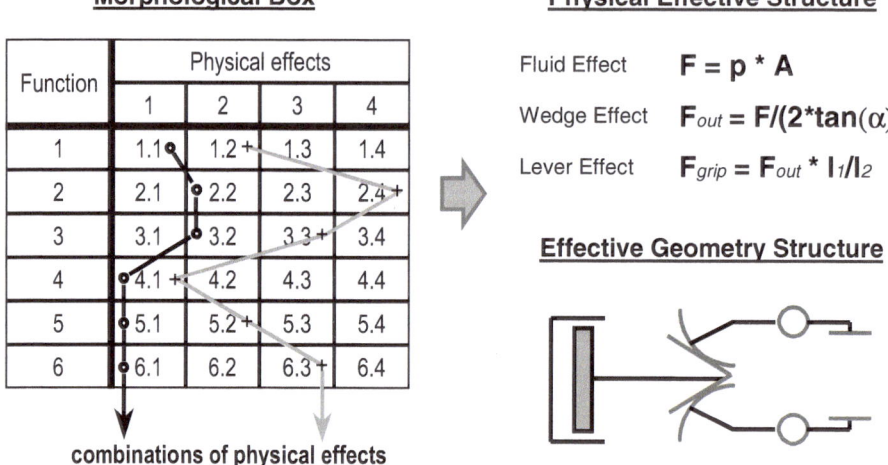

Fig. 2. Example for a morphological box and the derived effective geometry of a robot gripper (principle modelling)

simulation, by propagating the physical quantity input about the physical effective structure and its respective function flow. The result is the physical quantity output of the physical effective structure which is computed based on the mathematical equations of the used physical effects [3].

With the transition from effective geometry to *embodiment design* a designer assigns material to the effective geometry and gives the effective lines and surfaces a solid design. Thereby he has to satisfy the requirements from the requirement model. After the embodiment design the design process has finished.

The design activity "solution finding" absorbs a lot of time within the design process. This search process takes place at the transition from special function modeling to principle modeling. To support a designer within the design process by *finding solutions* a solution base is needed. In the solution base solutions are described by information from different stages of design and from different views, as mechanical engineering or chemical engineering [1]. In order to provide the product model with "knowledge" the whole information of the used solution principles has to be represented. Therefore, the solution base has to be part of an integrated product and production model (PPM) [2].

3   Approach for Low-Level CAD Data Interpretation

The recognition of product functions is not a straight forward interpretation process because the function of a product is determined by its use. For example, although the intended use of a simple pen is to write something, it can be used for several things like scratching yourself, you can use it as a weapon by throwing it at somebody or you can use it for supporting purposes like holding a camera in a specific position on an uneven surface.

The main idea behind our approach of low-level CAD data interpretation is to perform a kinematic analysis of the product and to reason about its physical effects and functions. This idea is motivated by the interpretation a human would do when he gets an unknown product in his hands. He would start to press and turn product components, look at the behavior and reason about its functionality. The same approach is used for our computerized product interpretation. The only information available for the interpretation of a product is its shape, i.e. its geometry and topology. The designer interacts with the product by selecting product components and applying forces and torques upon them. These forces and torques are propagated through the product geometry and the resulting motion is computed. This motion reflects the product behavior and is interpreted accordingly to gain knowledge about product functions and used physical principles.

Fig. 3 gives a quick overview of our approach for low-level CAD data interpretation. Starting with the geometry model of a product in the first step the products kinematic structure is created. The creation and meaning of the kinematic structure will be explained below in more detail as well as all other steps of interpretation. Based on the kinematic structure of a product the second

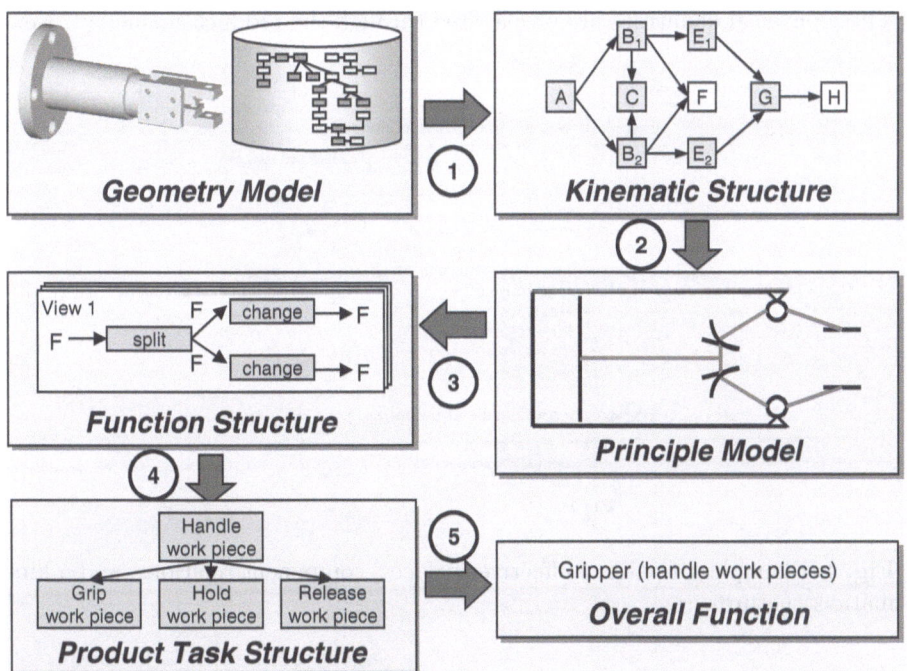

Fig. 3. Approach of low-level CAD data interpretation

step of our approach builds the principle model which is formed by the effective geometry structure and the physical effective structure, i.e. the physical effects behind the product solution. Taking the physical effects as the starting point for the next interpretation step the special function structure is created. As described in the previous section the special function structure is the input-output transformation flow of physical quantities, where a function verb specifies the kind of transformation. This function structure is already a very abstract view on the product. But the most abstract view on the product will be gained in steps 4 and 5 by the creation of a product task structure and specification of the overall product function. After fulfilling these last steps the product model is filled with all information of the early stages of product design. This information can then be used for further product analysis, product variation, or product optimization. In the following sections the single steps of low-level CAD data interpretation are described in more detail.

3.1   Step 1: Kinematic Structure

The kinematic structure is a flow structure of physical quantities between product components with extra information about pairs of effective surfaces and component relations (*Fig. 4*). For the creation of the kinematic structure first the flow of physical quantities through the product is determined. Therefore, product components are selected and forces and torques are applied upon them. These physical quantities are propagated through the product geometry[1]. Next,

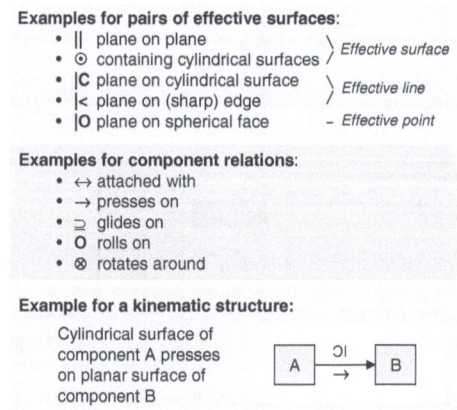

Fig. 4. Examples of pairs of effective surfaces, component relations, and a kinematic structure

---

[1] The methods for this propagation are already developed but they are not described in this paper because of complexity reasons.

contact surfaces between product components are classified and added to the kinematic structure as symbols. The last step is to determine the component relation according to the pair of effective surfaces, the relative movement of the product components, and the resulting degree of freedom. The component relations are also added to the kinematic structure as symbols.

3.2   Step 2: Principle Model

Within the second step the principle model which is formed by the effective geometry structure and the physical effective structure is created. First the physical effects are recognized by checking the kinematic structure against solution patterns as you can see in *Fig. 5*. Therefore, a set of solution patterns for all physical effects must be created. As the number of physical effects is not that big (far less than 1000) it is just a matter of time until the set of solution patterns for the recognition of physical effects is complete. For the recognition of an effect each product component in a kinematic structure is compared with all solution patterns. If several patterns are matching the user must interact and select the according pattern. A solution pattern, such as the lever effect in *Fig. 5*, is compared against a product component X by comparing its related components, the according component relations, pairs of effective surfaces and geometric constraints. If they match then the physical effect is determined which the component is fulfilling within this case of kinematic analysis.

After recognition of the physical effective structure the effective geometry structure is reconstructed. This is done by transforming the projected geometry of the pairs of effective surfaces between product components into the principle sketch. The effective surfaces of each part are connected by effective spaces which are represented by lines (see *Fig. 3*). If the related solution pattern of a physical effect contains special information the principle sketch can be completed accordingly.

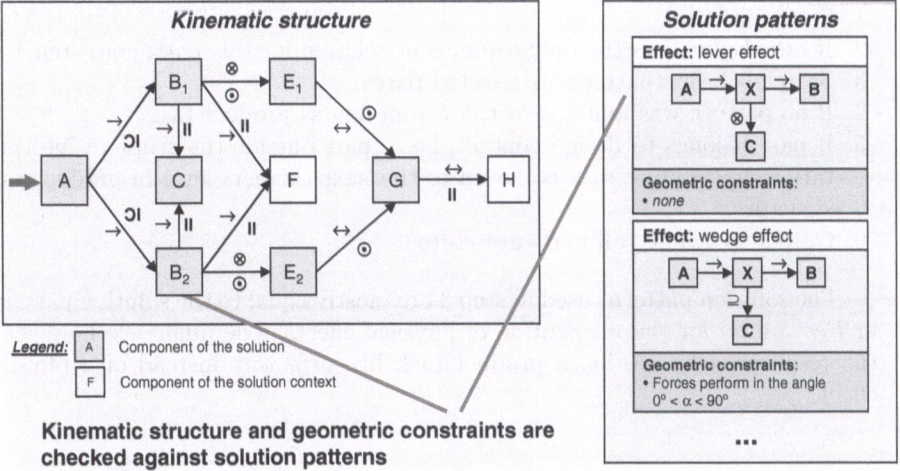

Fig. 5. Recognition of physical effects

3.3   Step 3: Function Structure

The function structure of a product is a graph which contains the flow of physical quantities and their transformation. This transformation is expressed by function verbs. The set of function verbs which is necessary for the description of the quantity transformation is very small, e.g. we are using less than 10 verbs at our institute, ROTH uses 5 verbs as described above. For the recognition of the function verbs a pattern approach can be used again. For each function verb a pattern is created which specifies the type and quantity of the inputs and outputs and additionally some rules the pattern must fulfill. For example the function verb *channel* expresses a displacement of a physical quantity without further transformation. The pattern to find this function has an input I_1, an output O_1 and the rule $O_1 = I_1 \wedge |O_1| = |I_1|$. Another example is the function verb *change* where a physical quantity is displaced and increased or decreased. The pattern to find this function has an input I_1, outputs $O_1, O_2, ..., O_n$ and the rule $n \geq 1 \wedge (\forall i = 1..n : O_i = I_i) \wedge (\forall i = 1..n : |O_i| \neq |I_i|)$. After the complete set of patterns is matched against the physical effects in the physical effective structure, the found functions of all effects are connected respectively in a graph and the reconstruction of the function structure is finished.

3.4   Step 4: Product Task Structure

The product task structure is another type of function structure but on a higher level of abstraction. Here the product functionality is described by a verb and a noun like *handle work piece*. The structure is normally created during the design process by a decomposition of product tasks. The decomposition of a task into sub-tasks is mostly motivated by a decomposition in time, i.e. the product has to fulfill the sub-tasks sequentially or concurrently. The algorithm to recognize the product task structure of a product is based on the kinematic structure and works as follows:

1. Select any part
2. If *attached with* is the only component relation for this part, goto step 1
3. Search solution pattern for selected part
4. If no pattern was found, select flow function as product task
5. If part belongs to design context, i.e. a part outside the scope of interpretation, its product task is moved to the next (higher) level in product task structure
6. Goto step 1, until all parts are visited

The solution patterns used in step 3 are mostly equal to the solution patterns in *Fig. 5* used for the recognition of physical effects. The difference however is the resulting solution, i.e. a product task like *grip part* instead of a physical effect.

3.5   Step 5: Overall Function

The determination of the overall function depends on the different cases of analysis, i.e. if the same product is analyzed with different user interactions it is very probable that different overall functions are recognized. This reflects the statement from the beginning of this section that the function of a product is determined by its use.

For the determination of the overall product function a library of products must be available. This library must include a classification of products with their related characteristics of functions and principles. Product tasks, functions and principles recognized in the previous steps are compared against the elements in this library. If several matching products are found the most specific solution is selected. In any case the found solutions must be presented and accepted by the user. If a solution is selected the different product task structures recognized in step 4 are merged and completed according to the product task structure of the solution.

The following section presents an example for the redesign of a product.

4   Example

Within this section the interpretation process is shown with the example of a robot gripper. *Fig. 6* shows the whole redesign from the geometry model to the product task structure. For the interpretation of the robot gripper geometry two different cases were analyzed.

The first case which is shown in the left column of *Fig. 6* is the interaction with the rod, i.e. a force pushing the rod in horizontal direction. The propagation of the initial force through the product geometry and the analysis of the contacting surfaces result in the shown kinematic structure. Components are represented by letters. Pairs of effective surfaces and component relations are depicted at the arrows between components. A (the rod) presses on B_1 and B_2 (the jaws). The jaws press on C (the spring) and F (the work piece, which is a design context). Additionally the jaws are rotating around the pins (E_1 and E_2). The pins are attached to the housing G and the housing is attached to the robot arm H which is a design context too. The effective structure is gained by matching subgraphs of the kinematic structure with solution patterns of physical effects. The projection of the pairs of effective surfaces, the connection of the surfaces by effective spaces and the additional knowledge of the solution patterns result in the principle sketch below the kinematic structure. The *wedge effect* is recognized for the rod, the *lever effect* is realized by the gripper jaws and the spring is abstracted by the *law of Hooke*. The analysis of the components and their related physical effects yield to the function structure below the principle sketch. The rod channels and splits the initial force into two forces. These forces are changed quantitatively by the jaws and their energy is stored by the spring. Finally the matching of the flow functions against solution patterns of product tasks result in the left branch of the product task structure at the bottom of *Fig. 6*.

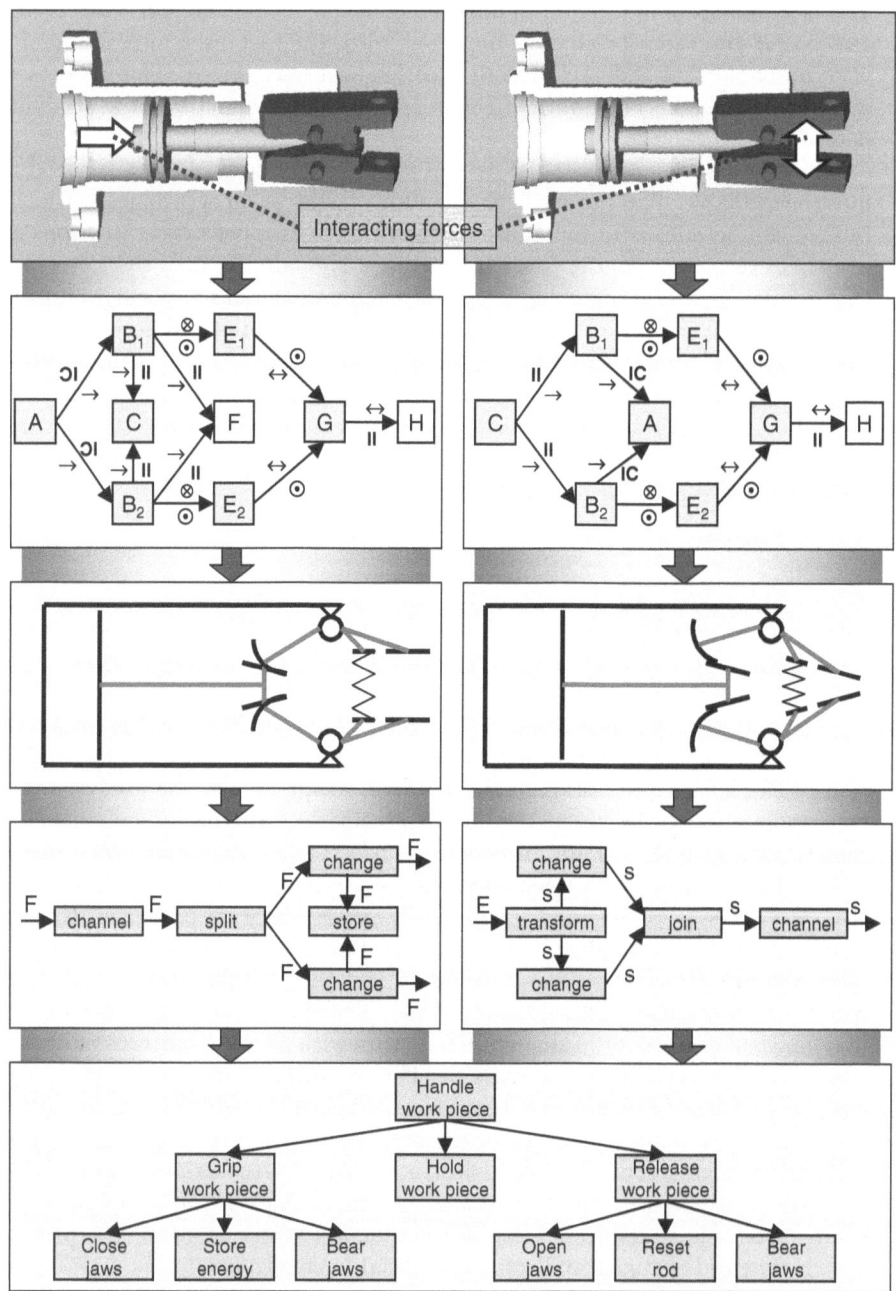

Fig. 6. Example for low-level CAD data interpretation

The analysis of the second case is performed in the same manner, but the initialization is different. Within this case the relaxing of the spring is simulated, i.e. forces are acting on the jaws. The resulting kinematic structure is very similar to the kinematic structure of the first case but traversed in the opposite direction. The effective structure is the same but with a different position of the effective surfaces. Similar to the function structure of case 1 is the resulting function structure of this case but again traversed in the opposite direction. Finally the right branch of the product task structure at the bottom of *Fig. 6* is found for this case.

The analysis of the robot gripper is stopped after two cases. The system was able to find the overall function *handle work pieces* and completes the product task structure.

5   Conclusion

We presented an approach for low-level CAD data interpretation with the objective to gain knowledge about the early design stages, mainly functional modeling and principle modeling. The main idea behind our approach of low-level CAD data interpretation is to perform a kinematic analysis of the product and to reason about its physical effects and functions. The detailed and explicit description of this abstract product knowledge helps to better understand a product, to discover dependencies between product components, and to find good starting points for product optimization.

Acknowledgment

This work is funded by the *Deutsche Forschungsgemeinschaft DFG* (German Research Community) in the *collaborative research center* (project number 346) *"Computer Integrated Design and Manufacturing of Single Parts"*.

References

1. Hans Grabowski, Harald Kunze, Ralf-Stefan Lossack, and Arno Michelis. Solution base for engineering design problems. In *CIRP Design Seminar*, Stockholm (Sweden), Juni 2001. CIRP. 16
2. Hans Grabowski, Stefan Rude, Markus Maier, and Arno Michelis. Integrated Product and Process Modelling. In *The European Conference on Integration in Manufacturing*, Dresden (Germany), September 1997. ESPRIT. 16
3. Hans Grabowski, Stefan Rude, Markus Maier, and Arno Michelis. Integrated Product and Process Modelling as a Basis for the Realization of modern PDT-Concepts. In *ProSTEP Science Days*, Darmstadt (Germany), 1998. 16
4. Rudolf Koller. *Konstruktionslehre für den Maschinenbau – Grundlagen zur Neu- und Weiterentwicklung technischer Produkte*. Springer-Verlag, Berlin, 1994. 13
5. Gerhard Pahl and Wolfgang Beitz. *Engineering Design: A Systematic Approach*. Springer-Verlag, Berlin, 1988. 13, 15

6. Karl-Heinz Roth. *Konstruieren mit Konstruktionskatalogen – Konstruktionslehre*, volume 1. Springer-Verlag, Berlin, 2. edition, 1994. 13, 15

7. F. Zwicky. *Entdecken, Erfinden, Forschen im Morphologischen Weltbild*. Droemer-Knaur, München, 1971. 15

Knowledge-Based Partial Matching: An Efficient Form Classification Method

Yungcheol Byun[1], Joongbae Kim[1], Yeongwoo Choi[2],
Gyeonghwan Kim[3], and Yillbyung Lee[4]

[1] Dept. of EC, Electronics and Telecommunications Research Institute, Korea
{ycb,jjkim}@etri.re.kr
[2] Dept. of Computer Science, Sookmyung Women's University, Korea
[3] Dept. of Electronic Engineering, Sogang University, Korea
[4] Dept. of Computer Science, Yonsei University, Korea

Abstract. An efficient method of classifying form is proposed in this paper. Our method identifies a small number of matching areas by their distinctive images with respect to their layout structure and then form classification is performed by matching only these local regions. The process is summarized as follows. First, the form is partitioned into rectangular regions along the locations of lines of the forms. The *disparity* in each partitioned region of the comparing form images is measured. The *penalty* for each partitioned area is computed by using the pre-printed text, filled-in data, and the size of a partitioned area. The *disparity* and *penalty* are considered to compute the *score* to select final matching areas. By using our approach, the redundant matching areas are not processed and a feature vector of good quality can be extracted.

1 Introduction

To develop a system of form processing applicable in a real environment, there are several issues to be resolved. First of all, various types of form document should be classified as a known form. Unfortunately, the form classification requires a substantial amount of processing time, especially with a large number of forms that need to be classified. That is because the increased number of registered form documents(or models) in a form model database means increased amount of time for analysis. A considerable span of time necessary to analyze the form due to the bulk of the images which is larger than that of individual characters should be considered. Even forms with complex structures ought to be processed at a high recognition rate. Also, the form documents with a similar form structure should be classified well. In addition, the system should be robust against noises and image degradation.

Many methods to process form documents have been proposed until now. Some methods are proficient at recognizing the form layout structure and identifying items, and the resulting information of a form is useful for interpretation[1, 2]. Various approaches have been introduced to classify form documents[3, 4, 5, 6]. Also, a desirable result was obtained from the viewpoint of recognition

D. Blostein and Y.-B. Kwon (Eds.): GREC 2002, LNCS 2390, pp. 25–35, 2002.
© Springer-Verlag Berlin Heidelberg 2002

rates [7, 8, 9]. However, all of this research treated all areas of a form document equally in general, which caused lengthy computation time. For various form documents with complex structures and noises, a new applicable system is also needed to overcome the problems related with the recognition rates. Hence, we propose a system of form classification that centers on the method of partial matching to resolve such problems. By performing structural recognition and form classification on only some areas of the input form, valuable time could be saved and an input form could be processed with a high recognition rate.

In the following section, we will discuss the outline of the form classification system framework. In section 3, we explain the form classification methods based on partial matching. We will discuss our experimental results in section 4. Finally, we offer closing remarks in section 5.

2   Outline of Form Classification

Our method will identify a small number of local regions by their distinctive images with respect to their layout structure and then match only these local regions to classify an input form. The process can be summarized as follows(Fig. 1): First, structures of the layout of all the forms to be processed are recognized. Each image is partitioned into rectangular-shaped local areas according to specific locations of horizontal and vertical lines(Sect. 3.1). Next, the *disparity* in each partitioned local area of the forms is measured(Sect. 3.2). The *penalty* for each local area is computed by using the pre-printed text, filled-in data and the size of a partitioned local area to prevent extracting erroneous lines(Sect. 3.3). The *disparity* and *penalty* are used to compute the *score* to select final matching areas(Sect. 3.4).

3   Form Classification Based on Partial Matching

3.1   Partition of Form and Feature Extraction

Form partition is performed by using the identified structures of all the types of forms to be processed. The form document must be partitioned so that the feature of the form can be extracted robustly. Partitioning must also meet the requirement of conferring stability to forms and matching areas that are in transition. The location and the starting/ending point of the lines in a form should be considered to solve these problems. The adjacent two line segments farthest away from each are bisected first. Then, the process is repeated with the resulting halves. This process of partitioning is performed recursively until the distance between any two adjacent line segments is smaller than a certain threshold value.

Fig. 2 shows an example of the partition. Fig. 2(c) indicates the form structure that is overlapped by two kinds of form documents to be processed, that is (a) and (b). The dotted lines shown by Fig. 2(d) indicate the horizontal $(a1, a2, a3)$ and vertical $(b1, b2, b3)$ separators. Fig. 2(e) shows the result of partition.

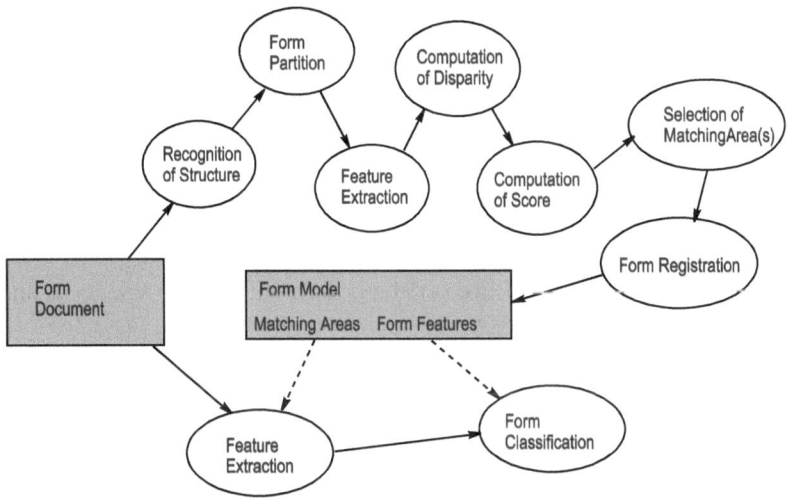

Fig. 1. A model-based system of form processing

A feature vector for each partitioned area is then extracted, which consists of the location and starting/ending position of horizontal and vertical lines. For example, a vertical line is represented as (v_l, v_s, v_e). In this case, v_l stands for the location of a vertical line(x coordinates), v_s and v_e indicate the starting/ending position of a vertical line(y coordinates) respectively. If m vertical lines exist in a partitioned area, a feature vector is constituted as follows.

$$((v_{l1}, v_{s1}, v_{e1}), (v_{l2}, v_{s2}, v_{e2}), \ldots, (v_{lm}, v_{sm}, v_{em}))$$

3.2   Calculation of *Disparity* Using DP Matching

To select the matching areas from the partitioned local areas the *disparity* is measured. The *disparity* represents the distance of a layout structure between two form documents in a partitioned area, which is calculated using DP matching. The DP matching algorithm is defined as follows:

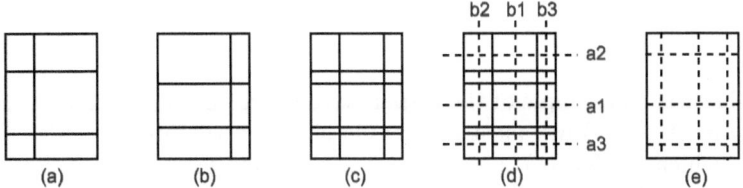

Fig. 2. The distribution of lines and a partition of forms

$$g(i,j) = \min \left\{ \begin{array}{c} g(i-1,j) + C \\ g(i-1,j-1) + d(i,j) \\ g(i,j-1) + C \end{array} \right\} \tag{1}$$

where i and j represent two indices of the respective vectors, and C is a constant of DP matching. By $g(i,j)$, the weight in a weighted graph is computed. In the cases that the numbers of elements in the two feature vectors are m and n, $1 \leq i \leq m$, $1 \leq j \leq n$ are satisfied. If the number of the form document is 2 and the two feature vectors for each area are $((a_{l1}, a_{s1}, a_{e1}), (a_{l2}, a_{s2}, a_{e2}), \ldots, (a_{lm}, a_{sm}, a_{em}))$ and $((b_{l1}, b_{s1}, b_{e1}), (b_{l2}, b_{s2}, b_{e2}), \ldots, (b_{ln}, b_{sn}, b_{en}))$ respectively, then $d(i,j)$ is defined as follows:

$$d(i,j) = |a_{li} - b_{lj}| + \alpha(|a_{s1} - b_{s2}| + |a_{e1} - b_{e2}|) \tag{2}$$

In the DP matching of line segments, the distance between two matching lines is added to the measure. In detail, the position and the length of a line would be determined from the the location and starting/ending points of the two lines and added to the measure(see Equ. 2). In this case, α represents the constant that indicates the extent to which the distance of starting/ending point is reflected in the equation. As a result of the DP matching, a weighted graph is generated. In the weighted graph, the shortest path, $k_1, k_2, \ldots, k_Q$ satisfying the following condition is found and the *disparity* is calculated.

$$disparity(d) = \min \left(\sum_{i=1}^{Q} w(i) \right) \tag{3}$$

where $w(i)$ is a function which returns the weight of the k_i node. The path has a minimal penalty which connects a $(0,0)$ node and a (m,n) node in the weighted graph.

If three form documents (for example A, B, and C) exist, the possible pairs are AB, BC, and CA(see Fig. 3). The disparity values for each pair of two forms called a disparity plane. That is, a disparity plane is generated by comparing two form documents. A disparity vector is constituted by all the values of disparity in the corresponding area of all disparity planes. In the case of Fig. 3, the following disparity vector for a partitioned area can be obtained.

$$\mathbf{d} = (d_1, d_2, d_3) \tag{4}$$

Table 1 is an example of the disparity values for each partitioned area where the forms A and B are most clearly classified by the area a_2. Also, a_1 is the area in which the forms B and C are classified well. For the forms denoted by C and A, a_4 is the most distinctive area. In this case, even if the disparity value for $a_2(0.20)$ is smaller than the disparity value for a_4 (0.80), both areas should be treated equally to classify all the form documents because they could most

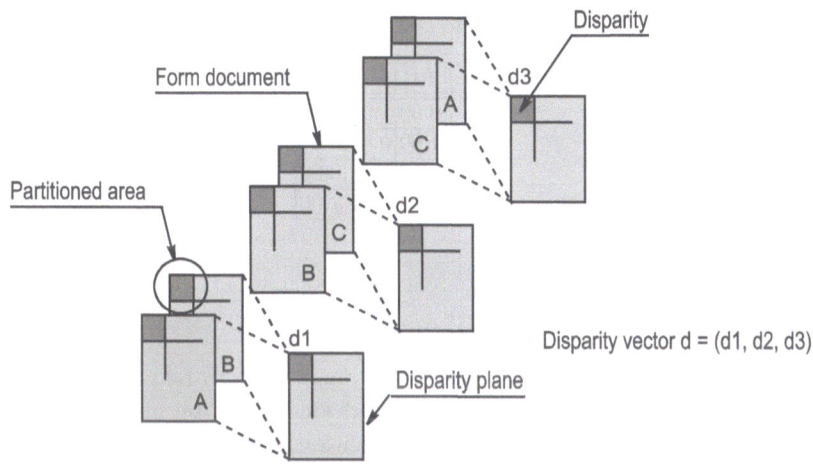

Fig. 3. Partitioned areas and disparity vectors for the local areas

clearly distinguish AB and CA respectively. To reflect this consideration the normalization for each vector element(*disparity*) is performed. In Table. 1, the process is performed column by column by using the following equation:

$$d_{Nij} = \frac{d_{ij}}{d_{Mj}}$$

$$(5)$$

$$1 \le i \le m, \quad 1 \le j \le_n C_2$$

where d_{ij} indicates the jth vector element of the disparity vector for the ith partitioned local area, and d_{Mj} indicates the maximal *disparity* among the jth vector element for all the area. The m and n represent the number of local area and the number of form documents to be processed respectively. As a result, the *disparity* for the local area where each pair of form documents is most clearly distinguished becomes 1, and the local area in which the layout structure is equal becomes 0. Table 2 shows the result of *disparity* normalization.

Table 1. An example of the disparity values for each local area

area	AB	BC	CA
a_1	0.10	0.50	0
a_2	0.20	0.40	0.10
a_3	0.10	0	0.70
a_4	0	0.10	0.80

Table 2. The normalized disparity values of the *disparity* for Table 1

area	AB	BC	CA
a_1	0.50	1.00	0
a_2	1.00	0.80	0.13
a_3	0.50	0	0.88
a_4	0	0.20	1.00

3.3   Computation of *Scores* Using Knowledge

Now, The matching areas where the form documents to be processed are classified well are selected from the partitioned local areas by using the *disparity*. The simplest method is to use the summation and average of all the vector element(*disparity*) for each disparity vector as a *score*, s_{avg_i}. We compute the *score* by using the equation as follows:

$$s_{avg_i} = \frac{1}{{}_n C_2} \sum_{j=1}^{{}_n C_2} d_{Nij} \tag{6}$$

$$1 \le i \le n, 1 \le j \le {}_n C_2$$

where s_{avg_i} is the *score* of the ith matching area and satisfies $0 \le s_{avg_i} \le 1$, and n indicates the number of form documents to be processed.

In this case, although the matching areas are selected by the *score*, erroneous lines can be extracted when the pre-printed texts and the filled-in data exist in the areas of a filled form. Therefore, it is desirable to select the local areas where the pre-printed texts and the filled-in data less exist relatively as the matching areas. As a result, the size of a matching area, the degree of existence of the pre-printed texts and the filled-in data are considered and measured as a *penalty*(p), which is used to adjust the s_{avg_i} and compute the *score*(s_i). The *score*(s_i) is measured by using the following equation:

$$s_i = s_{avg_i} - p_i \tag{7}$$

$$p_i = \beta_1 r_{p_i} + \beta_2 r_{t_i} + \beta_3 \left(1 - \frac{a_i}{a_M}\right)$$

where r_{p_i} is the overlapped ratio which means the degree of existence of the pre-printed texts in the ith partitioned local area. And a_i and a_M indicate the size of the ith area and the maximal size of area from all of the partitioned areas, respectively. The constants represented by β_1, β_2, and β_3 indicate the extent to which the three terms are reflected in the equation. When pre-printed texts exist in an area, the overlapped ratio of the pre-printed text, r_{p_i}, is computed by using the following equation:

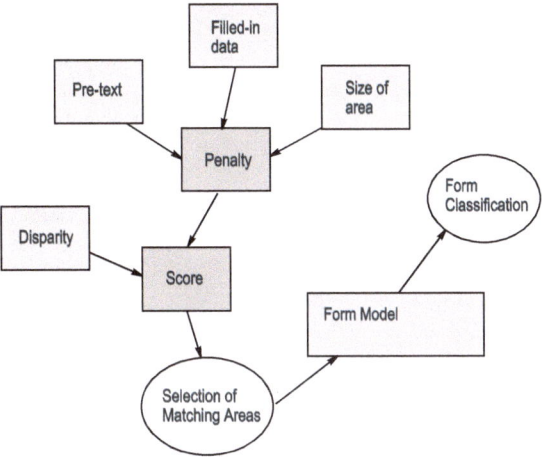

Fig. 4. The *penalty* and *score* to select matching areas

$$r = \frac{1}{N} \times \frac{a_o}{a_l} \tag{8}$$

In this case, N means the number of the form documents to be processed, and a_o represents the overlapped area of the partitioned local area and the pre-printed text area. Also a_l represents the size of the partitioned local area(Fig. 5). Fig. 6 (a) represents the pre-printed texts extracted from a blank credit card slip, and (b) represents the resulting information for the area of the pre-printed texts. (c) shows the overlapped representation of the partitioned local areas and the pre-printed text areas.

The overlapped ratio of the filled-in data can be measured by using the same method also. Therefore, according to the above equation, the *penalty*(p) is large when the number of the pre-printed texts and the filled-in data which exist in a partitioned local area is large. As a result, the value of *score*(s_i) is small, which decreases the possibility of selection of the ith local area as a matching area.

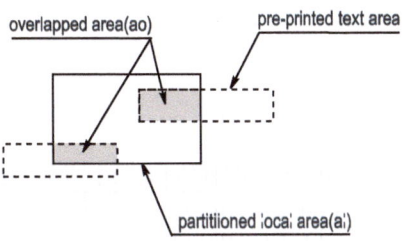

Fig. 5. The overlapped area of a local area and two pre-printed text areas

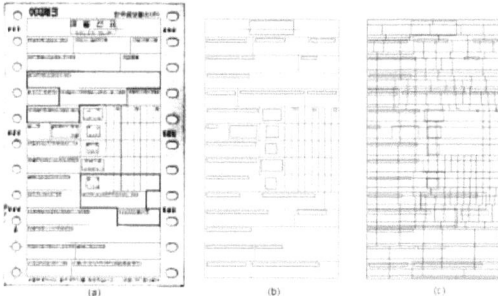

Fig. 6. An example of the pre-printed text areas extracted automatically

3.4   Selection of Matching Areas Using *Scores*

Finally, the matching areas which will be used to classify form documents are selected by referring the *score* measured before. For example, if the resulting scores are as illustrated in Table 3, the area denoted by a_2 is the area which will be selected firstly. If the number of matching areas is 2 then a_2 and a_1 are selected. By this method the order of selection is a_2, a_1, a_3, and a_4.

4   Experimental Results

We created a system in order to test the proposed method. Our system was implemented on Pentium PC(PII 366) using a C++ language. In this experiment, a total of six types of credit card slips were used. The slip images was very similar in their structure, but the pre-printed texts and the filled-in data to be recognized could be located at the different regions. Also, the image could be contaminated by the various kinds of noises. More specifically, the slips of Hankook, Union, E-Mart, Easy, Kiss, and Yukong were used to test. A total of 246 form images that include six unfilled training forms and 240 filled test forms were scanned in a 200 dpi mode. The average size of the image of the experimental form documents was 832×1141 pixels.

Before computing the *disparity* for the corresponding parts in each member of a pair, form documents were partitioned with a partition parameter, 13/13

Table 3. The *score* value which is computed by *disparity* and *penalty*

area	AB	BC	CA	s_{avg_i}	p	s_i
a_1	0.50	1.00	0	0.50	0.05	0.45
a_2	1.00	0.80	0.13	0.64	0.1	0.54
a_3	0.50	0	0.88	0.46	0.12	0.34
a_4	0	0.20	1.00	0.40	0.1	0.30

which was determined to yield a desirable result by experiments. The *disparity* was measured and the *score* was then computed. A total of 14 areas with the highest scores were selected as matching areas, and the form classification was performed according to the number of the matching areas. The test forms were classified with the increase of the number of the matching area. In this case, the *score* to select the matching areas was computed with various values of the β_1, β_2, β_3, and we analyzed the effect of the pre-printed text, filled-in data, and the size of local area upon the recognition rate.

First, the *score* was measured when the β_1, β_2, β_3 were 0, 0, 0, respectively. The *score* was then measured with various values of a β_1. We could know that the recognition rate increased more when the constant, β_1, was 0.2, 0.4, and 0.6 than the result when the constant was 0.0. However, the recognition rate decreased when the β_1 was 0,8 or 0.9 on the contrary because if the β_1 was large the effect of the *disparity*, that is, s_{avg_i} decreased when the $score(s_i)$ was computed.

The relation of the filled-in data and the recognition rate was examined. The experimental result was similar to that of the pre-printed texts. If the value of the β_2 increases the effect of *disparity* decreases, which makes the rate of recognition low relatively, and the number of the matching areas large to acquire a satisfactory result. Next, the relation of the size of area and the rate of recognition was examined by increasing the β_3. If relatively large local areas were selected as matching areas the rate of recognition increased more than the result when the size of area was not considered. Especially, desirable results were acquired when the β_3 was 0.2. And, if the β_3 was larger than 0.2, the results were then similar one another, which means that the algorithm of feature extraction extracts the feature vectors consistently for all of the local area larger than a specific area.

Finally, all of the factors, that is, the pre-printed text, the filled-in data, and the size of a local area were considered in the experiment. The resulting rate of recognition was then compared to the previous results. Fig. 7 shows the previous results and the result when the $\beta_1, \beta_2, \beta_3$ are $0.47, 0.57, 0.49$ respectively, which were decided by experiments. As a result, we could see that searching and matching only a small number of structurally distinctive areas yields a high

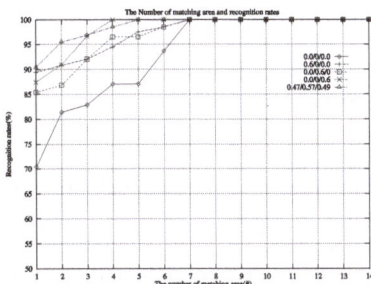

Fig. 7. The rate of recognition according to the all of the factors

rate of classification. 100% rate of classification was obtained by using only five matching areas, which is only 2.5% of the total size of an input form image. the average time to classify an input form was 0.78 seconds.

5   Conclusion

In this paper, we proposed a new method for processing form document efficiently using partial matching. The pre-printed texts, filled-in data, and the size of areas were considered to select matching areas. The redundant local areas containing noises were not selected so as to extract a good feature vector with respect to the recognition rate and the computation time. A feature vector of good quality could be extracted by avoiding the areas where erroneous lines could be extracted. From the experiments discussed in the previous section, we could see the following details: The rate of recognition increased when the pre-printed texts, filled-in data, and the size of areas were considered to select matching areas. By using the DP matching method for computing the *disparity* and the knowledge of a form which was represented by the *penalty*, the form document with the pre-printed texts, the filled-in data, and noises could be effectively handled. Even the form documents with a similar structure could be classified correctly by searching and matching only the structurally distinctive areas. The input form could be processed within a relatively short span of time compared to the time when the total image is processed. Only 2.5% of the total size of an input form was used to obtain a 100% rate of classification. As a result, an efficient form processing system applicable in real environment could be constructed.

References

[1] T. Watanabe, Document Analysis and Recogntion, IEICE Trans. Inf. & Syst., Vol.E82-D, No.3, pp.601-610, 1999. 25
[2] T. Sobue and T. Watanabe, Identification of Item Fields in Table-form Documents with/without Line Segments, MVA, pp.522-525, 1996. 25
[3] R. G. Casey, D. R. Ferguson, K. Mohiuddin and E. Walach, Intelligent forms processing system, MVA, Vol. 5, pp.511-529, 1992. 25
[4] J. Mao, M. Abayan, and K. Mohiuddin, A Model-Based Form Processing Sub-System, ICDAR, pp.691-695, 1996. 25
[5] S. W. Lam, L. Javanbakht, and S. N. Srihari, Anatomy of a form reader, ICDAR, pp.506-509, 1993. 25
[6] A. Ting, M. K. Leung, S.-C. H, and K.-Y. Chan, A Syntactic Business Form Classifier, ICDAR, pp.301-304, 1995. 25
[7] S. L. Taylor, R. Fritzson, and J. A. Pastor, Extraction of data from preprinted forms, MVA, Vol. 5, pp.211-222, 1992. 26
[8] Y. Ishitani, Model Matching Based on Association Graph for Form Image Understanding, ICDAR, pp.287-292, 1995. 26
[9] P. Heroux, S. Diana. A. Ribert, and E. Trupin, Classification Method Study for Automatic Form Class Identification, IWFHR, pp.926-928, 1998. 26
[10] C. L. Yu, Y. Y. Tang, and C. Y. Suen, Document Architecture Language Approach to Document Processing, ICDAR, pp.103-106, 1993.

[11] R. Lorie, A System for exploiting Syntactic and Semantic Knowledge, DAS, pp.277-294, 1994.

[12] Y. Byun and Y. Lee, Efficient Form Processing Methods for Various Kinds of Form Documents, DAS, pp.153-156, 1998.

Robust Frame Extraction and Removal
for Processing Form Documents

Daisuke Nishiwaki[1], Masato Hayashi[2], and Atsushi Sato[1]

[1] Multimedia Research Labs.
[2] Social Information Solution Division, NEC Corporation
1-1, 4-chome Miyazaki, Miyamae-ku, Kawasaki, Kanagawa 216-8555 Japan

Abstract. A new frame extraction and a removal method for processing form documents is proposed. The method robustly extracts scanned pre-printings such as frames and lines. It consists of a frame detection process and a frame removal process. In the frame detection process, the center coordinates are extracted using a Generalized Hough Transformation-based method. Then, using those coordinates, an inscribed rectangular image for each frame is produced. In the frame removal process, the detected frame image is removed along the outside of the rectangular edge. These processes are repeated to remove the target frames successfully by changing some pre-processings such as reducing and enhancing. The method was applied to some types of images. They are postal codes on mail and forms received by facsimiles. In both cases, there often can be seen low quailty pre-printings. For those low quality images, convetional approach such as model pattern maching was not well worked because of local distortion. Through experiments in frame detection and removal of the images, we demonstrated that all of the frames could be successfully removed.

1 Introduction

At GREC'97, Y. Choi and Y-B Kwon introduced trend research works for the processing of form document and described problem [1]. The restoration of defective character has been identified as one of the problems. Previously, we have also proposed a method for restoring defective handwritten-numeral characters written in a black ruled line form [2],[3]. At GREC'99, in his keynote speech Professor Nagy introduced us the future work of the form processing[4]. Achieving frame detection and removal using less format information is essential. In this paper, we focused on that.

In many conventional methods, projection profiles for the x- and y-axes are usually used to detect the frames pre-printed on a form documnet based on frame size and another format information. However, in some cases, the format information is not helpful for extracting pre-printings such as frames and lines, because of their distortions due to low resolution scanning or low quality printing. These distortions produce confusion between frames and character strokes. In this case, physical frames are not exist as estimated by format information.

D. Blostein and Y.-B. Kwon (Eds.): GREC 2002, LNCS 2390, pp. 36–45, 2002.
© Springer-Verlag Berlin Heidelberg 2002

To overcome this problem, we once suspected to detect frame image directly, and studied how to detect the center points for each frame. Because, once the center points are obtained, it should be easy to detect the frame by using size information even though the points do not stand at precisely regular intervals. Therefore, we propose a Generalized Hough transform(GHT)-based frame detection and removal method. The GHTs are well studied and widely used extension to the basic Hough transform [6]. However, the GHTs as well as the basic Hough transform often detects errneous peak in the voting area. For the problem, we also studied how to select the correct peak corresponding to the frame center points, and proposed a linear function to select them successfully. The function represents linearlity, regurality of vertical coordinates and intervals for the series of the selected ceter points as candidates for each frames. By using this function, we succeeded to select the ceter points. The following sections describe the method in detail and show how the effectiveness of the method was tested in some experiments by using images of postal codes and facsimile images.

2 Frame Detection and Removal

2.1 Process Overview

Figure 1 shows the outline of the frame extraction process. It consists of five steps, edge detection, voting, frame-center candidate selection, frame center detection, and frame removal. The second and the third step are focus of this paper.

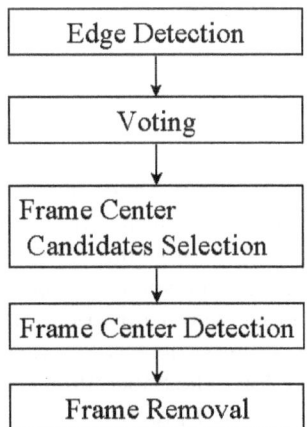

Fig. 1. Frame removal process

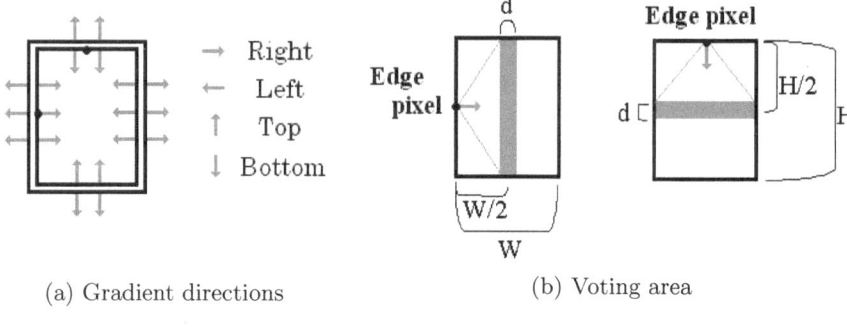

(a) Gradient directions   (b) Voting area

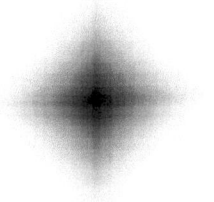

(c) Voting result

Fig. 2. Voting method

2.2   Edge Detection and Voting

In the edge detection step, edge pixels are obtained for each frame image by using a differential operation(Fig. 2(a)). It is expected that the loading of voing process can be reducted by using only edge pixels while keeping the precision of detecting the frame center point. Then the gradient direction is detected by using the operation result for each of the edge pixels. It is well known calculation cost is reduced pretty much to execute the vointing process along the direction [7]. The direction is classified into one of four directions, Right, Left, Top, and Bottom in every 90 degree. This is because detection target is rectangular. In the figure, the arrows show each direction. The Sobel filter is applied for the edge detection process and the gradient direction process.

The voting is performed for all edge pixels. Figure 2(b) shows examples of voting areas for two edge pixels which have the right and the bottom directions respectively. The width of the frame is W and H is the height. In the figure, a long and narrow rectangle colored in gray is produced as the voting area for each edge pixel. The length of the voting area is H for the pixels which have the right direction, and the length of the voting area is W for the pixels which have the bottom direction. The distance between the edge pixel and the major axis of the voting area is set to $W/2$ for the horizontal directions, $H/2$ for the vertical ones. The width of the voting area is set to d for both directions. Voting areas for the left and for the top direction respectively are obtained in the same way. In almost conventional OCR applications, the values of W and H can be obtained in advance as design data. In actuality, the gradient direction is not always obtained toward the center of each frame. Because, notches in the edge pixels produce the bad influence. Therefore, the voting area is set widely toward the gradient direction.

In addition, to achieve higher processing speed, scanned images are once reduced in half size. Therefore, this countermeasure is expected to be effective for images of low resolution as well as low quality. Figure 2(c) shows an example of the voting. The gray scale represents accumulated voting. The deepest point is the center of the frame. As shown in the figure, the voting is done on the same $x - y$ space as the target image. This is different from the well known the basic Hough transform[5] that uses the $\rho - \theta$ space.

If no character is written or printed in a frame, it is not so difficult to obtain the center point by using the method described in the above and shown in Fig 3. Figure 3(c) is the voting result for an original image(Fig.3(a)). The white circles represent peaks corresponding to the frame center points. The voting is applied to all edge pixels. Therefore, voting result also appears in the out of the frame image. In actuality, voting results are not uniform for each frame as described now because the voting is also applied to the character edge pixels as well as the frame edge pixels. Several peaks appear in each frame as a result. We, here, call these peaks center candidates.

We found that it could not be decided as the ceter point even though the peak had the maximum voting value. Because when pre-printed frame image is blurred and/or something symmetrical character like H or O is written in the frame, confusion is occured. Therefore, a method which selects the correct peak corresponding to the frame center is needed.

2.3   Frame Center Detection

To select the correct peaks from the center candidates for the frames, a regularity of intervals for the candidates, and a deviation of vertical coordinates are studied for a series of the selected center candidates. An estimation value E was used to do that as follows:

$$E = w_1 \times Sum - w_2 \times Reg - w_3 \times Dev, \qquad (1)$$

where Sum is the sum total of the voting for each selected center candidate, Reg is the regularity of intervals for the selected center candidates, Dev is the deviation of vertical coordinates for them, and w_1, w_2 and w_3 are weight values. Reg is the sum total of the squared error between a detected interval and its prescribed interval for each frames. When the selected center candidates stand as the same interval with the prescribed one, Reg is equal to zero. Dev is the

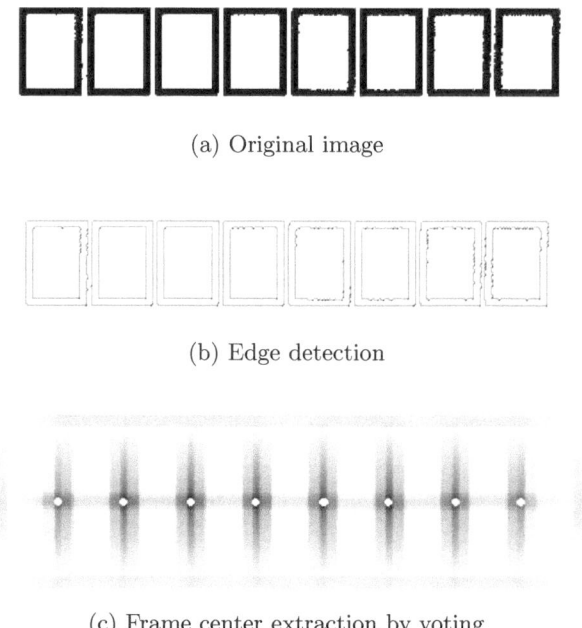

(a) Original image

(b) Edge detection

(c) Frame center extraction by voting

Fig. 3. Frame center detection

squared sum total of the y-coordinate position gap between a center candidate and the neighboring candidate on the right side. When the the selected center candidates stand horizontally in a line, Dev is equal to zero. The combination of the candidates that obtain the maximum value E are decided as the center points.

2.4   Frame Removal

In the frame removal step, each of the frame is removed by using detected center points. An inscribed rectangle for each frame is generated, and then several pixels of thickness removed along the inscribed rectangle of each frame.

As one of the other frame removal approaches, we can study a frame model maching method. This method generates an ideal frame image by using format information and matches this with a real scanned image. If the quality is good and no distortion appears after image scanning, this approach can be applied. However, these conditions are not satisfied in out targets which have local distortion. It would be difficult to detect a certain position which represents the maximum overlapping.

3   Experimental Results

3.1   System Overview

A prototype system was designed based on the proposed method. Figure 4 shows the process flow of the system. To obtain higher performance, speed and accuracy, an iterative process was used. In the first trial, the target image is reduced by half to obtain higher speed. The coordinates are also reduced at this time. If frame removal is impossible in this trial, then the original-sized image is used in the second trial. If frame removal is impossible in the second trial, then the original image is enhanced. The 3×3 majority filter is applied here. It is possible to check whether the trial was successfully finished or not by checking the estimation value E in each trial. In the formula 1, the same weight was applied for w_1, w_2 and w_3.

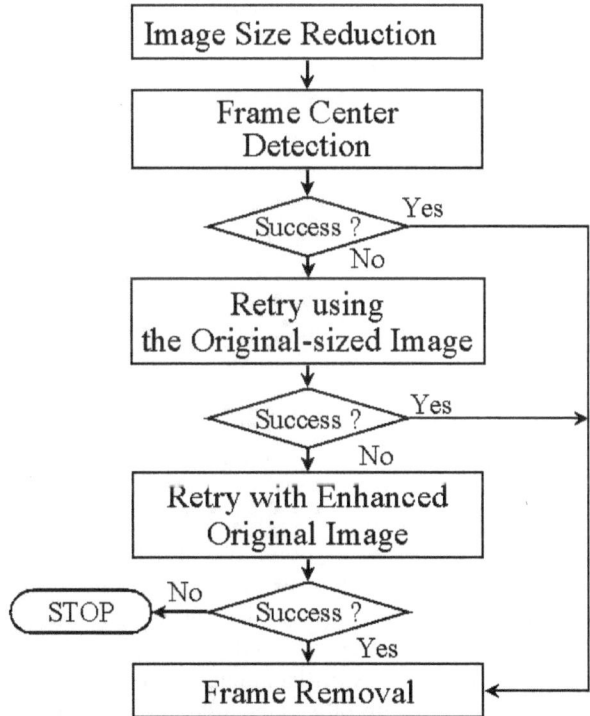

Fig. 4. Frame removal system overview

3.2   Results

To confirm the efficacy of the proposed method, we performed experiments for two types of images: postal code images, and facsimile images. Figure 5(a) shows an example of a postal code image. Eight-digit-number frames must be pre-printed in a certain size and position. However, the size and position are not always exactly the same as required by the regulation standard. The postal code frames are sometimes printed with intervals slightly different from the standard.

The voting result is shown in Figure 5(b). Several center candidates were obtained for each frame. Figure 5(c) shows the result after the frame removal. Even though the characters were not printed in the center of each frame, all frames were successfully removed. Figure 5(d) shows another example of a postal code image. The upper part of each frame is blurred. The voting result is shown Figure 5(e). As shown in Figure 5(f), all frames were successfully removed. In both cases, successful results were obtained in the first trial because the image sizes were large. In figure 5(a) and 5(d), destination parts are hidden with gray. We examined this method for 50 of mail images including these two cases. All frames in the test set were successfully removed. The processing time was between 50 and 60 ms when Pentium IV(1.5GHz) is used. If the original size image used, it was confirmed that the processing time became around 290 ms.

Figure 6(a) shows an example image that was sent by a G3-type facsimile. The small frame size is 7 mm × 5 mm, and the large one is 8 mm × 6 mm. The thicknesses of the frame lines are 0.5 mm and 0.1 mm respectively. In the form, black-ruled line-type frames are printed. They are filled in with handprinted capital letters. Handprinted capital letters include horizontal and vertical strokes. The 2nd column of the 5th line is one of the most remarkable field. One might expect a confusioned result in the frame center detection step. Therefore, it was interesting to see it the proposed method could successfully remove the frame.

Figure 6(b) shows the edge detection image for the second field of the fifth line. For this image, the first trial was a failure because the frame size was small and the received image was blurred. Detecting frames like this is difficult using the conventional method that uses a projection profile. This is because the several peaks of the characters obscure detection of the ture frame coordinates. Many candidates were produced as potential centers(Fig. 6(c)). However, the true center points were successfully selected. As a result, frame image was removed successfully by using our method(see Fig. 6(d)). In this case, the result was obtained in the final trial. We prepaered another forms they are fill in handwritten numerals and Japanese Kata-kana characters. And the received images were prepared by way of both an office facsimile and personal one. In total, 156 fields were examined, and confirmed that all of the frames were uccessfully removed. The total processing time was 35 ms for the the second field of the fifth line when the same processor used with the mail images. The processing time for the retry which uses the original size image was 24 ms.

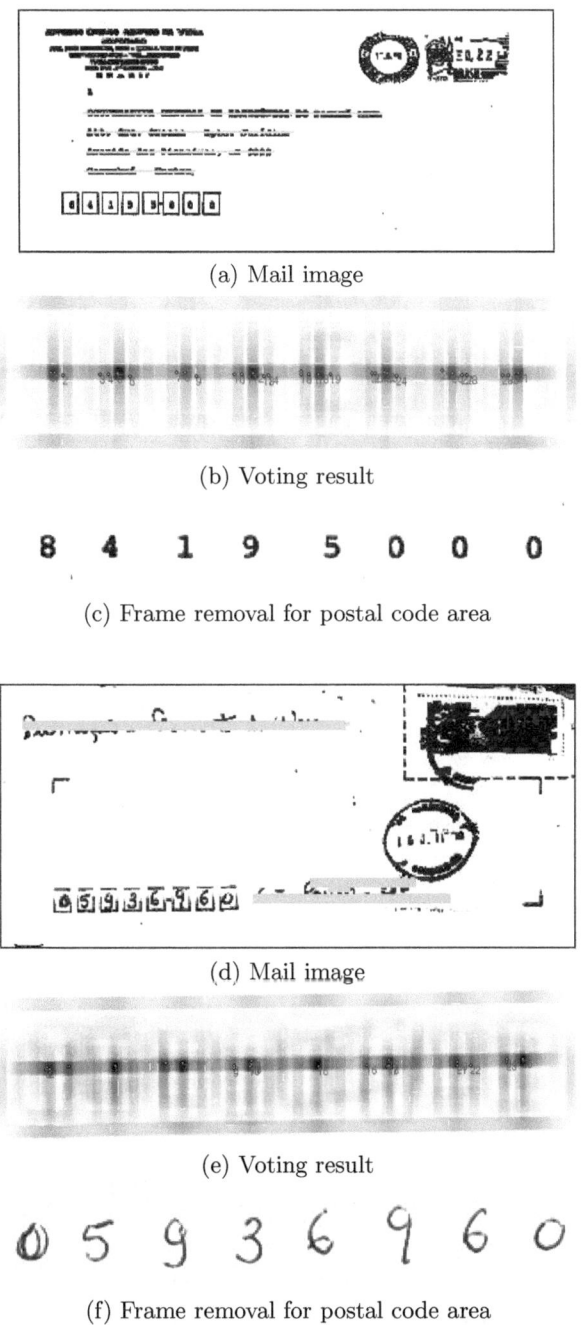

(a) Mail image

(b) Voting result

8 4   1 9   5   0   0   0

(c) Frame removal for postal code area

(d) Mail image

(e) Voting result

0 5 9 3 6 9 6 0

(f) Frame removal for postal code area

Fig. 5. Postal code frame removal

(a) Facsimile image

(b) Edge detection for a field

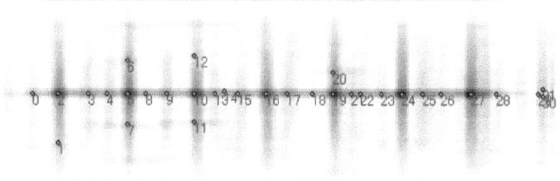

(c) Frame center candidates

H I J K L M N

(d) Frame removal

Fig. 6. Application for facsimile image

4   Conclusion

We proposed a robust frame extraction and removal method for the processing of form documents. The method first detects the center point for each frame. To detect the center point, a Hough Transform-based method was used. It produces several peaks as the candidate centers for each frame. The proposed method selects a plausible center point from center point candidates. To do this, an estimation function is used. A maximum value is obtained when the the center candidates stand in a straight line with the prescribed interval. By using our method, a prototype frame extraction and removal system was produced. The system processes target images iteratively by performing a scale reduction and an image enhancement. Experiments in frame extraction and removal for postal code images and facsimile images showed that all of the frames were successfully removed. In future work, we will combine our previously reported method for restoring handwritten characters[2] with this frame detection and removal method.

Acknowledgement

The authors would like to express their gratitude to their colleagues in the Pattern Analysis Technology Group for their helpful discussion and cooperation. Finally, the authors would like to thank the reviewers for their fruitful comments and suggestions.

References

[1]   Y. Choi and Y-B Kwon: Business Graphics and Form Processing: A Survey, Proc. of GREC'97, pp. 111-118 (1997).   36

[2]   D. Nishiwaki et al.: A New Recognition Method of Black Ruled Line Form including Connected Handwritten Numerals, Proc. of GREC'97, pp. 152-159 (1997).   36

[3]   D. Nishiwaki et al: An Improvement of Numeral String Recognition Performance on Black Ruled Line Forms using Character Touching Type Verification, Proc. of GREC'99, pp. 168-175 (1999).   36

[4]   D. Lopresti and G. Nagy: Automated Table Processing: An (Opinionated) Survey, Proc. of GREC'99, pp. 109-134 (1999).   36

[5]   R. O. Duda and P. E. Hart: Using the Hough transform to detect lines and curves in pictures, Commun. ACM, pp. 11-15 (1972).   39

[6]   D. H. Ballard and C. M. Brown: COMPUTER VISION, Prentice-Hall, Inc. (1982).   37

[7]   C. Kimme et al: Finding Circles by An Array of Accumulators, Comm. ACM 18, 2, pp. 120-122 (1975).   38

Issues in Ground-Truthing Graphic Documents

Daniel Lopresti[1] and George Nagy[2]

[1] Bell Labs, Lucent Technologies Inc.
600 Mountain Ave. Room 2D-447, Murray Hill, NJ 07974, USA
[2] Department of Electrical, Computer, and Systems Engineering
Rensselaer Polytechnic Institute Troy, NY 12180, USA

Abstract. We examine the nature of ground-truth: whether it is always well-defined for a given task, or only relative and approximate. In the conventional scenario, reference data is produced by recording the interpretation of each test document using a chosen data-entry platform. Looking a little more closely at this process, we study its constituents and their interrelations. We provide examples from the literature and from our own experiments where non-trivial problems with each of the components appear to preclude the possibility of real progress in evaluating automated graphics recognition systems, and propose possible solutions. More specifically, for documents with complex structure we recommend multi-valued, layered, weighted, functional ground-truth supported by model-guided reference data-entry systems and protocols. Mostly, however, we raise far more questions than we currently have answers for.

> And diff'ring judgements serve but to declare,
> That truth lies somewhere, if we knew but where.
> *William Cowper, 1731-1800*

1 Introduction

Is ground-truth indeed fixed, unique and static? Or is it, like beauty, in the eyes of the beholder, relative and approximate? In OCR test datasets, from Highleyman's hand-digitized numerals on punched cards to the UW, ISRI, NIST, and CEDAR CD-ROM's, the first point of view held sway.[1] But in our recent experiments on printed tables, we have been forced to the second [14]. This issue may arise more and more often as researchers attempt to evaluate recognition systems for increasingly complex graphics documents.

Strict validation with respect to reference data (*i.e.*, ground-truth) seems appropriate for pattern recognition systems designed for categorical objects from finite, well-defined classes where an appropriate set of samples is available. (The choice of sampling strategy is itself a recondite topic that we skirt here.) It is more problematic for structured entities, like tables and drawings, with dyadic or hierarchical relations. We examine the major components that seem to play

---

[1] A table defining the various acronyms used in this paper appears in Appendix A.

D. Blostein and Y.-B. Kwon (Eds.): GREC 2002, LNCS 2390, pp. 46–67, 2002.
© Springer-Verlag Berlin Heidelberg 2002

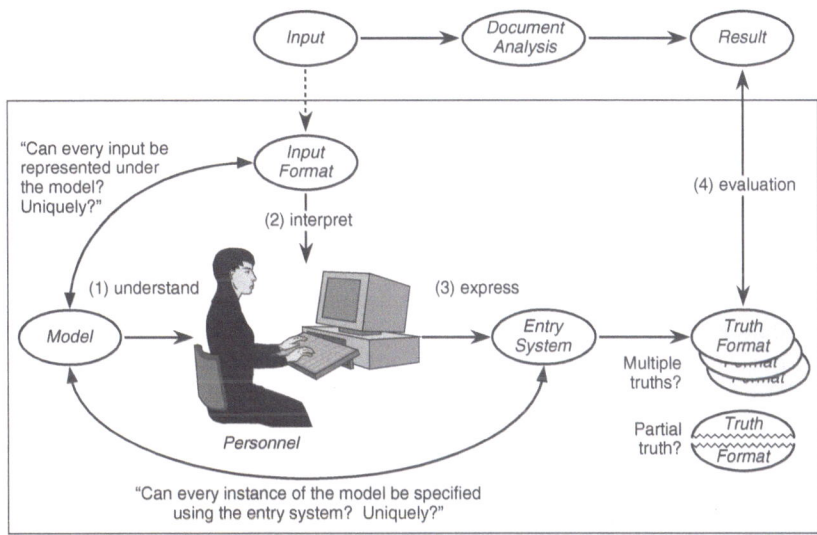

Fig. 1. Overview of the ground-truthing process

a part in determining the nature of reference data. In the conventional scenario, ground-truth is produced by recording the interpretation of each test document using a chosen data-entry platform. Looking a little more closely at this process, we study its constituents and their interrelations (see also Fig. 1):

Input Format. The input represents the data provided to both interpreters and the recognition system. It is often a pixel array of optical scans of selected documents or parts thereof. It could also be in a specialized format: ASCII for text and tables collected from email, RTF or LaTeX for partially processed mainly-text documents, chain codes or medial axis transforms for line drawings.

Model. The model is a formal specification of the appearance and content of a document. A particular interpretation of the document is an instance of the model, as is the output of a recognition system. What do we do if the desired interpretation cannot be accommodated by the chosen model?

Reference Entry System. This could be as simple as a standard text editor like Unix vi or Windows Notepad. For graphics documents, some 2-D interaction is required. DAFS-1.0 (Illuminator), entity graphs, X-Y trees, and rectangular zone definition systems have been used for text layout [8, 16, 21, 35]. We used our own system, Daffy, for tables (see Fig. 4) [14]. Because the reference data should be more reliable than the recognition system being evaluated, the data is usually entered more than once, preferably by different operators, or even by trusted automated systems. Questions that we will examine in greater detail are the conformance of the reference entry system to the model (Is it possible to enter reference data that does not correspond to a legal instance of the model? Are there instances of modeled reference data that cannot be entered?), and its bias (Does it favor some model instances over others?). To avoid discrepancies, should we expeditiously redefine the model as the set of representations that can be produced using the reference entry system?

Truth Format. Clearly, the output of the previous stage must carry with it more information than the input. To facilitate comparison, traditionally the ground-truth format is identical, or at least equivalent, to that of the output of the recognition system. In more difficult tasks it may be desirable to retain several versions of the truth. We may also be willing to accept partial reference information. For instance, we may be satisfied with line-end coordinates of a drawing even if the reference entry system allows differentiating between leaders, dimension lines, and part boundaries. Is not all reference data incomplete to a greater or lesser extent? More flexible truth formats may be needed.

Personnel. For typeset OCR, knowledge of the alphabet in question is usually considered sufficient. For handwriting, literacy in the language of the document may be necessary. For more complicated tasks, some domain expertise is desirable. We will discuss the effects of subject matter expertise versus specialized training (as arises, for instance, in remote postal address entry, medical forms processing, and archival engineering drawing conversion). How much training should be focused on the model and the reference entry system versus the topical domain? Is consistency more important than accuracy? Can training itself introduce a bias that favors one recognition system over another?

Although each of these constituents plays a significant role in most reported graphics recognition experiments, they are seldom described explicitly. Perhaps there is something to be gained by spotlighting them in a situation where they do not play a subordinate role to new models, algorithms, data sets, or verification procedures.

2   The Role of Ground-Truth in Evaluation

As mentioned, we are interested in the evaluation of recognition systems through quantitative measures based on comparing the output files of the system with a set of reference ("truth") files produced by (human) interpreters working from the same input. This is by no means the only possible scenario for evaluation. Several other methods have merit, including the following:

1. The interpreter uses a different input than the recognition system (for example, hardcopy instead of digitized images).
2. The patterns to be recognized are produced from the truth files [13, 15, 17], as in the case of the bitmaps generated from CAD files for the GREC dashed-line and circular curve recognition contests [6]. Do we lose something by accepting the source files as the unequivocal truth even if the output lends itself to plausible alternative interpretations?

3. The comparison is not automated: the output of the recognition system may be evaluated by inspection, or corrected by an operator in a timed session (a form of goal-directed evaluation).
4. Functional evaluation without reference data or ground-truth: in postal address reading, the number of undeliverable letters may be a good measure of the address reader's performance.
5. Double-blind evaluation is used in medical image interpretation [36]. The output of the automated system is included among outputs generated by human operators. The evaluators then rank the outputs (without reference to any specific ground-truth file).

Each of these methods is vulnerable to some form of ambiguity in interpretation. The notion of ambiguity is not, of course, unique to pattern recognition. Every linguist, soothsayer and politician has a repertoire of ambiguous statements. Perceptual psychologists delight in figure-ground illusions that change meaning in the blink of an eye (*e.g.*, the image on the left in Fig. 2). Motivation is notoriously ambiguous: "Is the Mona Lisa smiling?" We do not propose to investigate ambiguity for its own sake, but only insofar as it affects the practical aspects of evaluating document analysis systems.

Evaluation is not the only possible use for ground-truth, obviously. Labeled training data is central to the development of machine learning algorithms that are employed in many graphics recognition systems. While we focus our attention on evaluation in this paper, we note that some (but not necessarily all) of our conclusions have analogs in this other realm as well.

We provide examples, from the literature and from our own experiments, of non-trivial problems with each of the five major constituents of ground-truth. Unless and until they are addressed, these problems appear to preclude the possibility of real progress in evaluating automated graphics recognition systems. For some of them, we can propose potential solutions.

   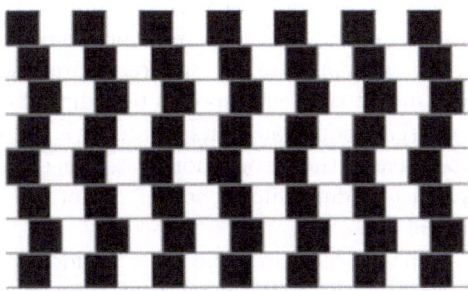

Fig. 2. Two well-known optical illusions. For the image on the left, how would you label figure/ground? For the image on the right, are the horizontal lines parallel or do they slope?

3   Input Format

The natural input format for scanned paper documents is a bilevel, gray-scale, or RGB pixel array at some specified sampling resolution, *e.g.*, 300 or 400 dpi. For electronic documents, there is ample choice. Documents intended for human consumption (printing or display) are often encoded in PostScript (PS) or Portable Document Format (PDF). At a given spatial sampling rate, these are lossless encodings and can be converted to bitmaps. PS and PDF are seldom used as input for document recognition experiments because the mapping to a pixel array is many-to-one.

How much formatting information Hypertext Markup Language (HTML) preserves depends on the specific encoding and on the browser. However, HTML can only represent layout and not semantic content, and is rapidly being replaced by XML. XML is only a meta-language: both the style of rendering and the semantic interpretation require ancillary information.

4   Model

What is a model? Who needs it?

The notion of models derives from the natural sciences, where they provide a concise explanation of observed facts and may predict yet unobserved phenomena. In DIA, we deal entirely with digital artifacts; therefore the models of interest are of a generative nature. Given an appropriate selection of input parameters, the model is expected to produce a digital object of the type being modeled. The model may be deterministic (blackboard systems, schema-based systems, syntactic methods, and procedurally coded rules [3]) or stochastic (Markov, Hidden Markov, stochastic context-free grammars).

Stochastic models can be used to generate synthetic data for experimentation. For instance, appropriately calibrated character defect models can produce multitudes of patterns [2], and relatively simple natural language models can produce multitudes of sentences. More complex models can generate valid addresses, tables [34], and even short stories.

In experimentation with real data, models are used differently. An instance of a model that corresponds to a digital object may be viewed as a compressed representation that omits whatever is not essential for the purpose at hand. The objective of the recognition system is either to recover the model parameters, or at least to determine whether or not the given object is a valid instance of the model.

The model is in some sense an idealized representation of the digital object. Therefore one could argue that imperfections in the object need not be represented in the model. So should we model, instead of the object, the intent of its maker? The perceptions of its observers?

For some automated data entry applications, a suitable model for a text region might be simply a string of character identities from a given alphabet.

This was the model used in the ISRI OCR benchmarks, with the alphabet corresponding to the printable characters of Latin-1 ASCII. This model could not represent bullets and Greek letters, nor two-dimensional symbol clusters like subscripts, superscripts, and mathematical formulas [27]. The Panasonic OCR experiments using the novel *Moby Dick* followed similar conventions [7], which are more appropriate for literary rather than technical material.

Even this simple model raises questions. For example, should typographic errors like "hamnered" or "W00D" be corrected? This would reduce the error rate on natural-language text if the OCR system used language context, but would be quite misleading if the system were intended for correcting galley proofs.

For information retrieval, a better model might be a collection of words from a pre-defined vocabulary. Most full-text search techniques can be implemented on such a model. NIST attempted to mount a large experiment to determine the degree to which "meta-data" (in this instance, additional format information) can improve query response [9, 10, 11], but the necessary experiments, combining OCR and IR, proved too difficult to execute.

Sequences of words are generally insufficient for postal addresses and forms. For such applications, it is desirable to model the types of acceptable fields (street address, delivery date), the permissible contents of fields (which are usually not enumerable), and the constraints between the field contents (*e.g.*, the presence of a city in a state). Such models can often be formalized as a specialized grammar. Note, however, that we would need a different model if (as is often the case) the 11-digit delivery-point zip-code was the recognition target [30].

For page reconstruction, symbol-based models are clearly inadequate. Knuth formulated the side-bearing model for typographic conventions based on pages composed from slugs and rulings inserted into a frame [18]. Alternatively, we may resort to a page-representation language like Rich Text Format (RTF) or proprietary word-processing code. However, the description of a page in any of these formats is not unique, and therefore it is difficult to use such models for page interpretation experiments. ScanSoft developed XDOC, which represents the location and size of the bounding box of each word, typeface, type size, and word transcription [29]. This model does not require character-level segmentation.

Such models allow checking the appearance of a reconstructed page more easily than a pixel-level model because they are insensitive to small changes in line or word location, and to subtle font substitutions. However, they correspond only to the lowest level of representation afforded by word processors, and neglect the logical labeling provided by style sheets. We may know that a one-line paragraph is in a large, bold font, but not that it is a section title. Modeling higher-level logical entities requires something like SGML or ODA, which separate form from function.

For tables, Wang developed an elegant model in the form of a formal data type [33]. The essence of this representation is that items in the cells of a table can be uniquely identified or addressed as the intersection of a set of categories (row and column headers). She found in her experiments that the greatest shortcoming of the model was its inability to deal with footnotes, which are often essential for correct table interpretation. Furthermore, the model does not take into account any ordering (*e.g.*, alphabetical or numerical) of items in a row or column. The model carries no semantics: for instance, in a table of birth and death dates, lifetimes do not have to be positive. Wang's model does, however, have a rendering component that is completely separate from the logical representation.

Early experiments in equation recognition used LATEX [22]. Note that this model is fundamentally different from languages designed for computer algebra such as Maple or Mathematica. The former describes the layout of mathematical symbols, while the latter is directly executable and allows for symbolic manipulations (*e.g.*, simplifying expressions). However, given a suitable parser, LATEX equations can be mapped into Maple or Mathematica, while the transformation in the other direction can be achieved by applying appropriate layout rules.

Models for engineering drawings and circuit diagrams generally have provisions for representing the geometry and connectivity of the primitive entities. The model may also incorporate predefined symbols, as well as line-types and configurations such as dimension sets, center-lines, and hash-marks. Constraints may be introduced for ensuring the compatibility of orthogonal projections. It is, however, risky to restrict the model to valid objects or circuits because there is no guarantee that the original drawing satisfies these constraints. The leftmost two views depicted in Fig. 3 are perfectly legal 2-D drawings, but taken together do they correspond to a real 3-D object? (If so, what is the side view?)

Modeling cartographic maps raises similar problems. Only when there is good reason to have faith in the validity of the documents, as in the case of previously verified cadastral maps [5], should the model include all applicable constraints. It would obviously be futile to use PSPICE to verify the integrity of hand-drawn circuits if the model for digitization did not allow invalid circuits. If, on the other hand, the hand-drawn circuits are believed valid, then PSPICE could improve the conversion accuracy.

According to Blostein and Haken, modeling notational conventions is difficult because such notations are not formally defined but evolve by common

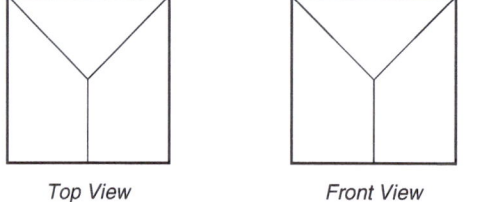

Top View   Front View   Side View

Fig. 3. A problem in line-drawing interpretation. What does the side view look like?

usage. They claim categorically that "there are no ground-truth models of ideal generator output" for diagram generation and recognition [3].

5   Reference Entry System

The data entry system should ideally be capable of supporting any valid instance of the specified model, and enforcing entry of only valid instances. If, for example, the model accommodates superscripts, then it is not enough to be able to enter the superscript above and to the right of the base symbol, but there must also be a way to indicate the asymmetric binary relation between the two symbols. If the model accommodates dot-dash lines, so must the reference entry system. If, in addition, the model is expressive enough to represent such a line as the center line of the projection of a cylinder, then the entry system must also be able to record this relationship.

Taking advantage of the full power of the reference entry system can reduce the ambiguity of having to contend with many different basic components that correspond to the same logical entity. For example, a dashed-line with specified beginning and end points is easier to match than an unordered set of collinear dashes.

In practice, the time required for entering the ground-truth is of paramount importance. With current computer speeds and storage capacities, the effort required to prepare ground-truth is almost always the limiting factor on the size of experiments. The fastest systems for entering ground-truth are seldom those developed for DIA experiments. For most common applications (*e.g.*, text, page layout, drawings, schematic diagrams, formulas, music notation), efficient manual data entry and editing systems were developed long before attempts at automating the process. It is worthwhile, whenever possible, to make use of such legacy systems. (Blostein and Haken recommend that such systems should also be exploited for automated recognition [3].)

Manual data entry systems fall into two categories. The first category, like the early word processors, was designed only to produce pleasing printed output. These systems had no need to represent higher-level logical relationships, and therefore seldom do. The second category, like spreadsheets, symbolic integrators, music editors, and circuit analysis packages, was intended from the start for computer analysis of the data. These systems, in addition to producing printable output, also offer means to represent logical relations and are therefore more suitable for advanced DIA.

Fig. 4 shows a screen snapshot of the Daffy tool we use for entering table ground-truth [14]. Daffy presents the same graphical interface to the user whether working with scanned document images or ASCII text. This particular feature is unusual; typically, the mechanisms used for selecting and manipulating two-dimensional image regions on a page (*e.g.*, in a drawing program) are very different from those used for selecting and manipulating text strings (*e.g.*, in a word processor), which are one-dimensional. Daffy treats all pages, whether encoded in TIF or ASCII, as two-dimensional entities. While this makes it easier

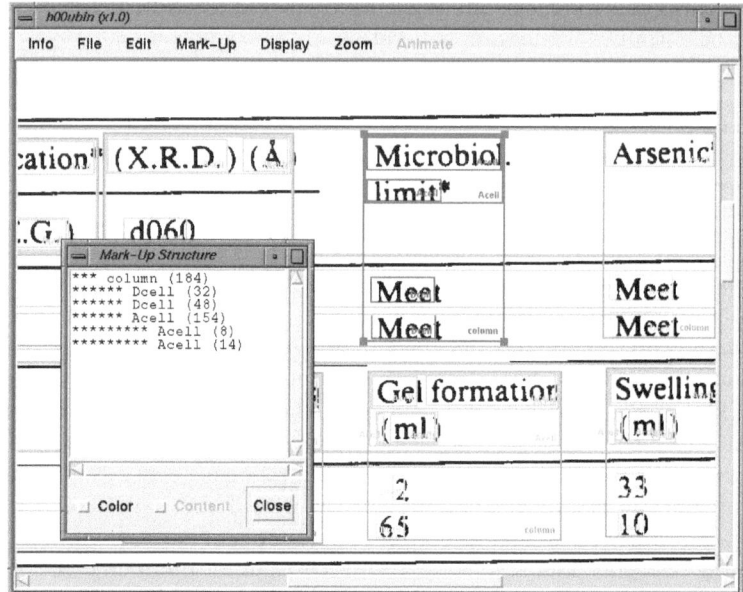

Fig. 4. Screen snapshot of the Daffy tool for ground-truthing tables

to learn the tool and transfer expertise acquired in one domain to the other, it may also tend to encourage a more consistent viewpoint in those preparing the ground-truth. Interpretation ought not to depend on the file format of the input.

Reference entry systems often incorporate automated tools to reduce data entry errors. Examples are spelling and syntax checkers, difference detectors, and consistency verifiers.

6 Truth Format

As we have noted, for basic OCR, strings of alphabetic symbols encoded in a standard format like ASCII or Unicode are adequate. To obtain a unique string representation for each text zone, in the ISRI benchmarks of OCR, blank lines and leading and trailing spaces on a line were eliminated, and consecutive spaces within a line were compressed to a single space [27]. The University of Washington OCR database used TeX escape codes [24], while Bettels used SGML and a European Patent Office (EPO) proprietary character set to encode strings for multilingual patents [32]. Text zones are generally used because the string representation of the text portion of a whole page may not be unique unless the reading order is specified or inferred. The string representation is often modified to carry minimal 2-D information by means of tabs and carriage returns.

There is no widely accepted format for page layout, and every article that we have seen on the subject proposes its own taxonomy and file format for entities above the word level. At DARPA's request, RAF Technology spent several years preparing a page representation language called Illuminator (originally "DAFS") for both OCR ground-truth and output [8], but it did not catch on widely.

Some researchers on tables have established their own format for representing specific aspects of tables of interest. Wang must have developed some file structures for her model. Abu-Tarif proposed using Microsoft Excel [1], while others have suggested a DBMS. However, none of these capture the graphic qualities of tables.

For line drawings, DXF (the Drawing eXchange Format developed by AutoDesk for its AutoCad drafting software) seems to be an appropriate intermediate-level format. DXF files are structured into HEADER, TABLES, BLOCKS, and ENTITIES sections, and are usually partitioned into layers. They contain lists of specifier-value pairs for coordinates, line types, line thicknesses, annotations, dimensioning, viewing angles and distances, methods of curve and surface generation, and coordinate systems. Several vendors offer programs (*e.g.*, CAD-Overlay) that allow interactive digitization of line drawings by overlaying the scanned drawing with the results of operator-issued AutoCAD commands. The fact that DXF and other CAD languages may not be able to capture some aspects of hand-drawn diagrams assumes less and less importance, as most engineering drawings are now prepared using a computer.

7   Personnel

For preparing ground-truth for ordinary OCR, rudimentary literacy, in the usually accepted sense of the term, may not even be necessary. The prevalence of offshore data entry companies suggests that knowledge of English is not a prerequisite for accurate preparation of textual ground-truth. While domain expertise on the part of ground-truthers would always seem to be desirable, there is a potential downside. Experts may indeed make fewer "simple" recognition errors, but at the same time they could introduce new types of discrepancies in higher-level interpretation. The "excess baggage" that comes from understanding a subject deeply could actually prove to be a disadvantage, as suggested in Fig. 5. In some cases, lack of familiarity may actually be preferable.

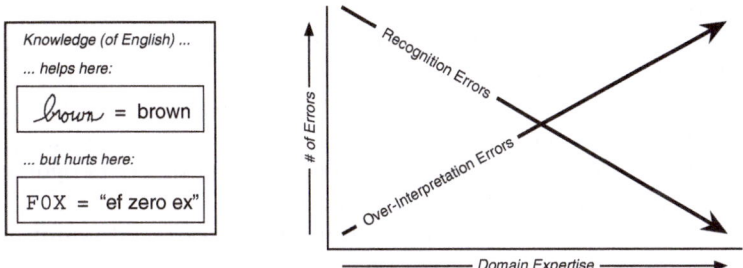

Fig. 5. The expertise effect

Page layout is another matter. Many DIA researchers ignore standard typographic nomenclature, and invent their own entities and terminology. The preparation of ground-truth couched in these terms must therefore be performed by the research team.

For more complex documents, some expertise is required. Even rows and columns are difficult to demarcate in foreign-language tables, and any interpretation is out of the question. The tractability of table interpretation ranges from that of relatively common material comprehensible to a large fraction of the public (*e.g.*, stock-market quotes, baseball statistics, railroad schedules), to complex financial statements, tabulations of physical and chemical properties and experimental results, and reports of statistical tests that can be understood only by specially trained personnel.

To create DXF-formatted ground-truth for engineering drawings, obviously a trained computer draftsman is required. However, an expert draftsman may capture the intent of the drawing rather than what is actually scanned. By way of analogy to the foreign-language data entry operators, perhaps someone familiar with the graphical data entry system, but entirely naive with respect to the specified objects, should enter drawing ground-truth.

8   Intrinsic Ambiguity

Many of the questions we have raised concerning ground-truth have been visited before, directly or indirectly, by researchers who have preceded us. Every researcher who reports experimental results must contend with the matter. In this section we survey briefly some of the larger-scale efforts where the development of ground-truth and, in particular, issues of ambiguity and/or errors in the data have played a primary role.

The well-known University of Washington UW1 CD-ROM database, for example, consists of over one thousand page images made available to the international community by Haralick and Phillips, *et al.* [12, 23, 24, 25]. For the scanned pages in the dataset (as opposed to the synthesized images), the ground-truth was carefully constructed using a double data entry, double reconciliation procedure. In other words, each page was interpreted independently by two different individuals. These data were then compared by two other individuals, each of whom attempted to correct any discrepancies between the two original interpretations. Then, in a final stage, a single person took the results from the two reconcilers and performed a final reconciliation. This care resulted in an error rate estimated to be extremely low (about 29 per million characters) and, as a result, UW1 has become one of the most-used resources in document analysis research. There was, however, a significant cost for this degree of rigor: each page required roughly six person-hours to produce.

It is interesting to note that the UW1 protocol is oriented towards preventing data entry errors and making sure there is a single, unique representation for every entity in the dataset. It does not allow for the possibility of multiple legitimate interpretations; the master reconciler is the final "arbiter." The zoning process, a task likely to lead to disagreement, was also handled by a single individual. Such policies are entirely appropriate for the kinds of applications UW1 was intended for, but may not be easily extensible to ground-truthing for more complex document analysis tasks.

METTREC, the Metadata Text Retrieval Conference, was planned by the National Institute of Standards and Technology (NIST) to evaluate document conversion using OCR and its impact on information retrieval technologies [9, 10, 11]. Over 67,000 pages were scanned from the *Federal Register* for 1994. Since the electronic text for each page was already available, the most labor-intensive parts of the ground-truthing process included scanning the page images, discarding blank pages, and ensuring the correct correspondence between image files and page numbers (so that images could be associated with the proper electronic text). This process was semi-automated by attempting to locate and read the page number field using OCR. The sort of ground-truth needed to evaluate text-based retrieval in this context is fairly basic; the page is treated as a "bag of words" with no real need to represent higher-level structure. Even so, and despite the fact that the basic steps seem relatively straightforward, the sheer quantity of pages made the task challenging.

In their paper describing the TRUEVIZ ground-truthing system, Kanungo *et al.* acknowledge that "a protocol for ground-truthing documents" is key, but this is not one of the issues they address in the work [16]. TRUEVIZ is a portable tool written in Java and uses XML to encode ground-truth markup. It offers multilingual support via Unicode and provides custom input methods to extend Java in this regard. It does not seem to address the potential for alternate interpretations, however.

Miyao and Haralick's paper on ground-truthing for music states that one of the requirements for ground-truth is "...whether or not the meaning of each piece of music can be interpreted correctly and whether or not the music can be correctly converted into playable data" [20]. Their current ground-truth database is small (eight sheets), and the paper does not consider the question of how to evaluate recognition results relative to the truth. Musical notation is, of course, rife with ambiguity – the artist's interpretation is central to a performance – so this would seem to be an application where the notion of a single, well-defined ground-truth may be hard to defend.

Fig. 6 shows a scanned fragment from the "Grosse Fuge" by Ludwig van Beethoven, one of the most demanding and best-known works in the string quartet literature. As every student of elementary music theory will know, two eighth notes of identical pitch that have been tied together are equivalent to a single quarter note of the same pitch. In nearly every instance, such a substitution would be considered acceptable for a ground-truther, data entry tool, or recognition algorithm (indeed, it may even be preferred because it simplifies the notation). Not here, however.

A discussion with members of the Guarneri String Quartet is telling in this regard, and highlights the complexity of the relationship between what is written on the page and interpretation ([4], page 70):

Fig. 6. A snippet from Beethoven's "Grosse Fuge" for String Quartet, Op. 133

> **Blum (interviewer):** "Another special notation in Beethoven, but this time far more enigmatic, is the slur he places over notes of identical pitch and of equal time value, as found, for instance, in the Grosse Fuge ... How do you interpret this marking?
>
> **Tree (violist):** Now we're getting into something very delicate. We all agree that something should be done, but not on what should be done. Interpreters of Beethoven have struggled with this question for more than a century and a half.
>
> **Dalley (2nd violinist):** Beethoven seems to want something between one note and two notes, possibly a double pulsation. I would prefer to make these pulsations with the vibrato, because it's hard to keep from doing too much if you make two impulses with the bow.
>
> **Steinhardt (1st violinist):** I've experimented with different possibilities. One idea is to make a kind of swell from the first note to the second ... Another idea ... is to make the second note an echo of the first ..."

None of these interpretations, however, are evident on the surface of the music; they only arise because we are aware of what Beethoven *could* have written (quarter notes) but chose not to. To the extent that ground-truthing is intended to support document understanding (as opposed to conversion) research, issues such as these are likely to pose a tremendous challenge.

We have uncovered a number of papers that are beginning to recognize intrinsic ambiguity and its impact on the truthing process. In explaining problems in ground-truthing for the classification of audio segments, Srinivasan *et al.* observe [31]: "Even the ground truth definition can be subjective, for example, a clear speech segment followed by mixed audio which includes speech tends to be classified as a single speech segment. Analogously, a clear music segment followed by mixed audio which includes music tends to be classified as a single music segment." Generally speaking, the indexing information that is required to group objects is more sensitive than content information that affects "only" the specific interpretation of an object.

In a paper describing the Pink Panther system for ground-truthing and benchmarking page segmentation, Yanikoglu and Vincent state [35]:

> "Creating segmentation ground-truth is not always straightforward. A ground-truthing handbook was therefore written in which potentially ambiguous

cases were described, in order to take as much subjectivity out of the ground-truthing process as possible. One of the decisions we have faced was the necessary level of detail; we have decided that, within limits, zones should be logically and structurally uniform, and that paragraphs were the smallest units of interest of regular text …

Furthermore, in order to remove the main source of ground-truthing ambiguity, which is deciding on the label (subtype) of a region, we have expanded GroundsKeeper to allow multiple region labels for a given region. However, the benchmarking algorithm currently uses only the first label of a region."

Even though they are addressing a problem from a fundamentally different domain, natural scene images versus scanned document pages, Martin *et al.* face exactly the same kinds of questions regarding interpretation, ambiguity, and perceptual refinement [19]. If the output from document analysis is intended to be what a human would perceive when viewing the document (regardless of the way it is encoded), then perhaps there is not such a large difference between documents and natural scenes.

So far as we are aware, Martin and his colleagues are among the first to examine variability in the interpretations registered by different ground-truthers. While there seems on the surface to be much disagreement, they argue that most of this is due to perceptual refinement and therefore not significant. They describe an evaluation measure that can factor out such discrepancies, and conclude that there is indeed consistency between human ground-truthers, at least for the application in question [19].

Recall, however, the inherent ambiguity of the image on the left in Fig. 2. While ground-truthers may very well agree on the locations of the edges in this image, could they ever agree on a labeling for the regions (without being coerced in some way)?

In yet another domain, medical image processing, Ratiu and Kikinis describe four different levels of "gold standard" for evaluating segmentation techniques for cross-section data from the Visible Human Project [26]. As do Martin *et al.*, they acknowledge inherent ambiguity; in their "8 karat" standard, multiple trained specialists label a given image. An algorithm is judged to have passed if its results "fall within the limits … either of the individual labelmaps, or within the average thereof." Their "14 karat" standard allows for different degrees of confidence on the part of the labelers, as well as negative annotations ("I don't know what this is, but I know what it isn't."). The remaining two standards are perhaps less interesting from our perspective as they deal with creating the ground-truth by working from a higher-resolution version of the data than that which is presented to the recognition system.

Returning to the "easiest" problem we face in document analysis, the interpretation of character patterns, some interesting (and entertaining) examples of ambiguity can be found in Chapter 3 of the book by Rice *et al.* [28].

9   Issues and Open Questions

In this section we attempt to collect together some of the questions we have raised throughout the earlier parts of this work, in addition to some new ones.[2] Our intention is that these might serve as the basis for future research investigations, or at least an interesting debate.

Is it sufficient to have just one version of the ground-truth when the input admits more than one interpretation? Provided the designated ground-truth is acceptable, this could never help a bad algorithm look better, but it could unduly penalize a good algorithm. If, however, there is debate as to whether the ground-truth is correct, that could be a problem. Who is to make this decision? (Do we need a fair and impartial "jury" for ground-truths?)

What kinds of experiments could be used to confirm that a proffered ground-truth is appropriate for the task at hand? In other words, how can one tell whether uncertainties in the ground-truth will have a significant effect on the evaluation of an automated recognition system?

What is the impact of the mismatch between human interpretation, what the entry tool can represent, and what the algorithm is capable of recognizing? Consider the case of circuit diagrams and whether or not electrical continuity between components is required. In attempting to ground-truth a circuit that appears to be disconnected (*e.g.*, Fig. 7), there are eight possible scenarios as enumerated in Table 1. Note that most of these introduce the potential for undesirable side-effects.

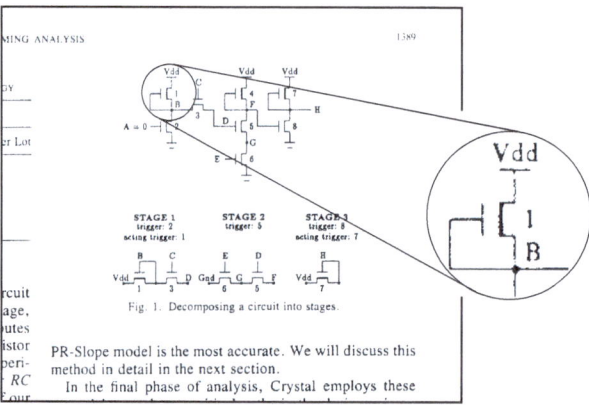

Fig. 7. A problem in circuit diagram recognition. Is the transistor connected to Vdd?

---

[2] We admit readily that some of the notions that have been presented are an attempt to generalize to graphics the lessons that we have learned from text and table interpretation.

Table 1. Interaction between interpretation, entry system capabilities, and algorithm

Electrical continuity required in circuit model employed by ...			
Human Truther	Entry System	Algorithm	Impact
no	no	no	no problem
no	no	yes	algorithm penalized
no	yes	no	human frustrated
no	yes	yes	human frustrated
yes	no	no	algorithm penalized
yes	no	yes	no problem
yes	yes	no	algorithm penalized
yes	yes	yes	no problem

Context clearly plays an important role in human interpretation. What is the minimal context necessary for a ground-truthing task? What is the relationship between context-dependent interpretation and domain expertise? How does expectation fit in? For example, consider showing a ground-truther the following sequence of words:

CAT ... MEW ... DOG ... ARF ... COW ... MOO

He/she will certainly record that the final word is "moo," even though it was written above using two zeros instead of "ohs."

On the other hand, if we show the same ground-truther the following sequence:

C65 ... M32 ... D03 ... A23 ... C77 ... MOO

there is a much better chance he/she will identify the zeros as zeros.

Is the following position, then, correct? The system is right when it arrives at an interpretation that some human might reasonably arrive at. It is wrong when it arrives at an interpretation that no human would arrive at. This is opposed to the current paradigm which states that the system is right when it arrives at the same interpretation that a specific individual arrived at when he/she created the ground-truth.

The issue of interpretation is surely central in ground-truth, since that is the final goal of DIA. Since DIA is generally less powerful than a human interpreter, it is likely to result in more ambiguity. One operational question, then, is whether the ground-truther should also allow for some degree of ambiguity, or whether the evaluation method should take into account ambiguous automated interpretation versus clear human interpretation. At the very least, perhaps the ground-truth, like the output of many automated recognition systems, should include some confidence measure that can be used in the evaluation criteria.

Do elements of the ground-truth have to be labeled with regard to their importance in some intrinsic hierarchy? Since the relative importance of errors depends on the ultimate use of the document analysis results, this suggests that ground-truth is application dependent. This will be increasingly the case as higher level information, containing less redundancy, is extracted from the document. For example, it may be less costly to mistake a dashed-line-crossing in a drawing for a plus sign than to label a cross-section as "drive-chain detail" instead of "brake assembly." We have barely touched on the issue of the metric used to compare the automated recognition result with the ground-truth(s), and on the cost-accounting for errors.

It appears that with ground-truth we have a similar situation to the recognition of unsegmented objects. We cannot do the segmentation (layout analysis) without interpreting the objects, and it may be hard to do the interpretation without segmentation. However, humans can interwine the two easily. Low-level ground-truth can be subject to ambiguities that may vanish when we consider high-level, database-type truth. In other words, creating ground-truth for intermediate results may be more difficult than for end-to-end evaluation.

10   Recommendations

In the absence of a theoretical framework, sound and objective evaluation of experimental evidence is essential for evaluating new (and old) approaches to the recognition of graphics documents. In evaluation, the nature and quality of the ground-truth are often the limiting factors. We offer the recommendations below in the hope of inviting further thought and research on the subject.

General-purpose data-entry tools are preferable to specialized systems that favor the experimenters' interpretations.

For structure recognition, and perhaps even for some instances of object recognition, it is better to have multiple versions of the ground-truth, with the algorithm allowed to match any version, than to require an arbiter to decide arbitrarily on a single "correct" interpretation.

If possible, consider a layered approach, where specific, low-level ground-truth is used to check intermediate results, but high-level functional ground-truth is used for overall system evaluation. The criteria for ground-truth used for evaluation can be markedly different from that of the labels necessary to build or train a model for classification.

The consistency of several versions of the ground-truth for a particular set of inputs could be determined by comparing them using the same metric as will be used to evaluate the algorithm. In the usual case, where human performance is superior to machine performance, we would expect the differences between human versions of the ground-truth to be less than those between the algorithm and any human version. If, however, there are multiple legitimate interpretations, such an analysis should reveal this fact.

It is important to avoid mismatched constraints between instructions to human ground-truthers, the capabilities of the data entry system, and what the algorithm is designed to do.

The role of context must be carefully controlled. Either the machine should be designed to exploit such context, or the context should be concealed from the human ground-truthers. The design of ground-truth collection requires as much care as the design of the automated parts of the evaluation.

Most applications tolerate some degree of ambiguity. The degree of latitude allowed for human ground-truthers should be reflected by a similar tolerance for the machine output.

The relative importance of different types of mistakes is often application-dependent. Bayesian analysis takes this into account through the formulation of a cost function. Such cost functions are more appropriate for evaluation than simple error counts. Ideally, the cost function must also take into account dependencies. (By analogy, any teacher knows that students protest vocally when they are penalized twice for a mistake made in a lower-level or antecedent task.)

Most importantly, perhaps, keep in mind that carefully collected data, with accurate and comprehensive ground-truth, has often proved more valuable to the research community than the specific method or algorithm that it was originally destined to support.

Acknowledgement

The authors would like to thank the anonymous reviewers as well as Asad Abu-Tarif for their careful reading of the initial draft of this paper and for their helpful suggestions for improving it.

References

[1] A. A. Abu-Tarif. Table processing and understanding. Master's thesis, Rensselaer Polytechnic Institute, 1998. 55

[2] H. S. Baird. Document image defect models. In H. S. Baird, H. Bunke, and K. Yamamoto, editors, *Structured Document Image Analysis*, pages 546–556. Springer-Verlag, New York, NY, 1992. 50

[3] D. Blostein and L. Haken. Using diagram generation software to improve diagram recognition: A case study of music notation. *IEEE Transactions on Pattern Analysis and Machine Intelligence*, PAMI-21(11):1121–1136, November 1999. 50, 53

[4] D. Blum. *The Art of Quartet Playing*. Cornell University Press, Ithaca, NY, 1986. 57

[5] L. Boatto, V. Consorti, M. D. Bueno, S. D. Zenzo, V. Eramo, A. Esposito, F. Melcarne, M. Meucci, A. Morelli, M. Mosciatti, S. Scarci, and M. Tucci. An interpretation system for land register maps. *IEEE Computer*, 25(7):25–33, July 1992. 52

[6] A. K. Chhabra and I. Phillips. The Second International Graphics Recognition Contest – raster to vector conversion: A report. In K. Tombre and A. K. Chhabra, editors, *Graphics Recognition: Algorithms and Systems*, volume 1389 of *Lecture Notes in Computer Science*, pages 390–410. Springer-Verlag, Berlin, Germany, 1998. 48

[7] J. Esakov, D. P. Lopresti, J. S. Sandberg, and J. Zhou. Issues in automatic OCR error classification. In *Proceedings of the Third Annual Symposium on Document Analysis and Information Retrieval*, pages 401–412, Las Vegas, NV, April 1994. 51

[8] T. Fruchterman. DAFS: A standard for document and image understanding. In *Proceedings of the Symposium on Document Image Understanding Technology*, pages 94–100, Bowie, MD, October 1995. 47, 55

[9] M. D. Garris. Document image recognition and retrieval: Where are we? In *Proceedings of Document Recognition and Retrieval VI (IS&T/SPIE Electronic Imaging)*, volume 3651, pages 141–150, San Jose, CA, January 1999. 51, 57

[10] M. D. Garris, S. A. Janet, and W. W. Klein. Federal Register document image database. In *Proceedings of Document Recognition and Retrieval VI (IS&T/SPIE Electronic Imaging)*, volume 3651, pages 97–108, San Jose, CA, January 1999. 51, 57

[11] M. D. Garris and W. W. Klein. Creating and validating a large image database for METTREC. Technical Report NISTIR 6090, National Institute of Standards and Technology, January 1998. 51, 57

[12] J. Ha, R. M. Haralick, S. Chen, and I. T. Phillips. Estimating errors in document databases. In *Proceedings of the Third Annual Symposium on Document Analysis and Information Retrieval*, pages 435–459, Las Vegas, NV, April 1994. 56

[13] J. D. Hobby. Matching document images with ground truth. *International Journal on Document Analysis and Recognition*, 1(1):52–61, 1998. 48

[14] J. Hu, R. Kashi, D. Lopresti, G. Nagy, and G. Wilfong. Why table ground-truthing is hard. In *Proceedings of the Sixth International Conference on Document Analysis and Recognition*, pages 129–133, Seattle, WA, September 2001. 46, 47, 53

[15] T. Kanungo and R. M. Haralick. An automatic closed-loop methodology for generating character groundtruth for scanned documents. *IEEE Transactions on Pattern Analysis and Machine Intelligence*, PAMI-21(2):179–183, February 1999. 48

[16] T. Kanungo, C. H. Lee, J. Czorapinski, and I. Bella. TRUEVIZ: a groundtruth / metadata editing and visualizing toolkit for OCR. In *Proceedings of Document Recognition and Retrieval VIII (IS&T/SPIE Electronic Imaging)*, volume 4307, pages 1–12, San Jose, CA, January 2001. 47, 57

[17] D.-W. Kim and T. Kanungo. A point matching algorithm for automatic generation of groundtruth for document images. In *Proceedings of the Fourth IAPR International Workshop on Document Analysis Systems*, pages 475–485, Rio de Janeiro, Brazil, December 2000. 48

[18] D. Knuth. *The Texbook*. Addison-Wesley, 1984. 51

[19] D. Martin, C. Fowlkes, D. Tal, and J. Malik. A database of human segmented natural images and its application to evaluating segmentation algorithms and measuring ecological statistics. In *Proceedings of the International Conference on Computer Vision (ICCV)*, pages II:416–421, Vancouver, Canada, July 2001. 59

[20] H. Miyao and R. M. Haralick. Format of ground truth data used in the evaluation of the results of an optical music recognition system. In *Proceedings of the Fourth IAPR International Workshop on Document Analysis Systems*, pages 497–506, Rio de Janeiro, Brazil, December 2000. 57

[21] G. Nagy and S. Seth. Hierarchical representation of optically scanned documents. In *Proceedings of the Seventh International Conference on Pattern Recognition*, pages 347–349, Montréal, Canada, July 1984. 47

[22] M. Okamoto and A. Miyazawa. An experimental implementation of a document recognition system for papers containing mathematical expressions. In H. S. Baird, H. Bunke, and K. Yamamoto, editors, *Structured Document Image Analysis*. Springer-Verlag, Berlin, Germany, 1992. 52

[23] I. Phillips, J. Ha, R. Haralick, and D. Dori. The implementation methodology for the CD-ROM English document database. In *Proceedings of Second International Conference on Document Analysis and Recognition*, pages 484–487, Tsukuba Science City, Japan, October 1993. 56

[24] I. T. Phillips, S. Chen, J. Ha, and R. M. Haralick. English document database design and implementation methodology. In *Proceedings of the Second Annual Symposium on Document Analysis and Information Retrieval*, pages 65–104, Las Vegas, NV, April 1993. 54, 56

[25] I. T. Phillips, J. Ha, S. Chen, and R. M. Haralick. Implementation methodology and error analysis for the CD-ROM English document database. In *Proceedings of the AIPR Workshop*, Washington DC, October 1993. 56

[26] P. Ratiu and R. Kikinis. Squaring the circle: Validation without ground truth. In *Proceedings of the Third Visible Human Project Conference*, Bethesda, MD, October 2000.
http://www.nlm.nih.gov/research/visible/vhpconf2000/AUTHORS/RATIU2/RATIU2.HTM. 59

[27] S. V. Rice, J. Kanai, and T. A. Nartker. Preparing OCR test data. Technical Report TR-93-08, UNLV Information Science Research Institute, Las Vegas, NV, June 1993. 51, 54

[28] S. V. Rice, G. Nagy, and T. A. Nartker. *Optical Character Recognition: An Illustrated Guide to the Frontier*. Kluwer Academic Publishers, Norwell, MA, 1999. 59

[29] ScanSoft, Inc., Peabody, MA. *XDOC Data Format, Technical Specification Version 3.0*, May 1997. 51

[30] S. Setlur, V. Govindaraju, and S. Srihari. Truthing, testing and evaluation issues in complex systems. In *Proceedings of the Symposium on Document Image Understanding Technology*, pages 131–140, Annapolis, MD, April 2001. 51

[31] S. Srinivasan, D. Petkovic, and D. Ponceleon. Towards robust features for classifying audio in the CueVideo system. In *Proceedings of ACM Multimedia '99*, pages 393–400, Orlando FL, November 1999. 58

[32] H. R. Stabler. Experiences with high-volume, high accuracy document capture. In A. L. Spitz and A. Dengel, editors, *Document Analysis Systems*, pages 38–51. World Scientific, Singapore, 1995. 54

[33] X. Wang. *Tabular abstraction, editing, and formatting*. PhD thesis, University of Waterloo, 1996. 51

[34] Y. Wang, I. T. Phillips, and R. Haralick. Automatic table ground truth generation and a background-analysis-based table structure extraction method. In *Proceedings of the Sixth International Conference on Document Analysis and Recognition*, pages 528–532, Seattle, WA, September 2001. 50

[35] B. A. Yanikoglu and L. Vincent. Pink Panther: a complete environment for ground-truthing and benchmarking document page segmentation. *Pattern Recognition*, 31(9):1191–1204, 1998. 47, 58

[36] T. S. Yoo, M. J. Ackerman, and M. Vannier. Towards a common validation methodology for segmentation and registration algorithms. In S. Delp, A. DiGioia, and B. Jaramaz, editors, *Medical Image Computing and Computer-Assisted Intervention*, volume 1935 of *Lecture Notes in Computer Science*, pages 422–431. Springer-Verlag, Berlin, Germany, 2000. 49

A   Table of Acronyms

Table 2. Acronyms used in this paper

Acronym	Interpretation
CAD	Computer-Aided Design (or Drafting)
CEDAR	Center of Excellence for Document Analysis and Recognition
DAFS	Document Format Attribute Specifications
DARPA	Defense Advanced Research Projects Agency
DBMS	DataBase Management System
DIA	Document Image Analysis
DXF	an AutoCad file format
EPO	European Patent Office
GREC	International Workshop on Graphics Recognition
HTML	HyperText Markup Language
IR	Information Retrieval
ISRI	Information Science Research Institute
METTREC	Metadata Text Retrieval Conference
NIST	National Institute of Standards and Technology
OCR	Optical Character Recognition
ODA	Office Document Architecture
PDF	Portable Document Format
PS	PostScript
(P)SPICE	Simulation Program with Integrated Circuit Emphasis
RTF	Rich Text Format
TIF	Tagged Image Format
SGML	Standard Generalized Markup Language
UW	University of Washington
XDOC	a ScanSoft text output format
XML	eXtensible Markup Language

Sketch-Based User Interface
for Inputting Graphic Objects
on Small Screen Devices

Liu Wenyin[1], Xiangyu Jin[2], and Zhengxing Sun[2]

[1] Department of Computer Science, City University of Hong Kong
Tat Chee Avenue, Kowloon, Hong Kong SAR, China
csliuwy@cityu.edu.hk
[2] State Key Laboratory for Novel Software Technology
Nanjing University, Nanjing 210093, PR China
jxy@graphics.nju.edu.cn

Abstract. For small screen devices, such as PDAs, which totally depend on a pen-based user interface, traditional menu-selection/button-clicking based user interface becomes inconvenient for graphics inputting. In this paper, a novel sketch-based graphics inputting user interface is presented. By sketching a few constituent primitive shapes of the user-intended graphic object, the candidate graphic objects in the shape database are guessed and displayed in a ranked list according to their partial structural similarity to what the user has drawn. The user can then choose the right one from the list and replace the sketchy strokes with the exact graphic object with proper parameters, such as position, size and angle. This user interface is natural for graphics input and is especially suitable for schematic design.

1   Introduction

Currently, most graphics input/editing systems, including Microsoft Office, Photo-Draw, Visio, and many CAD systems, ask users to input graphic objects using mouse/keyboard with lots of toolbar buttons or menu items for selection. This clumsy user interface is inconvenient when drawing graphic objects on small screen devices, such as PDAs, for there is no room to accommodate so many toolbar buttons and menu items. The most convenient and natural way for human beings to draw graphics is via a pen-based interface, that is, using pens to draw sketches, just like on a real sheet of paper. Moreover, it is even better to recognize and convert the sketchy curves drawn by a user to rigid and regular shapes immediately. This is because, with the on-line graphics recognition as an immediate and useful feedback, the user can realize errors or inappropriateness earlier and therefore draw diagrams more perfectly.

In this paper, we discuss the issues of inputting composite graphic objects using the sketch-based user interface. When drawing a pre-defined graphic object, the user

D. Blostein and Y.-B. Kwon (Eds.): GREC 2002, LNCS 2390, pp. 67–80, 2002.

tends to divide the graphic objects into several primitive shapes, which are usually convex polygons (users usually do not regard a concave shape as an entire shape and tend to input them in several different parts) or ellipses. The user can input a primitive shape in a single-stroke or in several consecutive strokes. In our proposed user interface, we discover latent primitive shapes among user-drawn strokes and show the regularized shape on the screen immediately. Users can then adjust the recognition and regularization results by the suggestion of the system. This immediate feedback strategy makes the user interaction smoother and humanistic. Moreover, it has an extra advantage to reduce inner-stroke and inter-stroke noises, which are usually introduced by the non-proficiency or un-professionalism of the user.

After being recognized and regularized, primitive shapes, which belong to one graphic object, are grouped together according to their spatial and temporal relationships. They are then segmented and combined to form an object skeleton. Partial structural similarities are calculated between the user-drawn graphic object and the candidates in the database. The graphic objects that are the most similar to what the user has drawn are suggested to the user in a ranked list for the user to choose and confirm. In this way, the user does not need to manually go to menus to find and select what he/she wants. Hence, it is a more efficient, convenient, and natural way to input composite graphic objects.

We have implemented this user interface for composite graphic object inputting in our on-line graphics recognition system—Smart Sketchpad [1]. Experiments have shown the effectiveness and efficiency of the shape similarity measurement and the naturalness and convenience of the user interface.

2   Related Works

Very few research works have been done on such on-line graphics recognition. Zeleznik et al. [2] have invented an interface to input 3D sketchy shapes by recognizing the defined patterns of some input 2D graphics that correspond to certain sketchy solid shapes. Some sketchy input interfaces such as the one developed by Gross and Do [3], are mainly for conceptual and creative layout designs. Fonseca and Jorge and their group [4][5] have implemented an on-line graphics recognition tool that can recognize several classes of simple shapes. But their recognition approach is based on global area calculation. This simple rule can hardly distinguish ambiguous shapes such as pentagon and hexagon and therefore cannot achieve high recognition precision generally. The sketch recognition work reported by Arvo and Novins [6] is mainly for 2D on-line graphics recognition. Their approach continuously morphs the sketchy curve to the predicted shape while the user is drawing the curve. However, their main purpose focuses on the user studies of such sketch recognition system. Moreover, their recognition approach only handles two simplest classes of shapes (circles and rectangles) drawn in single strokes. This is not adequate for a real software tool that can be used for inputting most classes of diagrams. Anyway, their preliminary conclusion shows that such interface is particularly effective for rapidly constructing diagrams consisting of simple shapes. Hence, in order to provide the capability to input more complex diagrams, it is necessary to extend the sketchy graphics recognition approach to handle more complex and composite shapes.

3   The Proposed Approach

The minimum input unit of a graphic object in an on-line graphics recognition system is a stroke, which is the trajectory of the pen movement on a tablet with the pen-tip touching the tablet between the time when the pen-tip begins to touch the tablet and the time when the pen-tip was lifted up from the tablet. Several consecutive strokes can constitute a primitive shape, which can be a straight-line segment, an arc segment, a polygon, or an ellipse, etc. Composite graphic objects are more complex shapes that consist of several primitive shapes. The objectivity of on-line graphic object recognition is to convert the input strokes into the user-intended graphic objects. In our approach, we first recognize the input strokes into primitive shapes. The latest input primitive shapes (based on the assumption that the component of the same object must be inputted consecutively) are then grouped together according to their location vicinity as a query example. A content-based graphics retrieval procedure then starts in the object database based on partial structural similarities between the query example and those predefined composite graphic objects in the database. The most similar composite graphic objects are displayed in a ranked list for the user to choose. If the user cannot find his/her intended objects, he/she can continue to draw other components until the intended shape appears in the list and can be dragged and dropped to the drawing area. In order to avoid too much intrusive suggestions of composite shapes of low similarity, only those candidates whose similarities are above a threshold will be displayed for suggestions. Relevance feedback based on the component's input sequence is also employed to raise the system's performance.

There are four major stages in our proposed approach: primitive shape recognition, primitive shapes grouping, partial structural similarity calculation, and relevance feedback. We will discuss them in detail in the following subsections.

3.1   Primitive Shape Recognition

The first stage of our approach is primitive shape recognition. In this stage, an input sketchy line is first recognized as a known primitive shape type (which can be a line segment, an arc segment, a triangle, a quadrangle, a pentagon, a hexagon, or an ellipse). The recognized shape is then regularized to the most similar rigid one that the user might intend to draw. The entire primitive shape recognition process is divided into four sub-processes: pre-processing, feature extraction, closed-shape recognition, and shape regularization.

3.1.1   Pre-processing

Due to non-proficiency or un-professionalism, the sketchy line for an intended shape input is usually very cursive/unshaped and free-formed. For example, without using a ruler, a straight line drawn by a drafter is not so straight if measured strictly no matter how much attention the drafter is paying to the drawing operation. More often, the sketchy line is not properly closed. Hence, the sketchy line is not suitable for feature extraction directly. Pre-processing is first done to reduce all kinds of noises. The pre-processing stage includes four sub-processes: polygonal approximation, agglomerate points filtering, end point refinement, and convex hull calculating. Many intermediate points on the sketchy line are redundant because they lie (approximately) on the straight-line segment formed by connecting their neighbours. These points can be removed from the chain so that the sketchy line can be approximately represented by a polyline (an open polygon) with much fewer critical vertices. We apply the algorithm developed by Sklansky and Gonzalez [7] to implement polygonal approximation in this paper.

Due to the shaky operations caused when the pen-tip touches the tablet and when it is lifted up, there are often some hooklet-like segments at the ends of the sketchy lines. There might also be some circlet at the turning corner of the sketchy line. These noises usually remain after polygonal approximation. Agglomerate points filtering process is introduced to reduce these noises. The main idea of this process lies in the difference of point density. Polyline segments, which have a hooklet or circlet, usually have much higher point density than the average value of the whole polyline. The task of agglomerate points filtering is to find such segments and use fewer points to represent the segment.

Because it is difficult for the user to draw a perfectly closed shape, the sketchy line is usually not closed or forms a cross near its endpoints. In other words, it has improper endpoints. These improper endpoints are great barriers for both correct shape recognition and well regularization. For a sketchy line that has cross endpoints, we delete its extra points to make it properly closed. For a sketchy line that is not closed, we extend its endpoints along its end directions and make it closed. After that it can undergo other processing as if it were previously closed.

The sketchy line the user draws is often very cursive, and might also be concave. These noises have strong impact on the later feature extraction stage. We employ the classical algorithm developed by Graham [8] to obtain the convex hull of the vertices set, which is used to represent its original line and therefore remove those noises. Experimental results show that using convex hull instead of the original input stroke helps to raise the precision of primitive shape recognition in general cases, although it might introduce extra noises.

After pre-processing, line segments and arc segments can be distinguished from other closed-shapes by intuitive rules. A closed-shape is then represented by a polygon and needs further classification into more elaborate types (including triangle, quadrangle, pentagon, hexagon, and ellipse). Feature extraction and closed-shape recognition are key processes to fulfill this objective.

3.1.2   Feature Extraction and Closed-Shape Recognition

We regard all polygons having the same vertex number as the same shape type. For instance, diamonds, squares, rectangles, trapezoids, and parallelograms are all regarded as quadrangles. Hence, the feature we used for recognition must be irrelevant to the polygon's size, position, and rotation. Thereby, we employ the turning function [9] to obtain the feature vector of the polygon representation of the sketchy stroke.

Turning function $\Theta_A(s)$ measures the angle of the counter-clockwise tangent as a function of the arc-length s, staring from a reference point O on a polygon A's boundary. Thus $\Theta_A(0)$ is the angle v of the tangent at O from the x-axis, as in Fig.1. $\Theta_A(s)$ accumulates the turning angles (which is positive if the turning is left-hand and negative otherwise) as s increases.

Our definition of turning function is a little different from the commonly used one [9]. We see that the previous definition is dependent on both the traversal direction of the polygon and the reference orientation. We scaled the polygon so that its perimeter length is 1. We use the tangent at O as the reference orientation such that $\Theta_A(0)=0$ and determine a traversing direction such that all turning angles are positive. Hence, our turning function Θ'_A is a monotonous increasing one from [0, 1] into [0, 2π].

If an m-dimensional feature vector is needed, we equally divide the boundary of the polygon into m pieces (m=20 in this paper). Each element of the feature vector is the turning degrees accumulated in a corresponding piece of the polygon, as is defined in Eq. (1).

$$V_i = \begin{cases} \varphi_i & if \quad \varphi_i \geq 0 \\ \varphi_i + 2\pi & if \quad \varphi_i < 0 \end{cases} \tag{1}$$

where $\varphi_i = \Theta'_A(\dfrac{i \bmod m}{m}) - \Theta'_A(\dfrac{i-1}{m})$.

After feature extraction, we classify closed-shapes into more elaborate primitive shape types. A multi-class classifier based on support vector machines [10] is used to assign each input stroke to its corresponding primitive shape type.

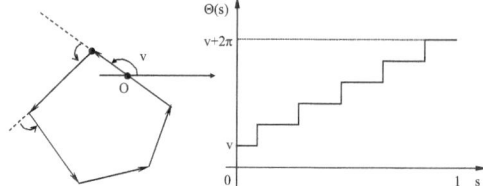

Fig.1. A polygon and its turning function

3.1.3 Shape Regularization

After shape type is known, fitting is employed to adjust the shape parameters (e.g., three vertices for a triangle) such that the recognized shape can best fit its original closed curve. We employ two basic types of fitting processes: ellipse fitting and polygonal fitting. For ellipse fitting, we first determine its axes orientations by finding the eigenvectors of the covariance matrix of the sampling points along the original curve at equi-length steps. The stroke's enclosing rectangle whose edges are parallel to the axes is used to determine the centre and axes lengths of the ellipse. For N-edge polygonal fitting, we first find its inscribed N-edge polygon that has the maximal area. As a result, we cut the original stroke into N pieces by the vertices of this inscribed polygon. By finding the linear regression result of each piece, the edges of the optimized N-edge polygon can be obtained.

The input shape that the user has drawn cannot precisely match the one he/her intends to input. Rectification process is employed to make the shape very similar to the one that the user has in mind. This process is currently rule based, including two sub-processes called vertical/horizontal rectification and rigid shape rectification. The shape is gradually regularized to a rigid shape as regular as possible following the arrowheads in Fig.2.

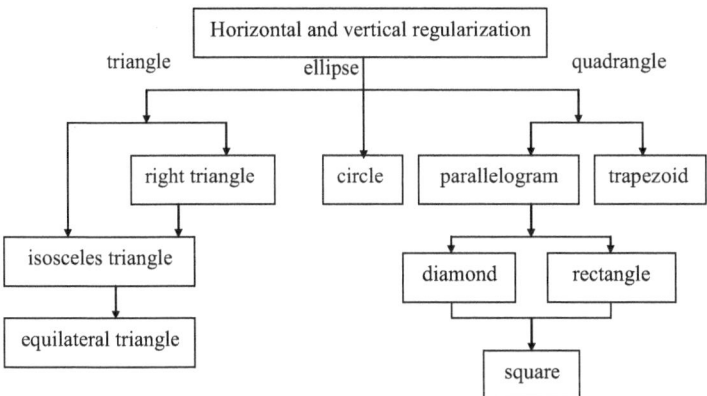

Fig.2. Shape rectification

3.2   Primitive Shapes Grouping

After primitive shape recognition, we group together the recognized and regularized primitive shapes that the user has already drawn to guess his/her intended composite graphic object. Based on the assumption that the user tends to draw an object in consecutive strokes, we only group together the latest drawn primitive shapes according to their location vicinity.

Denote a group as SG, which consists of k primitive shapes $(S_1, S_2, ... S_k)$. Denote ψ as a function to get the center point of the bounding box of a shape or a shape group. Define $Dis(x, y)$ as a function to get the distance between two points. We have the following definitions:

$$d = \max_i (Dis(\psi(SG), \psi(s_i)))$$ (2)

$$\sigma = \frac{d}{\sum_{i=1}^{k} Dis(\psi(SG), \psi(s_i))}$$ (3)

where d represents the compactness of components. The smaller d is, the more compact these components are. σ represents the imbalance between different components. The larger σ is, the larger the possibility that a component does not belong to this object. At initialization, the shape group $SG=\phi$. Then, we add primitive shapes from the latest drawn one and follow the reverse input sequence until one of the four conditions is met:

1. No ungrouped shapes available,
2. $d > d_{max}$,
3. $\sigma > \sigma_{max}$, or
4. The bounding box of SG is bigger than a given threshold.

Thus the ultimate SG is the shape group we guess for the user-intended shape. Fig.3 illustrates an example of primitive shapes grouping. The red line segment is the latest input shape, and it can be grouped together with two quadrangles. The circle cannot be grouped in because it would make the combined shape imbalanced (the σ value would exceed the threshold).

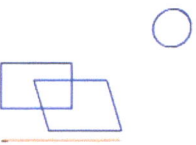

Fig.3. Primitive shapes grouping

3.3   Partial Structural Similarity Calculation

After primitive shape grouping, the combined shape group can also be regarded as a composite object. This object (the source object) and each candidate object (the destination object) in the database are then compared for similarity assessment. Since the inputted shape might be in an incomplete form, the similarity is not symmetric between the source and destination objects. It is a "partial" similarity measured from the source to the destination objects. For example, if the source object (an incomplete object) is a part of the destination object (a complete object), we think they are highly similar because the source object can be completed later. However, if the source object contains certain components that do not exist in the destination object, we think their similarity should be very low, no matter how similar other corresponding parts are.

First, we normalize the source object and the candidate object into a square of 100*100 pixels. Then we calculate the partial structural similarity in two stages. In the first stage, primitive similarity is only measured based on matched primitive shapes. In the second stage, these matched primitive shapes are removed from both the source and the designation objects. More elaborate shape similarity is then adjusted according to the similarity of the topological structures of the remained parts.

3.3.1   Similarity Between Primitive Shapes

An intuitive way to calculate the similarity between the source object and the destination object is to compare their components one-by-one based on identical shape types and relative positions. The overall similarity can be acquired through a weighted sum of similarity of their matched components. We define Average Point Shifting *(APS)* for two primitive shapes if their types are identical through the following rules:

1.   If both are line segments, denote them as P_1-P_2 and P_3-P_4. The *APS* of two line segments is defined as

$$APS_L = \frac{1}{2}(Min(Dis(P_1,P_3) + Dis(P_2,P_4), Dis(P_1,P_4) + Dis(P_2,P_3)) \qquad (4)$$

2.   If both are arc segments, denote them as P_1-P_{12}-P_2 and P_3-P_{34}-P_4, where P_{12} and P_{34} are the middle point of an arc. The *APS* of two arc segments is defined as

$$APS_A = \frac{1}{2}(APS_L + Dis(P_{12},P_{34})) \qquad (5)$$

3. If both are n-polygons (n is the number of vertexes), denote their vertex lists (in the same traversal order) as $P_0, P_1, \ldots P_{n-1}$ and $Q_0, Q_1, \ldots Q_{n-1}$ respectively. The *APS* between two polygons is defined as

$$APS_P = \underset{j=0}{\overset{n-1}{Min}}(\frac{1}{n}\sum_{i=0}^{n-1}Dis(P_i,Q_{(i+j)\bmod n})) \tag{6}$$

4. If both are ellipses, the *APS* is defined as the APS_P between their bounding boxes.

The partial structural similarity between two composite objects is then defined based on *APS*. Denote the component set of the source object as S, which has m components $S_1, S_2, \ldots S_m$. Denote the component set of the destination object as D, which has n components $D_1, D_2, \ldots D_n$. If $m=0$, the similarity is defined as 1. If $m>n$, the similarity between them is 0. Otherwise ($n \geq m$), we create a mapping j form $[1, m]$ to $[1, n]$, which satisfies the following two conditions:

    1. For each $i \in [1, m]$, S_i and $D_{j(i)}$ are of the same shape type.
    2. For $i_1, i_2 \in [1, m]$ and $i_1 \neq i_2, j(i_1) \neq j(i_2)$.

Enumerate all these possible mappings and denote them as a set J. The partial structural similarity between the source object and the destination object can be obtained by

$$Sim_{prim}(S,D) = \underset{j \in J}{Max}(1 - \frac{1}{m}\sum_{i=1}^{m}\frac{APS(S_i,D_{j(i)})}{100\sqrt{2}}) \tag{7}$$

where $100\sqrt{2}$ is the length of the diagonal of the normalized area.

3.3.2 Similarity between Object Skeletons

Admittedly, different users may have different opinions in decomposing a composite object into primitive shapes. Even the same user may change his ideas from time to time. E.g., for the composite graphic object called "envelope", a user may input it by drawing a pentagon and a triangle. However, he may also input it by drawing a pentagon and two line segments. See Fig.4 for illustrations of two ways to draw the "envelope". Although the ultimate combined objects look totally the same, they cannot match each other through the previous strategy presented in Sect. 3.3.1. Moreover, it is extremely difficult to enumerate all the possible combinatorial ways to draw a composite graphic object, in both spatial and temporal considerations. Hence, we also need to find other ways to measure similarity between two composite objects.

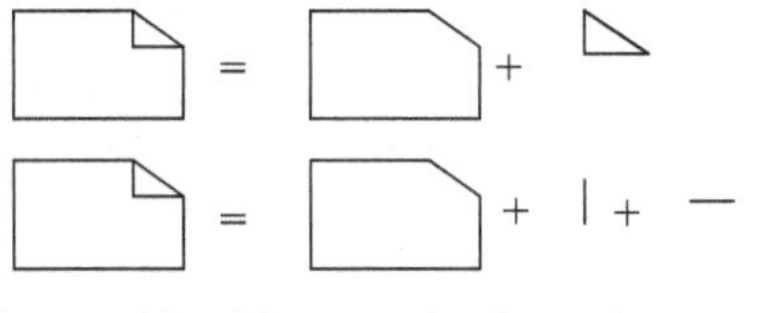

Fig.4. A composite graphic object may be drawn in several different ways

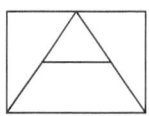

(a) the destination object in the database (some primitive shapes has been removed)

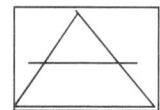

(b) the source object that the user has input (some primitive shapes has been removed)

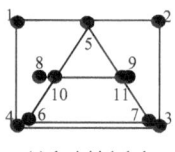
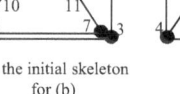

(c) the initial skeleton for (b)

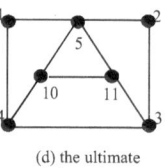

(d) the ultimate skeleton for (b)

Fig.5. Object skeleton acquisition

Object skeleton is a graph we use to represent the topological structure of the composite object. In order to acquire the skeleton of the object, we first generate an initial candidate skeleton and then merge redundant nodes and edges to obtain the ultimate one. See Fig.5 for illustration.

Node generation is first done based on the following rules:

1. All vertices of polygons are candidate nodes, such as Point 1~7.
2. All end points of line and arc segments are candidate nodes, such as Point 8 and 9.
3. The middle points of arc segment are candidate nodes.
4. All of the cross points (of two different primitive shapes) are candidate nodes, such as Point 10 and11.

After nodes generation, these candidate nodes divide the polygon and line segment into several pieces. Each piece can be logically represented by an edge (a line segment). If a node's distance to an edge is under a threshold, we move this node on the edge (choose the most nearest point on the line segment to replace the node) and split the edge into two pieces. E.g., in Fig.5 (c), node 5 splits the edge between node 1 and 2 into two pieces.

After getting the initial skeleton, those nodes that are close enough are merged. Edges must be correspondingly adjusted to adapt to the new nodes. All fully overlapping edges should be merged into one. E.g., in Fig.5 (c), node 4 and 6 are combined into node 4. And node 7and 3 are combined into node 3. Thus, the edge between node 6 and 7 is then eliminated. After these processes, we can get the ultimate object skeleton, as illustrated in Fig.5 (d). The object skeleton is represented by a graph consisting of a node set and an edge set.

The similarity between object skeletons is calculated based on their topological structures. Denote the node set of the source object skeleton as NS, which has m elements. Sort these nodes according to the sum of their x coordinate and y coordinate descendent, denote them as NS_1, NS_{12}, ... NS_m. Denote the node set of the destination object skeleton as ND, which has n elements. Use the similar way sort them in a list, denote them as ND_1, $ND1_2$, ... ND_n. Define the edge set of the source object as ES and the edge set of the destination object as ED. Define logical edge set of the destination object $LED=\{(i_1, i_2)| (ND_{i1}, ND_{i2}) \in ED, i_1, i_2 \in [1, n]\}$. If $m=0$, the similarity is defined as 1. If $m>n$, the similarity between these two skeletons is 0. Otherwise ($n \geq m$), create a mapping j from $[1, m]$ to $[1, n]$ which satisfies the following two conditions.

1. For each i_1, $i_2 \in [1, m]$ and $i_1 \neq i_2$, $j(i_1) \neq j(i_2)$.
2. For each i_1, $i_2 \in [1, m]$ and $i_1 < i_2$, $j(i_1)-j(i_2) < \varepsilon$, where ε is an integer threshold.

Define the logical mapped edge set under mapping j as $LME_j = \{(j(i_1), j(i_2)) | (ND_{i1}, ND_{i2}) \in ED$ and $i_1, i_2 \in [1, m]\}$. Thus we can define the following three metrics:

1.   Average Node Shifting (ANS)

$$ANS_j = \frac{1}{m} \sum_{i=1}^{m} \frac{Dis(NS_i, ND_{j(i)})}{100\sqrt{2}} \tag{8}$$

2.   Edge Precision (EP)

$$EP_j = \frac{\|LME_j \cap LED\|}{\|LME_j\|} \tag{9}$$

3.   Edge Recall (ER)

$$ER_j = \frac{\|LME_j \cap LED\|}{\|LED\|} \tag{10}$$

Enumerate all these possible mapping j to form a set denoted as J. The similarity between two object skeletons is defined as

$$Sim_k(S,D) = \underset{j \in J}{Max}(w_1(1 - ANS_j) + w_2 EP_j + w_3 ER_j) \tag{11}$$

Where w_1, w_2, $w_3 > 0$ and $w_1 + w_2 + w_3 = 1$. The computational complexity of this algorithm depends on ε. If $\varepsilon = 0$, the number of all possible mapping is C_n^m. If $\varepsilon \geq n$, the number of all possible mapping is P_n^m. Hence, the computational complexity of this algorithm is between C_n^m and P_n^m. The larger ε is, the larger the computational cost is, and more accurate result can be obtained.

3.3.3   The Overall Similarity Calculation

The computation complexity of similarity calculation between object skeletons is very high for a real-time system. Hence, we must reduce the node number to save calculation time. We find users have common senses to group some parts of a composite graphic object into one primitive shape. For instance, users usually draw some special primitive shape types (E.g., ellipse, hexagon, etc.) in a single stroke and seldom divide them into pieces (multiple strokes). The same situation is for some isolated primitive shapes, which do not have intersection with other parts of the object. These primitive shapes can be separated from other strokes of both the source and the destination objects and first compared by our primitive shape comparison strategy. Then, the remained parts of both objects can be compared through our object skeleton comparison strategy. Thus the node number of object skeleton can be much reduced. Denote the source object as S and the destination object as D. Denote the separated primitive shapes of S and D as S_{prim} and D_{prim} respectively. Denote the remained parts of S and

D are S_{rem} and D_{rem} respectively. The overall similarity between S and D is a linear combination of the similarities between S_{prim} and D_{prim} and that of S_{rem} and D_{rem}, as defined in Eq. (12).

$$Sim\ (S,D) = \begin{cases} 0, & if \quad Sim_{prim}(S_{prim}, D_{prim}) = 0 \quad or \quad Sim_k(S_{rem}, D_{rem}) = 0 \\ k_1 Sim_{prim}(S_{prim}, D_{prim}) + k_2 Sim_k(S_{rem}, D_{rem}), & otherwise \end{cases} \quad (12)$$

where k_1, $k_2 > 0$ and $k_1 + k_2 = 1$.

3.4   The User Interface and Relevance Feedback

We implement the proposed approach in our SmartSketchpad [1] system. After only a few components of a composite object are drawn, the similarities between the input source object and those candidates in the database are then calculated using the partial structural similarity assessment strategy. The most similar objects are displayed in a ranked candidate list (according to the descending order of similarities) in the smart toolbox for the user to choose. If the user cannot find his/her intended shape, he/she can continue to draw other components until the intended one appears in the list. In order to avoid too much intrusive suggestion of composite shapes of low similarity, only the first ten objects whose similarities are above a threshold (which is 0.3 currently) are displayed in the smart toolbox for suggestions. The smart toolbox will not give any suggestion if less than two components are drawn. E.g., if a user wants to input a graphic object called "bear", he/she can first draw its face and an ear by sketching two circles. But the intended object does not appear in the smart toolbox. Then the user continues to draw its right ears by sketching a smaller circle. This time, the wanted object appears and is ranked as the 5th in the smart toolbox for selection. If the user did not notice this, he/she can continue to draw its left eye by sketching a line segment.

This time, only three objects are listed and the wanted object is ranked as the first in the smart toolbox. See Fig.6 for illustration. However, the partial structural similarity calculation strategy we have proposed is not perfect yet. During the drawing processes, irrelevant objects may obtain higher ranks in the object list. Sometime, users have to draw the intended object completely so that it could be seen in the list. Or, even after the entire object is drawn, it still does not appear in the smart toolbox. In this case, the user has to click the "more" button continually to find his/her intended object. In order to avoid such situation, relevance feedback is employed to improve both input efficiency and accuracy. Because each user has his/her drawing style/preference, a specific user tends to input the same object through a list of components of fixed constitution (e.g., two ellipses and a triangle) and fixed input time sequence (e.g., ellipse-triangle-ellipse). Hence, the input primitive shape list offers much more information that we have not utilized in our partial structural similarity calculation. Each time when the user drags an object from the smart toolbox, the current grouped input primitive shape list will be saved as a "word" index for this intended object with a probability. E.g., a list rectangle-triangle may be used for index of "envelope" and "arrow box". The probability can be easily obtained from the relative times when the intended object is selected for this "word". Hence, for an inputted source object and a candidate object in the database, we not only can acquire their shape similarity using the partial structural similarity calculation strategy, but also can acquire the probability (initially, all possibilities are set to 0) that the candidate object is just the intend one. We linearly combine them with different weights to obtain a new metric for similarity comparison. Therefore, if a user always chooses the same candidate object for a specific input primitive shape sequence, the probability of this candidate object must be much higher than that of other candidate objects. Obviously, through this relevance feedback strategy, the intended object will obtain a higher rank than before.

#	Input	Sim	Rank	Candidates sorted in descending similarity
2		0.407362	12	
3		0.447857	5	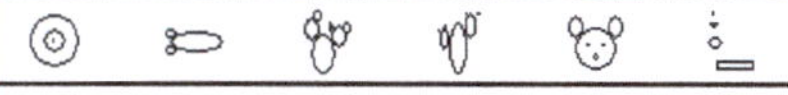
4		0.852426	1	

ID: # of input components   Input: The input components
Sim: Similarity of the intended object   Rank: Rank of the intended object

Fig.6. Input a composite graphic object by sketching a few constituent primitive shapes

4   Performance Evaluations

In this experiment, we created 97 composite graphic objects for experimentation in SmartSketchpad. All these objects are composed of less than ten primitive shapes, as shown in Fig.7. The weights and thresholds we used are $\varepsilon=20$, $w_1=0.4$, $w_2=0.3$, $w_3=0.3$, $k_1=k_2=0.5$. We randomly selected 10 objects (whose ID is 73, 65, 54, 88, 22, 5, 12, 81, 18, and 76) and draw these objects as queries. We recorded the intended object's rank variation together with the drawing steps. Results of four objects are shown in Fig.8. The horizontal coordinate is the input steps. In each step, only one component is drawn until the object is properly finished. The vertical coordinate is the rank of the intended object. In most cases, the intended object will appear in the smart toolbox (ranked in the first 10) after only a few components are drawn. Take Object 73 as an example. The user can input this object in 6 steps. After four components are drawn, the intended object's rank is 4. Thus the user can directly drag it form the smart toolbox. Averagely (of the ten objects we have tested), after 85.7% components of the intended object are drawn, it will be shown as the first one in the smart toolbox. In order to see it in the smart toolbox (top 10), only 67.0% of the intended object needs to be drawn. And the ratio is 68.7% for top 5 and 72.0 for top 3. The user does not need to draw the complete sketch for a composite graphic object. Only a few sketchy components are sufficient for the system to predict the user's intent. Hence, this user interface provides a natural, convenient, and efficient way to input complex graphic objects.

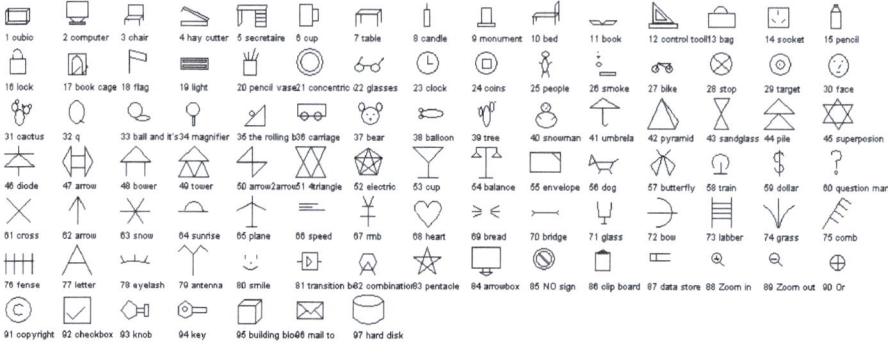

Fig.7. The composite graphic objects we have created for experimentation

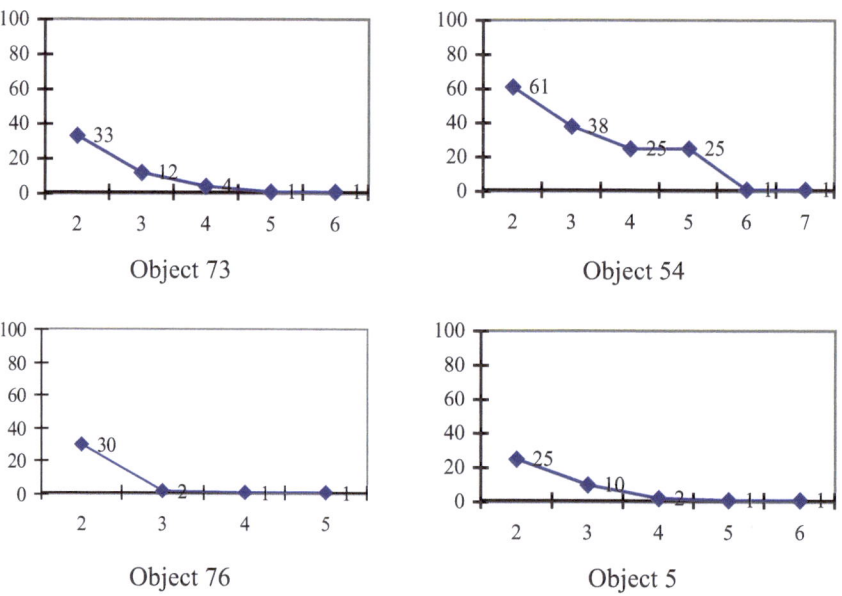

Fig.8. The intended object' ranks (vertical axes) and input steps (horizontal axes)

5   Summary

We proposed a novel user interface to input composite graphic objects by sketching only a few of their constituent simple/primitive shapes. The inputted shape's similarities to those objects in the database are updated in the drawing process and those promising composite objects are suggested to the user for selection and confirmation. By doing so, the user is provided with a natural, convenient, and efficient way to input composite graphic objects and can get rid of the traditional graphic input interface based on menu selection and this user interface is especially suitable for small screen devices.

Acknowledgement

The work described in this paper was fully supported by a grant from City University of Hong Kong (Project No. 7100247) and partially supported by a grant from National Natural Science Foundation of China (Project No. 69903006).

References

1. Liu, W., Qian, W., Xiao, R., and Jin, X.: Smart Sketchpad—An On-line Graphics Recognition System. In: Proc. of ICDAR2001, Seattle (2001) 1050-1054
2. Zeleznik, R.C., Herndon, K.P., Hughes, J.F.: SKETCH: An Interface for Sketching 3D Scenes. In: SIGGRAPH96, New Orleans (1996) 163-170
3. Gross, M.D., Do, E.Y.L.: Ambiguous Intentions: A Paper-like Interface for Creative Design. In: Proc. of the 9th Annual ACM Symposium on User Interface Software and Technology, Seattle, WA (1996) 183-192
4. Albuquerque, M.P., Fonseca, M.J., Jorge, J.A.: Visual Language for Sketching Document. In: Proc. IEEE Sym. on Visual Languages (2000) 225-232
5. Fonseca, M.J., Jorge, J.A.: Using Fuzzy Logic to Recognize Geometric Shapes Interactively. In: Proc. 9th IEEE Conf. on Fuzzy Systems, Vol. 1 (2000) 291-296
6. Arvo, J., Novins, K.: Fluid Sketches: Continuous Recognition and Morphing of Simple Hand-Drawn Shapes. In: Proc. of the 13th Annual ACM Symposium on User Interface Software and Technology, San Diego, California (2000)
7. Sklansky, J., Gonzalez,V.: Fast Polygonal Approximation of Digitized Curves. Pattern Recognition 12 (1980) 327-331
8. Graham, R.L.: An Efficient Algorithm for Determining the Convex hull of A Finite Planar Set. Information Processing Letters 1(4) (1972) 132-133
9. Esther, M.A., et al.: An Efficient Computable Metric for Comparing Polygonal Shapes. IEEE Trans. on PAMI 13(3) (1999) 209-216
10. Vapnik, V.: The Nature of Statistical Learning Theory. Springer-Verlag, New York (1995)

Experimental Evaluation of a Trainable Scribble Recognizer for Calligraphic Interfaces

César F. Pimentel, Manuel J. da Fonseca, and Joaquim A. Jorge

Departamento de Engenharia Informática, IST/UTL
Av. Rovisco Pais, 1049-001 Lisboa, Portugal
pimentelcesar@hotmail.com
mjf@inesc.pt
jaj@inesc.pt

Abstract. This paper describes a trainable recognizer for hand-drawn sketches using geometric features. We compare three different learning algorithms and select the best approach in terms of cost-performance ratio. The algorithms employ classic machine-learning techniques using a clustering approach. Experimental results show competing performance (95.1%) with the non-trainable recognizer (95.8%) previously developed, with obvious gains in flexibility and expandability. In addition, we study both their classification and learning performance with increasing number of examples per class.

1 Introduction

The increased interest in pen computers has attracted considerable attention to handwriting and sketch recognition. This paper describes a trainable recognizer for hand-drawn sketches and geometric figures, which expands on our previous work. While the simple approach we had developed earlier achieved fairly high recognition rates, it does not allow users to specify new gestures, requiring hand-coding and careful tuning to add new primitives, which is a tedious and cumbersome task. This considerably limits the flexibility of our previous approach in accommodating new gestures, strokes and interaction primitives. The recognizer we present here is a considerable advance from the one presented in [10]. Instead of limiting ourselves to using the pre-defined fuzzy-logic based method for a fixed number of different stroke classes, it allows us to use one of several learning algorithms to teach the recognizer new classes.

While this paper describes some preliminary work, our trainable recognizer exhibits respectable recognition rates (95%) achieved at a moderate training cost using a simple Naïve Bayes approach. The paper studies three basic tried-and-true techniques, Naïve Bayes, KNN and ID3 and a weighted majority combination of the three. Surprisingly enough, the simplest method outperformed the weighted majority technique not only in training costs but also in recognition performance.

D. Blostein and Y.-B. Kwon (Eds.): GREC 2002, LNCS 2390, pp. 81-91, 2002.

2   Related Work

In [10] we presented a fast, simple and compact approach to recognize scribbles (multi-stroke geometric shapes) drawn with a stylus on a digitizing tablet. Regardless of size, rotation and number of strokes, this method identifies the most common shapes used in drawings, allowing for dashed, continuous or overlapping strokes. The method described in [10] combines temporal adjacency, Fuzzy Logic and geometric features to classify scribbles with measured recognition rates over 95.8%. However, the algorithm devised, for its simplicity and robustness, makes it difficult to add new shapes. Each new primitive class added requires hand-coding and testing of new features in a slow, tedious and error-prone process.

Previous work on trainable gesture recognizers can be found in [23]. The author describes a trainable single-stroke gesture recognizer using a classic linear discriminator-training algorithm [7]. The goal of our work is somewhat different than Rubine's. Our purpose is to study the overall performance of several clustering algorithms in distinguishing between a large, dynamic and user-defined set of scribble classes. We have also focused our attention on choosing features that allow the recognizer to maintain a good recognition rate in case the user adds a new "complex" and "unexpected" scribble to the training set.

One notable difference from Rubine's recognizer is that our method allows the user to draw scribbles without any restriction and as close as possible to the intended shapes. This allows us to recognize virtually any complex shape we can think of, without the limitation of being single-stroked (i.e., without having to rule out crosses or arrows, contrarily to what happens in [23]). For these reasons, we sometimes say that our recognizer classifies scribbles instead of gestures. In order to accomplish this, we need to choose a richer feature set and more sophisticated learning algorithms as compared to the classic linear discriminator with thirteen features presented in [23].

Our work is based on a first approach to recognize schematic drawings developed at the University of Washington [12], which recognized a small number of non-rotated shapes and did not distinguish open/closed, dashed or bold shapes. Tappert [24] provides an excellent survey of the work developed in the area of non-interactive graphics recognition for the introduction of data and diagrams in vectorial CAD systems.

2.1   Geometric Features

The main features used by our recognizer are ratios between perimeters and areas of special polygons as the ones presented in Fig.1. To achieve this, we start by computing the convex hull of each scribble's points, using Graham's scan [17]. Using this convex hull, we compute three special polygons. The first two are the largest area triangle and quadrilateral inscribed in the convex hull [2]. The third is the smallest area-enclosing rectangle [11]. Finally we compute the area and perimeter for each special polygon to estimate features for the training and recognition process.

Our choice of features is based on those which have proven useful for our previous recognizer. They allow the recognition of shapes with different sizes and rotated at arbitrary angles, drawn with dashed, continuous strokes or overlapping lines.

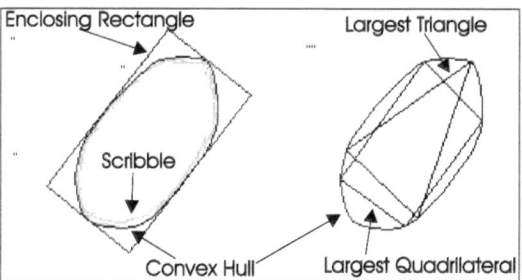

Fig. 1. Special polygons

3   Learning Algorithms

We have chosen three general algorithms for their simplicity, flexibility and robustness to implement our trainable recognizer. Each of these algorithms has different training and classification procedures. For training we need to gather a number of gestures as examples, and we need to have a class associated to each of these gestures. The class is, in other words, the name of the type of gesture (for example: Ellipse, Triangle, Delete), which is, of course, defined by the user. Training is therefore, an inductive procedure for it consists in learning a certain concept based on examples. It is obvious that some concepts are easier to learn than others are and that the more training examples we have, the more effective is our training.

Since the training set is completely user-defined, the algorithms are able to identify, with more or less performance, any type of gesture we can think of. Because the learning algorithms see a gesture as a set of geometric features (and, therefore, continuous numeric features) the training procedure can be seen as supervised clustering.

In the three following sections we present a brief explanation for each of the algorithms used. We also analyze the performance and training costs of a weighted majority combination of the three.

3.1   K-Nearest Neighbors (KNN)

The K-Nearest Neighbors is an old and simple machine-learning algorithm, which has proven to be very effective in most situations. Initial theoretical results can be found in [4] and an extensive overview in [7]. Furthermore, Bishop [1], studies on how the algorithm is related to the method of estimating probability densities.

The conceptual idea of the K-Nearest Neighbors algorithm is simple and intuitive. The estimated class for a given gesture instance (feature vector) is the most abundant class within the k nearest neighbor examples of that instance, where k is a set parameter. Neighborliness is measured wrt the vector distance between two instances. This explains why this algorithm works better with numeric features.

An essential step in the KNN algorithm is to normalize all feature values in training and classification instances. This normalization step prevents features with usually large values from having a disproportionately superior weight in the calculation of distances.

In summary, KNN training consists merely in normalizing and organizing the sample instances in structures that may allow a fast computation of distances. It is, as such, a very fast training procedure. KNN evaluation, on the other hand, consists in finding which are the closest k neighbors. The algorithm has good classification efficiency, however, large training sets place considerable demands in both computer memory and classification time required.

3.2   Inductive Decision Tree (ID3)

ID3 (Inductive Decision Tree) is a decision tree learning algorithm thoroughly described in [18], [19] and [20]. Later, Quinlan [22] discussed details and implementation issues concerning decision tree learning in general. Our version of ID3 also includes reduced-error pruning, as described in [21], to avoid over-fitting. Our approach can also handle continuous-valued attributes, using a method described in [8] and [9], and incorporates a pruning strategy discussed in [16] and [15].

ID3 is one of the most famous algorithms based on decision trees. Once a decision tree has been constructed during the training procedure, it will then be kept in memory and used for classification. Each non-leaf node is labeled with the name of the feature whose value must be tested, and the branches that originate from that node are labeled with the values (or intervals) that the feature can take. Given a gesture instance to be classified, we start from the tree's root, testing the respective feature and moving down through its respective branch. Eventually we will arrive at a leaf node labeled by the class name for that instance.

This algorithm is better suited to handle symbolic features or, at least, discrete-valued ones. Since our features consist mostly of continuous numeric values, it is necessary to transform them all into discrete values before training. This transformation discretizes each feature domain into several intervals and substitutes the feature values for the correspondent interval index. We also perform this process for the Naïve Bayes algorithm, as explained below.

In summary, while ID3 training consists in the construction of the tree based on the examples at hand, ID3 evaluation is the process of running an instance down the tree until it reaches a leaf node. ID3 presents average classification efficiency for continuous features. As is typical of decision-tree classifier methods the recognizer is very fast due to its low computational complexity.

3.3   Naïve Bayes (NB)

Documentation on Naïve Bayes' basic notions as well as other Bayesian Classifiers' can be found in [7]. Additionally, Domingos [6] presents an analysis on Naïve Bayes' performance with situations where the independence assumption does not hold (which is the most common situation in real world problems). A discussion concerning the method of estimating probabilities using the *m-estimate* can be found in [3].

The algorithm uses probabilities estimated from the training set. For a given instance (gesture), it computes the probability of each feature value belonging to each class. Then it combines these probabilities to discover which class is most likely the instance. This procedure assumes that there are no statistical dependencies between features. While generally this is not the case, the simplification is often acceptable and frequently we do not have enough information about the dependencies between features. This algorithm also requires discretizing features in the same way as ID3.

In summary, Naïve Bayes training consists in discretizing the samples and computing intermediate probabilities, which is a fairly fast process. Subsequently, Naïve Bayes evaluation consists in the computation of the final probabilities. As a result, the algorithm is also very fast in classifying. Even though we are dealing with continuous features, Naïve Bayes' classification efficiency in recognizing our gestures was proven to be better than any other algorithm we have tried.

3.4   Weighted-Majority (WM)

A description of the Weighted-Majority algorithm can be found in [13] and [14]. This algorithm is very commonly used in Machine Learning. It is adopted in situations where our chosen algorithm often performs a wrong classification while most alternative algorithms would succeed. Therefore, Weighted-Majority algorithm combines several algorithms to classify an instance and returns the mostly voted class. Some algorithms' votes carry more weight than others'. These weights are determined by the algorithms' classification efficiency.

We implemented a Weighted-Majority using the three previous algorithms (K-Nearest Neighbors, ID3 and Naïve Bayes). This way, whenever a class receives at least two votes, it will be returned as the result of classification. If, otherwise, each algorithm votes in a different class, the result will be the class voted by Naïve Bayes because it was proven to be the most truthful algorithm in this domain.

4   Experimental Assessment of the Algorithms

We will now evaluate the four learning algorithms described in the previous section. The first section describes the data collection procedure. Next, we describe the training process and the learning evaluation of each algorithm.   Finally we present and discuss the learning evolution of the different algorithms.

4.1   Data Collection

To train and evaluate the different learning algorithms described above we used a set of sample data collected using a Wacom PL-300 LCD digitizing tablet and a cordless stylus.

We asked 22 subjects to draw each multi-stroke shape twenty times using solid lines, ten times using dashed and ten times using bold lines. We also asked them to draw 30 times each uni-stroke shape. All these drawings yield a total of 9944 shapes. Fig. 2 shows the different classes of shapes drawn.

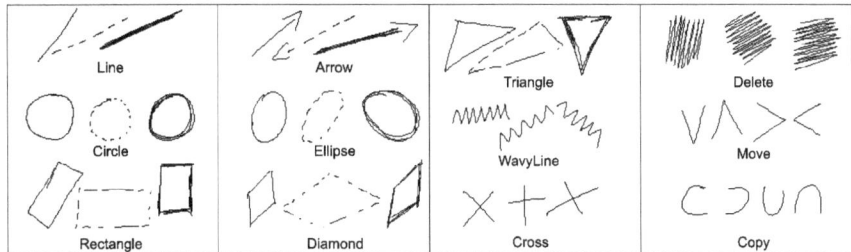

Fig. 2. Classes of shapes used in the trainable recognizer

We gave a brief description of the recognizer to the users, including the set of recognizable shapes. We also told them about the multi-stroke shape recognition capabilities and the independence of changes with rotation or size. Novice subjects had a short practice session in order to become acquainted with the stylus/tablet combination.

4.2   Training Process

We performed many test runs using the different algorithms described above. In most cases, we used twelve different gestures classes, namely, `Arrow`, `Circle`, `Cross`, `Copy`, `Delete`, `Diamond`, `Ellipse`, `Line`, `Move`, `Rectangle`, `Triangle` and `WavyLine`, as depicted in Fig.2. In all test runs, none of the gestures used in training were used for evaluating the algorithms. Furthermore, training and evaluation gestures were drawn by different subjects.

Every gesture instance is represented by a vector of 24 geometric features, as explained in section 2.1. Since all recognition algorithms used perform clustering techniques, working with twelve classes implies that we will need many examples per class in order to be able to distinguish them well. If, for instance, we only needed to distinguish between three or four classes, the algorithms would be able to perform with better recognition efficiencies at fewer examples per class. For these reasons, the training sets we have created are relatively large (2610 examples).

Each training set includes several instances and their correspondent classes, whilst an evaluation includes only unclassified instances. After a given algorithm classifies an evaluation set, the estimated classes are compared with the real classes that generated the instances (the gesture that the individual intended to draw). This is the method we have used to extract all results presented in this paper.

4.3   Learning Evaluation

We quickly noticed that the Naïve Bayes algorithm presents the highest classification efficiency (around 95% when the number of training examples is high). This combined with very acceptable training and classification complexity, makes this algorithm the best suited for the task in hand.

The K-Nearest Neighbors algorithm can also yield good classification efficiency (around 92%), but classification times become unacceptable for large training sets. This is because the algorithm must scan all training examples, to calculate their distances from the given gesture. Much work has been done on preprocessing the training samples so that KNN does not need to scan them all during classification. Some papers on this matter can be found in [5]. This type of optimization, however, is not present in our implementation of KNN.

ID3 was the worst performer yielding classification efficiencies lower than 77%. Also, ID3's training complexity is high and classification complexity, while acceptable is higher than Naïve Bayes'. We think that the reason behind such low classification efficiencies is due to using continuous numeric features, which are not well suited for decision trees. Although there are more sophisticated ways to coalesce continuous features into intervals (in contrast to the method devised), it is not clear how this algorithm will ever yield high classification efficiency in our domain.

Finally, the three-algorithm Weighted-Majority requires, the sum of their individual training and classification times. As expected, it incurs fewer mistakes than either algorithm alone, but it is not significantly better than the best of the three (Naïve Bayes). Despite making fewer mistakes, we observed that Weighted-Majority makes some mistakes that Naïve Bayes doesn't, when both KNN and ID3 are wrong.

4.4   Learning Evolution

We wanted to know how many examples of a gesture are sufficient to "teach" that gesture. To this end, we plotted recognition performance against training set size for the twelve solid shapes mentioned above, for each algorithm as shown in Fig.3.

We can see that, in order to distinguish between twelve different classes, all of the algorithms show the greatest classification efficiency increase up to 30 examples per class. This is the typical shape of what we call "the learning curve". Relatively speaking, we would say that Naïve Bayes and KNN exhibit acceptable classification efficiency with 50 examples per class, and tend to stabilize around 100 examples per class. We did not find ID3 to perform at an acceptable level for the reasons explained before.

We point out that the efficiency in recognizing a given gesture does not solely depend on its training instances. It also depends on how many distinct classes appear in the training set and how the given class differs from them. Based on the observations presented here we conclude that Naïve Bayes algorithm is the better suited for our domain. From now on all the results and recognition rates presented pertain to this algorithm.

5   Comparison with the Previous Recognizer

We will now make an experimental evaluation of the trainable recognizer using a set of data collected as described before, comparing its performance to the recognizer described in [10].

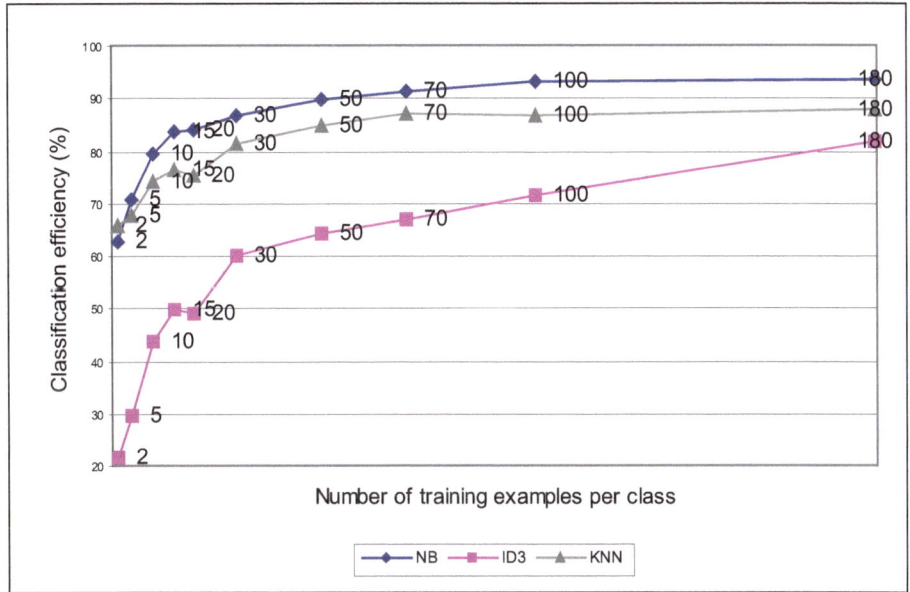

Fig. 3. Learning evolution based on the number of examples per class

5.1   Performance Evaluation

We selected a subset of the data described before. The values in the confusion matrix of Fig.4 describe the recognition results by the Naïve Bayes algorithm for the twelve solid shape classes, described in the previous sections. We used 2610 scribbles to train the recognizer under the conditions explained above and 959 other scribbles (from a different set of users) to evaluate it. There were 895 correct classifications out of 959 scribbles, corresponding to a global classification efficiency of 93.3%.

Before interpreting these results we point out that the distribution of correct and incorrect classifications is mostly a consequence of the selected features. Confusion matrices obtained from the other algorithms outline similar distributions. Focus, however, is now only on the selected algorithm: Naïve Bayes.

As shown by the confusion matrix, sometimes arrows are classified as crosses and vice-versa. This happens because these scribbles have two strokes. The number of strokes is one of the most important features for distinguishing arrows and crosses, provided that we are now only speaking of solid shapes.

Arrows are also sometimes misclassified as lines due to their geometrically similar shape. Copies are seen as ellipses when they are drawn as a nearly closed shape. If deletes are made a bit longer than usual, there may be some confusion with wavylines. Many of the ellipses are classified as circles, a very intuitive failure mode. Eight percent of the rectangles are taken as wavylines (perhaps the worst misclassification) because of the similarities between their enclosing rectangles.

	Arrow	Circle	Cross	Copy	Delete	Diamond	Ellipse	Line	Move	Rectangle	Triangle	WavyLine
Arrow	84.8		5.6			1.6		5.6		0.8		1.6
Circle		94.0			2.0		4.0					
Cross	5.9		94.1									
Copy		1.2		91.7			6.0			0.6		0.6
Delete					89.1			2.2	2.2			6.5
Diamond						98.7					1.3	
Ellipse		14.0					86.0					
Line								99.2				0.8
Move									100			
Rectangle			2.0			2.0		2.0		86.0		8.0
Triangle											96.1	3.9
WavyLine								1.7				98.3

Fig.4. Confusion matrix for solid shapes (values in percentage)

Finally, we wish to explain that "strange" misclassifications such as mistaking a rectangle for a wavyline can be easily solved by adding appropriate extra features. Such features should present distant values on each gesture class. The learning algorithm will automatically handle the task of using that feature to better distinguish the classes at stake. The features we have chosen have proven useful for these twelve classes, and since these are relatively diverse, we believe the recognizer will behave well for a broad variety of classes. In summary, our recognizer performs considerably well, presenting an overall classification efficiency of 93.3% in this experiment.

5.2   Performance Comparison

We will now present the result of a different experiment specifically intended to establish a comparison with the (non-trainable) fuzzy recognizer described in [10]. While the fuzzy recognizer presents 95.8% classification efficiency with all 12 gestures, we must point out that this result was obtained considering that some of the gestures called "shapes" (Arrow, Circle, Diamond, Ellipse, Line, Rectangle and Triangle) may exist in three different versions (Solid, Dashed and Bold). This performance, however, does not consider distinguishing line-style, which make for non-homogenous classes, thus impairing performance. To establish a "realistic" assessment, we have performed a similar experience with our trainable recognizer, described as follows.

We have trained the recognizer with 6022 gesture instances of all 12 classes, this time, considering the three different line-styles (Solid, Dashed and Bold). After, we ran the recognizer through an evaluation set of 3869 gesture instances. Again, training and evaluation sets were generated by different sets of people. The recognizer performed with a classification efficiency of 95.1%, higher than the previous case.

6   Conclusions

We have presented a simple trainable scribble recognizer for multiple stroke geometric shapes. While some of the work is still preliminary, the Naïve Bayes approach provides for good performance and classification efficiency, making it our method of choice. The classification rates measured under realistic conditions (95.1%) are promising. However, the trainable classifier requires many samples to achieve this value. While this can be perfected, the amount of work required is considerably less than the hand tuning previously required to add new shapes, for a performance comparable to that of the non-trainable recognizer.

References

1.   Bishop C. M. *Neural Networks for pattern recognition.* Oxford, England: Oxford University Press, 1995.
2.   Boyce J. E. and Dobkin D. P., *Finding Extremal Polygons,* SIAM Journal on Computing 14 (1), pp.134-147. Feb. 1985.
3.   Cestnik B. Estimating probabilities: A crucial task in machine learning. *Proceedings of the Ninth European Conference on Artificial Intelligence* (pp. 147-149). London: Pitman, 1990.
4.   Cover T. and Hart P. Nearest neighbor pattern classification. *IEEE Transactions on Information Theory*, 13, 21-27, 1967.
5.   Dasarathy B. V. *Nearest Neighbor (NN) Norms: NN Pattern Classification Techniques*, Los Alamitos, CA, IEEE Computer Society Press, 1991.
6.   Domingos P. and Pazzani M. Beyond independence: Conditions for the optimality of the simple Bayesian classifier. *Proceedings of the 13th International Joint Conference on Machine Learning* (pp. 105-112), 1996.
7.   Duda R. and Hart P. *Pattern classification and scene analysis.* New York: John Wiley & Sons, 1973.
8.   Fayyad U. M., *On the induction of decision trees for multiple concept learning,* (Ph.D. dissertation). EECS Department, University of Michigan, 1991.
9.   Fayyad U. M. and Irani K. B. Multi-interval discretization of continuous valued attributes for classification learning. In R. Bajcsy (Ed.), *Proceedings of the 13th International Joint Conference on Artificial Intelligence* (pp.1022-1027). Morgan Kaufmann, 1993.
10. Fonseca M. J. and Jorge J. A. *Experimental Evaluation of an On-line Scribble Recognizer*, Pattern Recognition Letters 22(12): 1311-1319, 2001.
11. Freeman H. and Shapira R., *Determining the minimum-area encasing rectangle for an arbitrary closed curve,* Communications of the ACM 18 (7), 409-413, July 1975.
12. Apte, A., Vo, V., Kimura, T. D. *Recognizing Multistroke Geometric Shapes: An Experimental Evaluation.* In Proceedings of UIST'93. Atlanta, GA, 1993.
13. Littlestone N. and Warmuth M. *The weighted majority algorithm* (Technical report UCSC-CRL-91-28). Univ. of California Santa Cruz, Computer Engineering and Information Sciences Dept., Santa Cruz, CA, 1991.

14. Littlestone N. and Warmuth M. The weighted majority algorithm. *Information and Computation* (108), 212-261, 1994.
15. Malerba D., Floriana E. and Semeraro G. A further comparison of simplification methods for decision tree induction. In D. Fisher & H. Lenz (Eds.), *Learning from data: AI and statistics*. Springer-Verlag, 1995.
16. Mingers J. An Empirical comparison of pruning methods for decision-tree induction. *Machine Learning*, 4(2), 227-243, 1989.
17. O'Rourke J. *Computational geometry in C*, 2nd edition, Cambridge University Press, 1998.
18. Quinlan J. R. Discovering rules by induction from large collections of examples. In D. Michie (Ed.), *Expert systems in the micro electronic age*. Edinburgh Univ. Press, 1979.
19. Quinlan J. R. Learning efficient classification procedures and their application to chess end games. R. S. Michalski, J. G. Carbonell and T. M. Mitchell (Eds.), *Machine learning: An artificial intelligence approach*. San Mateo, CA: Morgan Kaufmann, 1983.
20. Quinlan J. R. Induction of decision trees. *Machine Learning*, 1(1), 81-106, 1986.
21. Quinlan J. R. Rule induction with statistical data – a comparison with multiple regression. *Journal of the Operational Research Society*, 38, 347-352, 1987.
22. Quinlan J. R. *C4.5: Programs for Machine Learning*. San Mateo, CA: Morgan Kaufmann, 1993.
23. Rubine D. H. *Specifying Gestures by Example*, SIGGRAPH'91 Conference Proceedings, ACM, 1991.
24. Tappert, C. C., Suen, C. Y., Wakahara, T. *The state of the art in on-line handwriting recognition*. IEEE Transactions on Pattern Analysis and Machine Intelligence 12~(8), 787—807, 1990.

User Interfaces for On-Line Diagram Recognition

Dorothea Blostein, Ed Lank, Arlis Rose, and Richard Zanibbi

Department Computing and Information Science, Queen's University
Kingston Ontario, Canada, K7L 3N6
{Blostein,zanibbi}@cs.queensu.ca
lank@cs.sfsu.edu

Abstract. The user interface is critical to the success of a diagram recognition system. It is difficult to define precise goals for a user interface, and even more difficult to quantify performance of a user interface. In this paper, we discuss some of the many research questions related to user interfaces in diagram recognition systems. We relate experiences we have gathered during the construction of two on-line diagram recognition systems, one for UML (Unified Modeling Language) notation and the other for mathematical notation. The goal of this paper is to encourage discussion. The graphics recognition community needs strategies and criteria for designing, implementing, and evaluating user interfaces.

Keywords: diagram recognition, graphics recognition, user interfaces

1 Introduction

User interfaces are challenging to design and construct. The success of the user interface has an enormous influence on the success of recognition software. In this paper we discuss issues that arise in the construction of user interfaces. We draw on our experience in implementing two on-line diagram recognition systems, one for UML[1] notation [17] and the other for mathematics notation [29] [30]. Both of these build on the character recognizer and user interface of Smithies, Novins and Arvo [26].

1.1 Existing Research into User Interfaces

Human Computer Interaction (HCI) has been studied extensively [8]. This literature provides insights relevant to the construction of user interfaces for diagram recognition systems. Selected papers are discussed in the following sections.

---

[1]UML, the Unified Modeling Language, is a diagram notation for describing the structure of software systems. Early in the software design cycle, software engineers informally sketch UML diagrams on paper or whiteboards. UML recognition software can make this diagram information available to Computer Assisted Software Engineering (CASE) tools.

D. Blostein and Y.-B. Kwon (Eds.): GREC 2002, LNCS 2390, pp. 92–103, 2002.

Software tools that support user-interface construction are reviewed by Myers [21]. Application of these tools to diagram recognition systems should be investigated. However, generic tools may be difficult to apply: diagram recognition requires close interaction between the user interface and the recognition algorithms. The user interface should provide the user with the highest-quality recognition results for a given amount of user input time. The user's input time can be spent before recognition (e.g., training a recognizer), during recognition (e.g., inspecting and correcting intermediate results), or after recognition (proofreading and correcting recognition errors).

HCI researchers have found that system design requires a deep understanding of the user's tasks [11] and the user's physical context [12]. Scenarios of use [7] can be developed to show the broader context of diagram use. For example, a diagram may be drawn as part of a brainstorming session. This requires a fluid drawing environment, in which users are not inhibited by requirements of syntactic correctness or complex editing techniques [27]. In contrast, the creation of a precise diagram (e.g., an engineering schematic, or a blueprint for a building) may be best served by tools that do not use on-line document recognition, but instead provide structured document entry with checking and analysis facilities.

If an on-line diagram recognition tool is used to support brainstorming, the tool must support collaborative work. This means that the system must be able to sort out whether strokes drawn by a pair of users are contributing to a single drawing element or to separate elements. While a single person can work with a small drawing surface, such as a piece of paper or a small computer monitor, group work requires a large physical drawing surface, such as a whiteboard or flip chart. Various activities accompany the drawing of a diagram, including gesturing, pointing, annotating and list-making. It is challenging for the user interface of an on-line diagram recognition tool to permit all of these activities; in addition the user must be able to easily specify what parts of the drawing are candidates for recognition.

1.2 On-Line versus Off-Line Recognition

Our focus is on user interfaces for on-line diagram recognition systems. Here we briefly contrast on-line and off-line systems. The user interface is pervasive in an on-line system, responding to the strokes the user draws. In contrast, an *ideal* off-line system has a minimal user interface: a user presents a stack of paper to the scanner, and the recognition system processes this fully automatically, with no recognition errors. In reality the recognition results are not perfect, so a user interface is provided to support proofreading and error correction.

On-line systems process modest amounts of data, whereas off-line systems can be applied to large-scale data acquisition tasks. This difference in data volume has a significant impact on user interface requirements. In the on-line case, the user draws diagrams while connected to the computer, and is available for proofreading and correcting the recognition results. In the off-line case, the large volume of data may mean that the user cannot invest the time to proofread and correct the recognition results. This problem can be addressed both by developing highly accurate recognition algorithms, and by finding ways for subsequent software to use noisy recognition results. For example, noisy OCR results are used in the text editing system of [2]: the scanned document image is presented to the user, with the noisy OCR result hidden from sight but used for content searches.

1.3 The Interaction between Editing, Generation and Recognition

Diagram recognition has traditionally been well-separated from diagram generation and editing. This arises naturally in off-line diagram recognition, where the goal is to get a one-time transfer of information from paper documents to an electronic form. As discussed below, on-line diagram recognition can lead to a closer relationship between recognition, editing and generation.

Figure 1 illustrates the image and information aspects of a diagram [4]. Diagram *generation* is a translation from information to image, whereas diagram *recognition* is a translation from image to information. In traditional off-line diagram recognition systems, a recognizer is applied once, to transform the image into information. This is followed by an *edit/generate cycle*, in which the user iteratively edits the information and views a re-generated diagram. The user edits both to correct recognition errors, and to change the diagram. For example, recognized music notation may be transposed; this creates new sheet music to fit the vocal range of a singer.

Diagram editors provide either a batch or WYSIWYG (What You See Is What You Get) user interface. In a batch system, such as LaTeX, the user directly edits a textual representation of the information, and requests image generation when desired. In a WYSIWYG system, the user views the diagram image, issues editing commands to update the information, and views an updated image created by the diagram generator [3]. The WYSIWYG interface does not require the user to learn a textual encoding of the diagram information.

On-line diagram recognition offers the possibility of more closely integrating the recognizer into the edit/generate cycle. A user begins by drawing a portion of a diagram, which is recognized and perhaps rectified. The user continues to edit: correcting recognition errors, moving and resizing diagram elements, and drawing new parts of the diagram. This scenario puts complex requirements on the user interface for the diagram recognition system. It must be possible to apply recognition algorithms to the new parts of the diagram, while preserving the recognition results (and user corrections) made to the old parts of the diagram. Ideally, the user should have one integrated piece of software which provides diagram recognition, diagram editing and diagram generation capabilities.

1.4 The User Interface Provided by Paper

Paper is popular and widely used. Before designing a user interface for document recognition, it is worth reviewing the properties of paper that make it so attractive: ergonomics, contrast, resolution, weight, viewing angle, durability, cost, life expectancy, and editorial quality [15]. Paper also has limitations: erasing is difficult, it is not possible to "make extra room" to expand parts of a diagram, it is hard to find information in a large stack of paper, and so on. A goal for future computer interfaces is to retain the advantages of paper while also providing the editing and search capabilities lacking in paper.

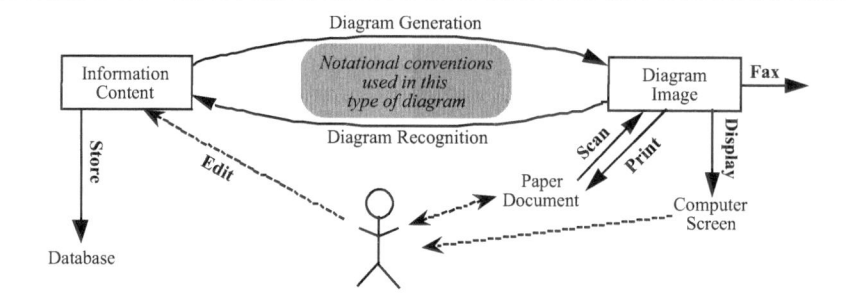

Fig.1. A diagram must be treated as both image and information. The image is displayed to the user, and supports image-level operations such as faxing. The information supports intelligent editing operations, content-based searches, and many other operations which require an understanding of the diagram

Gross and Do have observed that designers prefer to use paper and pencil [13]. In particular, designers reject the use of computers in the early, conceptual, creative phases of designing.   Paper and pencil permits ambiguity, imprecision, and incremental formalization of ideas. With paper and pencil, you draw what you want, where you want it, and how you want it to look.   In contrast, computer based tools force designers into premature commitment, demand inappropriate precision, and are often tedious to use when compared with pencil and paper.

The rest of this document discusses user interface issues related to quantifying the performance of a user interface (Section 2), supporting diagram input (Section 3), executing the recognizer (Section 4), displaying recognition results (Section 5), and supporting user correction of recognition errors (Section 6).   Important research problems exist in all of these areas.

2   Quantifying the Performance of a User Interface

The performance of a user interfaces is difficult to measure. One challenge is to separate the performance of the user interface from the performance of the recognition algorithms. (Evaluation of graphics recognition algorithms is discussed in [25]). The user interface plays a big role in determining how much time the user spends finding and correcting recognition errors. It is particularly difficult to develop a quantitative performance measure for incremental recognition, where the user can intervene at different steps of the design [16].

2.1 Comparing Automated versus Unautomated Diagram Entry

One approach to performance measurement is to compare automated diagram entry (using diagram recognition software) to unautomated diagram entry (where the user redraws the diagram using a structure-based editor, or the user directly types the information, as in LaTeX). Analogous measurements have been made for OCR, comparing the time needed to type and proofread a page of text to the time needed to proofread and correct OCR results produced for this page of text [9]. The conclusion was that OCR must be at least 98% correct to match the data entry time and accuracy (number of residual errors after proofreading) achieved by human typists. Unfortunately, there are no comparable results for diagram recognition. We would like to compare diagram entry (user time taken, residual errors) using two methods:

1. *Automated diagram entry:* On-line or off-line diagrams are processed by recognition software, producing information which is proofread and corrected by the user.
2. *Unautomated diagram entry:* A user enters diagram information via editing commands (using a batch or WYSIWYG editor, as discussed in Section 1.3).

The user time taken for method 1. depends both on the recognition algorithm (how many recognition errors there are for the user to correct) and on the user interface (how quickly the user can find and correct these errors). The performance baseline provided by method 2. is well-defined for text entry, but may change for diagram entry: the speed of skilled human typists is not likely to increase much in the next decade, whereas the speed of skilled human entry of diagrams (using method 2.) could increase due to advances in diagram editors. Since many users are interested in finding the fastest and most accurate method of entering diagram data, there is a real need for quantified comparisons of automated and unautomated diagram entry.

2.2 Other Performance Measures for User Interfaces

The previous section discusses a performance measure based on user entry time. That is one useful measure of performance, but additional measures are needed to capture other performance aspects. Can a user draw a diagram in a comfortable, freehand, unconstrained way? This is particularly important when the user does not start with a paper diagram, but instead is creating a diagram interactively. Also, how frustrating does a user find the recognition system? Currently, diagram editing and generation software tends to be less frustrating (and more predictable) than diagram recognition software. Quantifiable performance metrics are needed for these aspects of a user interface.

Nielsen has defined usability as consisting of five attributes: learnability, efficiency, memorability, errors and satisfaction [23]. Measuring these attributes is difficult. Nielsen and Levy discuss two categories of measurable usability parameters [24]: subjective user preference measures, which assess how much the users like the system, and objective performance measures, which measure the speed and accuracy with which users perform tasks on the system. Although users sometimes prefer the system on which they perform more poorly, in most cases preference and performance are positively associated [24]. Desirable usability principles listed in [14] include control, predictability, transparency, trust and privacy.

The *usefulness* of a design refers to the possibilities for action in the design, whereas *usability* refers to the user's ability to notice these possibilities for action [20]. Usability depends on the user's experience level. For example, a standard GUI provides extensive information about the set of actions that are available to the user. This is of great help to novice users. Expert users tend to prefer command-line interfaces: they have memorized the commands, and find it faster to enter a short keyboard command than to move the hand to the mouse, position the pointer, and click [20].

3   Supporting Diagram Input

Pens are commonly used as input devices for drawing diagrams. A pen is natural and easy to use on small surfaces (data tablets) and large surfaces (electronic whiteboards). A two-ended pen supports fast mode switching (Section 6). Perhaps the use of a programmable data glove could be investigated: all fingers can be involved, providing new ways of drawing, selecting objects, and correcting recognition results.

Diagram entry can occur through unconstrained drawing, or via gestures. Gestures are stylized pen strokes which invoke a command. In gesture-based entry, there is no need for diagram recognition – only gesture recognition is required. As an example, the Knight Project supports gesture-based creation of UML diagrams [10]. Unconstrained diagram entry is more natural (more like drawing on paper) than gesture-based diagram entry. However, recognition performance must be high. Otherwise users are likely to prefer the more constrained but more reliable entry method based on gestures.

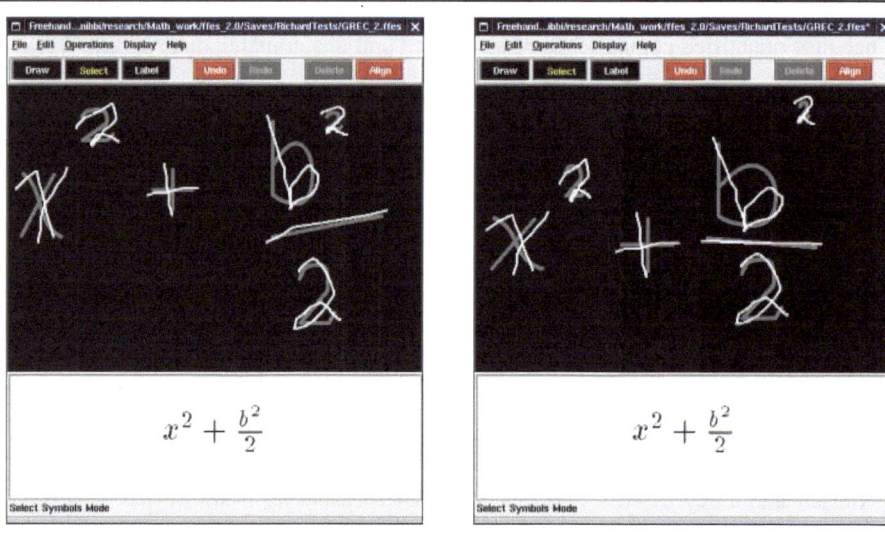

Fig.2. These screen shots from our mathematics recognition system [29] show three different methods of providing feedback about recognition results. (1) The lower panel displays the LaTeX output produced by the recognizer. Eager recognition is used, updating this panel every time the user enters or moves a character. (2) The upper panel shows character-recognition feedback provided by superimposed drawing annotations. (3) The left screen shot shows the original input, as written by the user. The right screen shot shows the result after a style-preserving morph (Section 5.3). The morph gradually moves and resizes symbols to display the baselines perceived by the recognizer

In [16], gestures used for drawing correction include erase (back-and-forth scribble), move (arrow) and select (enclosing circle). In trials on nine novice users, only 74% of 269 erasing gestures were effective; surprisingly, this did not discourage the users, who always chose to gesture when they had a choice between gesture and menu for erasing. There is need for standardization of gestures for common operations; existing software uses various gestures for the same operation [18].

4   Executing the Recognizer

Recognition can be *eager* (occurring while the user draws), or *lazy* (occurring only when explicitly requested by the user, or when recognition results are required by another operation). Lazy recognition prevents the recognition process from intruding on the user's creative phase [22]. In contrast, when eager recognition is combined with immediate changes to the display, the user may be distracted. For example, eager rectification of a user's strokes can be disorienting, because items that the user just drew suddenly change appearance.

Eager recognition is used in our mathematics recognition system (Figure 2). Every time the user draws a new math symbol in the drawing panel, the recognizer produces an updated LaTeX interpretation. This relies on the ability of the recognition algorithms to process incomplete expressions [29].

Lazy recognition is used in the Burlap system [19]. The meaning of a sketch is not recognized until the user actually tries to interact with the sketch. In other systems, recognition occurs after a sufficiently long drawing pause. For example, in the Electronic Cocktail Napkin, configuration recognizers are run when the user pauses more than five seconds. These recognizers search the drawing for instances of their patterns [13].

Many different types of diagrams are used in society, so it is useful if a diagram recognition system is able to accept multiple diagram types. The user can be asked to identify the diagram type, or the system can infer this information. Our UML recognition system currently handles Class, Use Case, and Sequence diagrams, with the user indicating ahead of time which type of diagram is being drawn [17]. The Electronic Cocktail Napkin provides definition and inference of diagram types, called *contexts* [13]. The user gives examples to define graphical rewrite rules (patterns and replacements). These are used to recognize configurations in a drawing. In one example, the pattern is a large rectangle surrounded by smaller squares, and the replacement is the configuration *table surrounded by chairs*. This configuration is defined in the Room context, and does not apply in another context such as Circuits. The system identifies the context when the user draws a symbol or configuration unique to one context. For example, a resistor symbol is found only in the Circuits context.

5   Displaying Recognition Results

Diagram recognition systems need to display a variety of information. Our UML recognition system [17] directly or indirectly displays: the drawing made by the user, the segmentation of strokes into glyphs, the classification of glyphs into "UML glyph" versus "alpha-numeric character", the results of character recognition and UML glyph recognition, and a rectified version of the drawing. Recognized glyph relationships are not explicitly displayed, but are used in glyph moving (e.g., if the user moves a box, attached arrows move as well). As another example, our mathematics recognition system [26] [29] [30] displays: the drawing made by the user, the segmentation of strokes into glyphs, character recognition results, a partially rectified version of the drawing, the recognition result as raw LaTeX and as typeset math notation produced from the LaTeX. The partial rectification uses a style-preserving morph [30]: characters and symbols are moved and resized, but retain the shape of their handwritten strokes. This gives the user feedback on the recognition of superscript, inline and subscript relations.

As discussed below, mechanisms for displaying recognition results include multi-view displays, superimposed drawing annotations, rectification, and morphing. Extra thought must go into the display of alternative or ambiguous recognition results.

5.1 Multi-view Displays

Multi-view displays [5] [6] show several views of related data, such as the drawn diagram and the recognition result. A two-view display can also be used to reduce input ambiguity, where user-drawn strokes can be categorized either as a command or as graphic input [16]. For example, if the user draws a circle around an object, the intent may be to select the object (a command) or to add the circle to the drawing (graphic input). A two-view display avoids this ambiguity by providing a drawing window for graphic input, and a correction window (which shows recognition results as a rectified drawing) for commands.

5.2 Superimposed Drawing Annotations

Superimposed drawing annotations (displayed in a contrasting colour) can be used in various ways. The results of glyph recognition can be presented via superimposed bounding boxes and labels. However, this gives the display a cluttered appearance. We have developed a neater, more readable way of displaying character recognition results, as shown in Figure 2. Lightly-coloured typeset symbols are placed behind the user's handwritten symbols. The typeset symbols are sized and scaled to fit in the bounding box of their associated handwritten symbol. Initial results are encouraging, with both handwritten and typeset symbols clearly visible.

The Electronic Cocktail Napkin uses superimposed drawing annotations to provide recognition and constraint feedback [13]. For example, a constraint [28] can cause one node to move whenever the user moves another node. The Electronic Cocktail Napkin displays constraints as superimposed drawing annotations, analogous to dimensioning annotations in mechanical engineering drawings. In conclusion after user trials, Gross and Do recommend that The Electronic Cocktail Napkin be run with low-level echoing and rectification turned off, to avoid distracting the user.

5.3 Rectification and Morphing

Rectification replaces drawing elements by stylized versions of their shapes. This provides useful feedback, but a user can be disoriented by the sudden change in drawing appearance. *Morphing* reduces this problem by displaying a series of images that show a smooth transition from the input to the rectified feedback [1]. Our mathematics recognition system uses a style-preserving morph in which the shape of the handwritten strokes is preserved; the symbols are gradually moved and resized to display the baselines that make up the structural interpretation of the mathematical expression (Figure 2). An experiment compared user performance with morphing feedback to user performance with typeset feedback shown in a separate window [30]. Expression-entry (and correction) time was equally good under both conditions. This is despite the fact that the style-preserving morph does not provide explicit symbol recognition feedback, and that the morphing itself introduces a delay of about one second. The success of the style-preserving morph may be due to reduced disruption of the participants' mental map; the typeset feedback is more disruptive because participants have to shift focus between the drawing window and the window that gives feedback via a typeset expression.

5.4 Treatment of Ambiguity

Mankoff et al. provide an interesting discussion of user-interface issues arising from recognition ambiguity [19]. Their goal is to create a toolkit supporting creation of effective user interfaces for software which involves recognition. Recognition errors easily confuse the user, and result in brittle interaction dialogues. Existing user interface toolkits have no way to model ambiguity. Mankoff et al. address symbol recognition ambiguity, segmentation ambiguity, and target ambiguity. (Target ambiguity occurs when it is unclear which drawing elements are the target of a user's selection operation.) Two design heuristics for increasing the usability of mediators that handle recognition ambiguity are described. The first heuristic is to provide sensible defaults, for example highlighting the top choice in an N-best list that is displayed to the user. The second heuristic is to be lazy. Put off making choices; these can result in errors. Later input may provide information that helps with disambiguation, and some input (such as comments and annotations) may never need to be disambiguated.

6   Supporting User Correction of Recognition Errors

Good user interface support is needed to allow the user to quickly locate and correct recognition errors. In some cases, domain-specific methods can be applied to help locate errors. For example, a recognition system for music notation can use audio feedback to allow the user to listen for recognition errors. A mathematics recognition system can use numerical evaluation to check the correctness of recognized equalities.

Modes are commonly used to support error correction. Our UML and mathematics recognition systems use diagram-entry and diagram-correction modes. The user clicks a button to change modes. More natural mode selection might be achieved by a two-ended pen. People are accustomed to using the two ends of a pencil for writing and erasing. The two pen ends can be interpreted in a variety of ways, perhaps using one end for drawing the diagram and the other end for commands and corrections. The user need to know which mode is currently selected. Mouse-based systems can use cursor shape to indicate mode. This does not carry over to our systems, because the electronic whiteboard does not provide cursor tracking when the pen is away from the drawing surface. We continue to investigate ways to handle modes in our recognition systems.

Overtracing can be used to correct glyph recognition errors [13]. If the user draws a blob that could be either 'circle' or 'box', both interpretations are retained. The user can later identify the element by drawing over it, this time more clearly as a box.

N-best lists have been widely used to allow a user to select from a set of recognition alternatives. Uses of N-best lists include the following [19]: a voice recognition system uses an N-best list to display candidates for a misrecognized word; a drawing system displays an N-best list for a misrecognized glyph; a drawing system uses an N-best lists to display various ways of grouping similar glyphs.

7   Conclusion

The user interface plays a significant role in a diagram recognition system. Unfortunately, it is difficult to define precise goals for a user interface, to find clear guidelines for user interface construction, or to quantify the performance of a user interface. We have reviewed the interaction between diagram editing, generation and recognition and have discussed five categories of user interface issues: quantifying the performance of a user interface, supporting diagram input, executing the recognizer, displaying recognition results, and supporting user correction of recognition errors. Many open problems remain in all of these areas. We look forward to discussions and future work which will lead to improved strategies and criteria for designing, implementing, and evaluating user interfaces for on-line diagram recognition systems.

Acknowledgments

The work reported in this paper benefited greatly from discussions with and software implementation by Sean Chen, T. C. Nicholas Graham, Jeremy Hussell, Alvin Jugoon, David Tausky, Jeb Thorley, Randall Tuesday, and James Wu. Funding by NSERC (Canada's Natural Sciences and Engineering Research Council) and CITO (Communications and Information Technology Ontario) is gratefully acknowledged.

References

[1]   J. Arvo, K. Novins, "Smart Text: A Synthesis of Recognition and Morphing," *Proc. AAAI Spring Symposium on Smart Graphics*, Stanford, Cal., March 2000, 140–147.

[2]   S. Bagley, G. Kopec, "Editing Images of Text," *Communications of the ACM*, **37**(4), Dec. 1994, 63–72.

[3]   D. Blostein, L. Haken, "The Lime Music Editor: A Diagram Editor Involving Complex Translations," *Software – Practice and Experience*, **24**(3), March 1994, 289–306.

[4]   D. Blostein, L. Haken, "Using Diagram Generation Software to Improve Diagram Recognition: A Case Study of Music Notation," *IEEE Trans. Pattern Analysis and Machine Intelligence* , **21**(11), Nov. 1999, 1121–1136.

[5]   K. Brooks, "Lilac: A Two-View Document Editor," *IEEE Computer*, June 1991, 7–19.

[6]   M. Brown, "Zeus: A System for Algorithm Animation and Multi-View Editing," *Proc. IEEE Workshop on Visual Languages*, Kobe, Japan, Oct. 1991, 4–9.

[7]   J. Carroll, editor, Scenario-Based Design: Envisioning Work and Technology in System Development, John Wiley, 1995.

[8]   HCI journals: ACM Transactions on Computer-Human Interaction; International Journal of Man-Machine Studies; Human-Computer Interaction; Behavior and Information Technology; Human Factors.

[9]   W. Cushman, P. Ojha, C. Daniels, "Usable OCR: What are the Minimum Performance Requirements?," *Proc. ACM SIGCHI 1990 Conf. Human Factors in Computing Systems*, Seattle, Washington, April 1990, 145–151.

[10] C. Damm, K. Hansen, M. Thomsen, "Tool Support for Cooperative Object-Oriented Design: Gesture Based Modeling on an Electronic Whiteboard," *Proc. CHI 2000*, The Hague, Netherlands, April 2000, 518–525 See also www.daimi.au.dk/~knight/.

[11] D. Diaper, "Task Observation for Human-Computer Interaction," in D. Diaper, editor, *Task Analysis for Human-Computer Interaction*, Ellis Horwood, 1989, 210–237.

[12] T. C. N. Graham, L. Watts, G. Calvary, J. Coutaz, E. Dubois, L. Nigay, "A Dimension Space for the Design of Interactive Systems Within their Physical Environments," *Proc. Designing Interactive Systems (DIS 2000)*, 2000, 406–416.

[13] M. Gross, E. Do, "Ambiguous Intentions: a Paper-like Interface for Creative Design," *Proc. Ninth Annual Symposium on User Interface Software and Technology (UIST'96)*, Seattle, Washington, Nov. 1996, 183–192.

[14] K. Höök, "Designing and Evaluating Intelligent User Interfaces," *Proc. 1998 ACM Int'l Conf. Intelligent User Interfaces*, San Francisco, California, 1998, 5–6.

[15] R. Hsu, W. Mitchell, "After 400 Years, Print is Still Superior," *Communications of the ACM*, **40**(10), Oct. 1997, 27–28.

[16] L. Julia, C. Faure, "Pattern Recognition and Beautification for a Pen Based Interface," *Proc. Third Int'l Conf. Document Analysis and Recognition*, Montreal, Canada, August 1995, 58–63.

[17] E. Lank, J. Thorley, S. Chen, D. Blostein, "On-line Recognition of UML Diagrams," *Proc. Sixth Int'l Conf. Document Analysis and Recognition*, Seattle, Washington, Sept. 2001, 356–360.

[18] A. Long, J. Landay, L. Rowe, J. Michiels, "Visual Similarity of Pen Gestures," *Proc. CHI 2000 Human Factors in Computing Systems*, The Hague, April 2000, 360–367.

[19] J. Mankoff, S. Hudson, G. Abowd, "Providing Integrated Toolkit-Level Support for Ambiguity in Recognition-Based Interfaces," *Proc. CHI 2000 Human Factors in Computing Systems*, The Hague, Netherlands, April 2000, 368–375.

[20] J. McGrenere, W. Ho, "Affordances: Clarifying and Evolving a Concept," *Proc. Graphics Interface 2000*, 179–186.

[21] B. Myers, "User Interface Software Tools," *ACM Trans. Computer-Human Interaction*, **2**(1), 1995, 64–103.

[22] M. Nakagawa, K. Machii, N. Kato, T. Souya, "Lazy Recognition as a Principle of Pen Interfaces," *Proc. INTERACT '93 and CHI '93 Conference on Human Factors in Computing Systems*, Amsterdam, 1993, 89–90.

[23] J. Nielsen, *Usability Engineering*, Academic Press, San Diego, California, 1993.

[24] J. Nielsen, J. Levy, "Measuring Usability: Preference vs. Performance," *Communications of the ACM*, **37**(4), April 1994, 66–75.

[25] I. Phillips, A. Chhabra, "Empirical Performance Evaluation of Graphics Recognition Systems," *IEEE Trans. Pattern Analysis and Machine Intelligence*, **21** (9), Sept. 1999, 849–870.

[26] S. Smithies, K. Novins, J. Arvo, "A Handwriting-Based Equation Editor," *Proc. Graphics Interface '99*, Kingston, Ontario, June 1999, 84–91.

[27] J. Tang, S. Minneman, "VideoWhiteboard: Video Shadows to Support Remote Collaboration," *Proc. Human Factors in Computing Systems*, 1991, 315–322.

[28] B. Vander Zanden, S. Venckus, "An Empirical Study of Constraint Usage in Graphical Applications," *Proc UIST'96 – Ninth Annual Symposium on User Interface Software and Technology*, Seattle, Washington, Nov. 1996, 137–146.

[29] R. Zanibbi, D. Blostein, J. Cordy, "Baseline Structure Analysis of Handwritten Mathematics Notation," *Proc. Sixth Int'l Conf. Document Analysis and Recognition (ICDAR 2001)*, Seattle, Washington, Sept. 2001, 768–773.

[30] R. Zanibbi, K. Novins, J. Arvo, K. Zanibbi, "Aiding Manipulation of Handwritten Mathematical Expressions through Style-Preserving Morphs," *Proc. Graphics Interface 2001*, Ottawa, Ontario, June 2001, 127–134.

Symbol Recognition:
Current Advances and Perspectives

Josep Lladós, Ernest Valveny, Gemma Sánchez, and Enric Martí

Computer Vision Center, Dept. Informàtica, Universitat Autònoma de Barcelona
08193 Bellaterra (Barcelona), Spain
{josep,ernest,gemma,enric}@cvc.uab.es

Abstract. The recognition of symbols in graphic documents is an intensive research activity in the community of pattern recognition and document analysis. A key issue in the interpretation of maps, engineering drawings, diagrams, etc. is the recognition of domain dependent symbols according to a symbol database. In this work we first review the most outstanding symbol recognition methods from two different points of view: application domains and pattern recognition methods. In the second part of the paper, open and unaddressed problems involved in symbol recognition are described, analyzing their current state of art and discussing future research challenges. Thus, issues such as symbol representation, matching, segmentation, learning, scalability of recognition methods and performance evaluation are addressed in this work. Finally, we discuss the perspectives of symbol recognition concerning to new paradigms such as user interfaces in handheld computers or document database and WWW indexing by graphical content.

1 Introduction

Symbol recognition is one of the significant applications within the area of pattern recognition. Fields like architecture, cartography, electronics, engineering etc. use domain-dependent graphic notations to develop their designs. The automatic interpretation of such documents, requires processes able to recognize the corresponding alphabets of symbols. Because of this wide range of graphic documents, each one with its own characteristic set of symbols, it is not easy to find a precise definition of a symbol. In a very general way, a symbol can be defined as a graphical entity with a particular meaning in the context of an specific application domain. Thus, and depending on the application, we can find different kinds of symbols according to their visual properties: simple 2D binary shapes composed of line segments (engineering, electronics, utility maps, architecture), a combination of line segments and solid shapes (musical scores), complex gray level or color shapes (logos), silhouettes (geographic symbols), etc. In this paper, we have taken the general definition stated above and hence, we have focused our attention on applications and methods developed to identify any meaningful entity in graphic documents. Good reviews on the state-of-the-art about symbol recognition were reported in previous Graphics Recognition Workshops and

D. Blostein and Y.-B. Kwon (Eds.): GREC 2002, LNCS 2390, pp. 104–128, 2002.
© Springer-Verlag Berlin Heidelberg 2002

related conferences [1, 2, 3, 4]. The main goals of this new overview are: first, to update the literature review on symbol recognition; secondly, to give a systematic and structured overview of methods, providing a double and related classification from two points of view, namely, applications and techniques. Finally, to address the set of challenges and open issues which can be derived from the analysis of current approaches.

From the point of view of applications, much of the research in graphics recognition has been addressed to the automatic conversion of graphic documents to a format able to be understood by CAD systems. Many efforts have been focused on the development of efficient raster-to-vector converters (e.g. [5, 6, 7, 8]). Moreover, performance evaluation methods have been developed [9] and contests on raster-to-vector conversion have been held in past editions of the Graphics Recognition workshop [10, 11]. However, raster-to-vector conversion should not be the final step. Complete raster-to-CAD conversion should also provide a semantic description and interpretation of the drawing. In this context, symbol recognition is required to identify graphic entities and it has been applied to many applications, including interpretation of logic circuit diagrams, engineering and architectural drawings and any kind of maps. Apart form raster-to-CAD conversion, other significant applications where symbol recognition plays an important role are interpretation of musical scores for conversion to MIDI and logo recognition. In section 2, we will review these application domains, pointing out the most relevant properties of each domain for symbol recognition. From the point of view of methods, symbol recognition is a particular application of pattern recognition. In section 3 we will describe symbol recognition according to the classical classification in statistical and structural approaches. In table 1, we summarize symbol recognition literature according to this double classification: application domains and methods for recognition.

From the analysis of all these methods, we can conclude that although many different approaches have been reported, it is difficult to find a general and robust method covering different domains with a well-established and evaluated performance. Authors tend to develop ad-hoc techniques which are difficult to be reused in other domains. Moreover, there is no way to compare different methods operating in the same domain or to validate the performance of any approach with a significant set of symbols and drawings. In section 4, we address and identify some open problems involved in the development of general, robust and efficient symbol recognition methods. We draw conclusions from the current state of art and outline some challenges for future research. Among these issues we have included symbol segmentation and representation, matching, learning, scalability and performance evaluation.

In the last section we discuss future perspectives for symbol recognition. New paradigms in the information society technology, such as keyboardless user interfaces for handheld computers, internet search engines based on graphical queries, on-line symbol recognition in graphics tablets and interactive pen displays, etc. also demand for symbol recognition capabilities, but with different requirements than classic applications.

Table 1. State of the art of symbol recognition in a twofold point of view: application domains and techniques

	logic diagrams	engineering drawings	maps	musical scores	architectural drawings	logo recognition	formula recognition	other applications
Structural matching	[12]–[16]	[17, 18]			[19, 20]	[21, 22]		[23]
Syntactic approaches	[24]–[26]	[27]–[30]		[31]	[32]		[33, 34]	[24]
constraint satisfaction	[35, 36]	[37]	[38, 39]		[40, 41]			[42]
Neural networks	[43]		[44]	[45]–[48]		[49, 50]		
Statistical classifiers	[51, 52]		[53]–[58]	[59]	[41]	[60]–[62]	[63]	[64]
decision trees	[65]							[66, 67]
Heuristic/ad-hoc techniques		[68]	[69, 55, 70]	[71]			[72]	[73]
Other	[13]	[74]						[75]

2 Application Domains of Symbol Recognition

2.1 Logic Circuit Diagrams

Electrical and logic circuit diagrams is one of the earliest application domains that focused its attention on graphical symbol recognition. A lot of contributions can be found in the literature [12, 36, 13, 51, 26, 14, 15]. The understanding and validation of electrical schematics and its conversion to an electronic format has become through the years a prototypical graphics recognition application. Circuit diagrams offer two advantages that have probably contributed to that. First, they have a standardized notation which is based on loop structures that characterize the symbols, and rectilinear connections between them. Such representational dichotomy between symbols and interconnections leads to the second advantage: symbols belonging to logic diagrams can be segmented in a reasonably easy way by distinguishing between lines and loops or background areas, in addition to small connected components, likely representing textual annotations.

2.2 Engineering Drawings

The first difficulty of engineering drawings is that we can not assume a standardized diagrammatic notation. Actually we can distinguish two levels of symbols. The first level consists of graphical entities that can have a different meaning depending on the context where they appear. Ablameyko [76] distinguished four types of graphical entities: arcs and straight lines, dashed lines, crosshatched areas, and dimensions. These primitives can represent an angular information, a section of a mechanical part, a symmetry axis, etc. Symbols at the second level are formed by an assembly of the low level primitives. The recognition of these elements combined with domain-dependent knowledge, gives meaning to the document and allows it to be converted to a GIS or CAD format. The problem of arc detection was recently studied in [77] proposing a method that combines two of the most reliable techniques in the literature. Hatched pattern detection is an important concern in the field of document analysis [78, 69, 73] and is usually solved by clustering parallel lines having the same slope angle and sorting them along a normal direction. Dimensions usually follow strict standards. Their interpretation and validation is very important, not only to fully understand the document but also to assist in the segmentation of other graphical entities. Since they are usually based in combinations of arrowheads, lines and text in particular configurations, syntactic approaches, usually based in graph grammars, are the most usually employed techniques [27, 28, 30]. Concerning to higher level symbols that usually represent mechanical parts, since they are very domain-dependent and even document-dependent, the automatic recognition cannot be made fully automatic and requires special interactive techniques and knowledge.

2.3   Maps

The conversion of maps to a GIS format has several challenges as the combination between cartographic information and satellite images, or the integration and conversion of maps from different areas (cadastral, telephone, water, etc.). From a general point of view, we could define three types of maps which have their own notational conventions. At the lowest level of difficulty, we could place cadastral city maps [69, 55, 79, 70]. In such type of documents, symbols have a polygonal shape often filled by a hatching pattern. These polygonal shapes represent parcels and the surrounding streets and their meaning is completed with text and annotations. Thus, symbol recognition is usually formulated in terms of detection of polygonal shapes and hatched patterns. Another subdomain of map interpretation focus on utility maps [53, 54, 39]. Utility maps contain information on network facilities of companies such as water, telephone, gas, etc. They are usually binary images consisting of lines and small symbols composed of geometric basic primitives (squares, circles, arrowheads, etc.). The recognition is usually very domain dependent. Finally, geographic maps [58, 38, 44, 57] are probably the most difficult class of documents of this domain. In this kind of maps, the graphical entities are associated with line objects that usually represent isolines and roads and, on the other hand, small symbols whose meaning is given by a legend. Color information plays an important role. Thus, a layer segmentation process is usually performed in terms of color quantization. Line objects are characterized by regular structures that can be represented by a linear grammar, i.e. they are composed of regular combinations of lines, points and gaps. Thus, to extract roads and isolines, lines are followed under a rule-based criterion. The detection of symbols usually is legend-driven, i.e. the legend is first detected and then, symbol patterns are extracted from it. The meaning of each prototype symbol is also captured by an OCR procedure. Variations of pattern matching are the most used techniques to recognize such symbols.

2.4   Musical Scores

The recognition of musical scores [45, 59, 80, 46, 71, 47] can be considered a graphics recognition domain not only from the point of view of the application but also from the point of view of the proposed procedural solutions. The particular structure of a musical score and its standardized notation have resulted in the development of a set of very specific techniques, only applicable to this family of documents. The interpretation process is organized in three stages. First, the extraction of staff lines, that allows to segment individual symbols, and can be performed by projection or run analysis techniques. Second, the recognition of individual notes that, since there is a finite set of standard symbols, are robustly recognized by neural networks or feature vector distances. The third stage is the interpretation of the whole musical score. The great part of the literature solve this task by using different variations of graph grammars.

2.5   Architectural Drawings

The interpretation of architectural plans is one of the most recent activities [40, 41, 19, 20]. In architectural drawings, the recognition of higher level entities such as walls, doors, windows, furniture, stairs, etc. allows the interpretation of the document and, hence, its conversion to a CAD environment to perform actions as design edition, validation, 3D visualization and virtual navigation inside the building. Two major symbol structures can be categorized: prototype-based symbols and texture-based symbols. Examples of symbols characterized by a prototype are doors and windows. On the other hand, symbols characterized by a structured texture as hatching or tiling represent walls, floors or stairs. Two problems make the recognition of architectural symbols difficult. First, there is no standardized notation and hence, a general framework for the interpretation of the documents is not a solved issue. Second, since symbols appear embedded in the document, its segmentation is difficult to be separated from the recognition. Due to that, recognition has to be done by searching throughout all the document and, hence it is an expensive process.

2.6   Logo Recognition

Logo and trademark recognition [49, 21, 22, 60, 50, 61] can be considered a symbol recognition application that differ from the other categories. While the purpose in the other subdomains is the interpretation of a certain diagram in which symbols are constituent graphical entities following a particular notation, logo recognition is devoted to clustering documents in terms of the originating institution and retrieval by content from document databases. Thus, while classical symbol recognition methods assume that the set of symbols to be recognized in a particular application are "similar" in terms of the constituent features, the recognition of logos requires a more general framework. Due to the unrestricted variety of instances, logo recognition is usually based on extracting signatures from the image in terms of contour codification, invariant moments, connected component labeling, etc. and match the unknown logo with the database models using different types of distance or neural networks. Since logos often combine text and graphics, the recognition in some cases also includes OCR processes.

2.7   Other Applications

In addition to a number of symbol recognition works that has been performed on other types of flow charts and diagrams [24, 73, 64] let us briefly describe three particular applications, namely formula recognition, on-line symbol recognition for pen-based user interfaces and WWW graphic indexing and querying. Mathematical formula recognition is at the frontier between OCR and symbol recognition. Actually, symbols in mathematical formulas can be considered as belonging to a particular font of characters. However, from the point of view of the structure of the formula and its interpretation, the problem falls out the classical OCR approaches. The existing literature [34, 63, 72] uses feature vectors to recognize individual symbols and syntactic approaches to validate the structure of the formula. Symbol recognition procedures are also used as a tool for man-machine interfaces [66, 81, 75, 82, 83]. This is not only an specific application domain but it also requires specific techniques because the recognition is performed on-line. The general goal is to use symbolic shortcuts in a pen-based environment that allow the user to perform operations such as select, delete, copy, or interactively draw in graphical design applications. Finally, a recent application area in which symbol recognition may play an important role is indexing by content on WWW documents. The number of WWW documents in Internet is growing very fast and the ability to make queries by graphical content would allow a more efficient search of information into the Web site. In the last years the problem of locating text in Web images has been addressed [84]. The definition of new XML-based vectorial formats as SVG makes symbol recognition techniques as useful tools to implement search engines based on graphical content.

3   Symbol Recognition Methods

Symbol recognition is one of the multiple fields of application of pattern recognition, where an unknown input pattern (i.e. input image) is classified as belonging to one of the predefined classes (i.e. predefined symbols) in a particular domain. We will take the traditional classification of pattern recognition into statistical and structural approaches to give a systematic and structured overview of symbol recognition methods. The goal is only to describe which methods have been used in symbol recognition, relating them to general pattern recognition strategies. For a more general and detailed discussion of pattern recognition, many excellent surveys and books can be found in the literature [85, 86, 87, 88].

3.1   Statistical Symbol Recognition

In statistical pattern recognition, each pattern is represented as an n-dimensional feature vector extracted from the image. Classification is carried out by partitioning the feature space into different classes, one for each symbol. Therefore, two issues are especially relevant for the performance of this kind of methods:

the selection of the features and the selection of the method for partitioning the feature space. In table 2, we have classified symbol recognition approaches according to these two issues.

The selection of the feature space depends on the properties of the patterns to be classified. The main criterion must be to minimize the distance among patterns belonging to the same class and to maximize the distance among patterns belonging to different classes. Additional interesting properties of feature space are invariance to affine transformations and robustness to noise and distortion. An interesting survey of feature extraction methods, applied to the related area of character recognition can be found in [89]. In symbol recognition, only a subset of all these features have been employed. We will classify them into four groups: those based on the pixels of the image, geometric features, geometric moments and image transformations.

The simplest feature space is the image space itself. The feature vector is composed of one feature for each pixel value. Usually, the image is first normalized to a fixed size. The main advantages are simplicity, low complexity and direct correspondence with visual appearance. However, the representation is not rotation invariant and it is very sensitive to noise and distortion. Another set of methods use geometric features: centroids, axes of inertia, circularity, area, line intersections, holes, projection profiles, etc. In relation with image space, the size of the feature vector can be reduced. A good selection of relevant features is critical to achieve high discrimination power, invariance to affine transformations. Feature extraction must be robust enough to reduce feature variability due to noise and distortion. Moment invariants are another kind of features which have also been applied to symbol recognition, Both the regular moments [51, 43] and the moments of Zernike [53] have been used. Moment invariants are easy to compute, they have relation with geometric properties, such as the center of gravity, the axes of inertia, etc, and they can be made invariant to affine transformations. Finally, features can also be defined through the application of some kind of transformation of the image. Features are taken from the representation of the image in the transformation space. Image transforms which have been used in symbol recognition include Fourier transform [51, 67], Fourier-Mellin transform [53] or special transforms to get signatures from the image [60].

Once a set of features have been chosen, classification consists in selecting a method to partition the feature space and to assign each feature vector to one of the predefined classes. In symbol recognition literature, we can find methods based on the concept of similarity, on neural networks and on decision trees. The

Table 2. Statistical symbol recognition approaches: crossing of features and classification methods

	Image	Geometric features	Moments	Image transformations
Distance-based	[58, 62, 64]	[54, 59, 90, 63, 91, 52, 61]	[51]	[60, 51]
Nearest neighbors		[57]	[53]	[53]
Decision trees		[66, 65]		[67]
Neural networks	[45, 49, 44, 47, 48]	[50, 46]	[43]	

simplest way to partition the feature space consists in defining a distance function among feature vectors and assigning each input image to the class with the closest representative. A slight variation is the *k-nearest neighbors rule*, where several representatives are taken for each class and, for each input pattern, the set of the k closest representatives is built. The pattern is assigned to the class having more representatives in this set. Neural networks have showed to have good classification rates in many different domains. One of their advantages is their learning ability to adapt themselves to the properties of the training set. Learning is performed automatically, providing the optimal parameters of the network to recognize the symbols in the training set. In decision trees, each node of the tree corresponds to an specific condition about the value of a particular feature. Classification is carried out by following the branches in the tree according to the result of condition testing, until one of the leaves is reached. The leaves of the tree correspond to recognized symbols.

3.2   Structural Symbol Recognition

In structural pattern recognition symbols are represented with a description of their shape using some suitable set of geometric primitives and relationships among them. For each symbol, a model of its ideal shape is built using these primitives. An input image is classified as belonging to the symbol giving the best matching between the representation of the image and the model of the symbol. Usually, straight lines and arcs are the primitives used to describe the shape of the symbols although sometimes, other geometric primitives, such as loops, contours or simple shapes (circles, rectangles, etc.) have also been used. Therefore, a previous vectorization step is required. Vectorization can introduce noise and distortion in the representation of images and thus many times, error-tolerant matching must be used.

A large class of structural approaches are based on a graph representation of the symbols [12, 36, 13, 14, 15, 92, 17, 23]. Nodes and edges of the graph correspond, respectively, to points and lines of the image, providing a very natural and intuitive description of symbols. Matching consists in finding the best subgraph isomorphism between the input image and the models of the symbols. With this approach, symbols can be found as subgraphs of the whole image allowing to perform segmentation and recognition at the same time. Distortion is handled using error-tolerant subgraph isomorphism graph edit operations to define an error model. The main drawback of graph matching is computational complexity. Some ways of reducing computation time have been explored.

Formal grammars - usually graph grammars because of bidimensional structure of symbols - are used in another family of structural approaches [24, 80, 25, 26, 34]. A grammar stores in a compact way all valid instances of a symbol or a class of symbols. The recognition of an input image consists in parsing its representation to test if it can be generated by the grammar. To handle distortion, differnt types of error-correcting parsers are proposed. Grammars are useful in applications where the shape of the symbols can be accurately defined by a set of rules, for instance, the recognition of dimension symbols in technical drawings [27, 28, 30] and the recognition of symbols composed of textured areas [32]. Joseph and Pridmore [29] show an alternative use of grammars in which the grammar not only describes the structure of the symbols, but also guides the interpretation of the whole drawing.

Another set of methods uses a set of rules to define geometric constraints among the primitives composing the symbol. Then, these rules are applied or propagated to find symbols in the input image [35, 38, 37]. In [42] a kernel based on a *blackboard architecture* guides the application of the rules, selecting and activating a set of procedures for searching new elements in the drawing when any primitive or symbol is recognized. A similar approach is used in [39]. In [40], the rules are organized in a constraint network. Symbols are identified by traversing the network and testing the rules at every node.

Some other approaches use deformable template matching and a structural representation of the symbols [93, 20] to handle distortion in hand-drawn symbols. The goal is to find a deformation of the model of the symbol resembling the input image. This goal is achieved through the minimization of an energy function, composed of an internal energy measuring the degree of deformation of the model and an external energy measuring the degree of similarity between the deformation of the model and the input image.

Hidden Markov Models can also be seen as structural methods since the structure of the symbol can be described by the sequence of states generating the image. Recognition consists in finding the sequence of states with higher probability. Features used to represent the symbols in HMM approaches include discrete cosine transformation [18], log-polar mapping [21] and image pixels [33]. HMMs are able to segment the symbols and to recognize distorted symbols.

Finally, there is a set of methods [69, 55, 73, 70, 72, 71, 68, 37], also based on a structural representation of the image, in which symbol representation and symbol recognition are not independent tasks. Symbol recognition is carried out by a set of specific procedures for each symbol, and the knowledge about the shape of the symbol is encoded in the procedure itself.

4   Open Issues

From the analysis of existing approaches, we can identify some unsolved or unaddressed issues, concerning the development of general, robust and efficient symbol recognition strategies. In this section, we will discuss the most significant of them, outlining open questions and future perspectives and challenges.

4.1   Segmentation

Symbol segmentation is a very domain-dependent task. A number of symbol recognition contributions assume that symbols have been previously segmented (see appendix A). However, it is not always feasible to break the drawing up into unique constituent components. In certain cases, only a partial or approximate segmentation can be done and domain-dependent knowledge or user assistance is required. The methods that separate segmentation and recognition in different stages, usually base the segmentation on features such as connected components, loops, color layers, long lines, etc. As we have seen in section 2, there are some domains were symbols can be separated from the rest of the drawing in terms of the knowledge of the notation. Thus, the easiest domain is probably musical scores where symbols can be segmented in terms of connected components after removing the staff lines by projections or run length smearing. Other applications such as maps support the segmentation on the presence of a legend. Thus, symbols in the legend are first segmented and then, this information is used as signature to index in the whole document. In electronic diagrams, symbol segmentation is achieved by removing long lines that connect loop-based graphical entities. In other domains such as logo or formula recognition, symbols can be segmented in terms of connected components. Difficulties arise when symbols appear embedded in the drawing. This is usual where symbols consist of an assembly of low level primitives such as arcs, lines, loops and crosshatched or solid areas. Efficient techniques have been proposed to detect each type of these low level primitives. However, at a higher level, considering that a symbol is an assembly of the above primitives, to find a part of the diagram that is likely to represent a symbol is not a trivial task. In such cases, assuming that the knowledge about the domain is required, the trend is to define symbol signatures based on simple features that allow to locate image areas with high evidence to contain a symbol. Doermann in [94] reviews different techniques for indexing of document images that can be used for symbol segmentation.

4.2   Symbol Representation

The selection of an structure for symbol representation can have strong influence in the performance of symbol recognition. It also has a strong relationship with the selection of a matching method, although both issues should be clearly distinguished. A suitable structure should be compact, complete, i.e. general enough to represent symbols from different domains, discriminant, i.e. able to maximize the intra-class similarity and the inter-class dissimilarity, computationally manageable, extensible and able to support distortion models.

In statistical symbol recognition, feature vectors are a very simple representation with low computational cost. However, discrimination power and robustness to distortion strongly depends on the selection of an optimal set of features for each specific application. There is no comparative study on the performance of symbol recognition with different sets of features. Moreover, the number of features must be small and sometimes methods for reduction of dimensionality must be applied. Finally, these methods need a previous segmentation step, which it is not always an easy task because of the embedding of symbols in the drawing.

In structural representations, feature selection is not so critical because they usually rely on a vectorial representation, although vectorization introduces some errors in the representation. The main advantages of structural representations are generality, extensibility and ability to include distortion models.

In conclusion, there is no optimal and general structure for symbol representation and it seems not easy to define a representation powerful enough and general enough to perform well in different domains and applications. Probably, a comparative study on the performance of different representations on several problems and applications is required. Moreover, further research can be done to explore the feasibility of mixed representations combining both approaches and being able to represent symbols in a more general and complete way. This approach can allow the application and combination of several classifiers - see section 4.3 - to get better performance. Signatures have emerged in pattern recognition applications as a simple, flexible and general structure to represent relevant shape properties. Their application to symbol representation could be an interesting approach for general segmentation approaches and applications such as indexing and retrieval of graphic documents.

4.3   Matching

Matching is the procedure to decide to which symbol corresponds an unknown input image. It is the core of any symbol recognition approach. Some desirable properties for matching are the following:

- **Generality:** the ability of matching to be applied to a wide number of different applications.
- **Extensibility:** the ability of matching to work if we add new symbols to be recognized, without need for being changed or rewritten.
- **Scalability:** it refers to the performance of matching when the number of symbols is significantly high (hundreds or thousands), in terms of recognition rates and computation time.
- **Robustness to distortion:** matching must be able to recognize distortions due to noise, hand-drawing or feature extraction errors, without increasing confusions among different symbols.
- **Low computational complexity**.

Concerning to these properties, in statistical methods, generality, scalability and robustness to distortion will depend on the selection of an optimal set of features, although there are no evaluation studies on these subjects. Adding new symbols can require re-learning classification parameters since the feature space is modified. They are usually computationally efficient methods. On the other hand, structural approaches are easier to generalize to different domains and to extend with new symbols to recognize because all graphic symbols share a similar representation using lines and points, the usual low-level primitives. Errors produced by vectorization can increase confusions among different symbols, making matching difficult to scale to a great number of symbols. Most of these methods are able to recognize distorted symbols with different degrees of distortion allowed. Finally, they are methods having high computational complexity, increasing it significantly with the number of symbols to be recognized.

As we can see, many approaches using both statistical and structural approaches have already been developed, and both strategies have attractive properties for symbol recognition. Therefore, maybe research efforts should be concentrated, not in finding new general methods, but in combining several existing methods following some kind of combination scheme. This strategy could lead to significant improvements on performance and could provide new perspectives to the scalability problem. Another interesting issue is that of parallelization of algorithms to reduce computational complexity in structural approaches.

4.4   Learning

Symbol recognition requires the selection of some representative for each symbol type. When patterns are represented by numeric feature vectors, the representative is easy to be computed by the mean or the median of the feature vectors. However, when objects are represented in terms of symbolic structures, the inference of a representative is no longer clear and it is not an obvious task. The computation of a mean symbol is useful in applications that require the learning of a prototypical pattern from a set of noisy samples. Two categories for symbol representative computation can be stated [95]. The fist approach looks for a representative among the set of samples. Thus, given a set of symbols $S = \{s_1, \ldots, s_n\}$, the *set median symbol* $\hat{s}$ is defined as a symbol $s_i \in S$ that has the minimum combined distance to all other elements in the set. Formally,

$$\hat{s} = arg \min_{s_i \in S} \sum_{s_j \in S} d(s_i, s_j), \tag{1}$$

where d denotes the distance function between symbols. In general, the inference of a representative is not constrained to the set of samples but has to be searched among the set of all possible symbols of the particular context. Formally, let $S = \{s_1, \ldots, s_n\}$ be a set of symbols and let Ω be the alphabet of all possible symbols, the *generalized median symbol* or just the *mean symbol* $\bar{s}$ is defined as:

$$\bar{s} = arg \min_{s \in \Omega} \sum_{s_i \in S} d(s, s_i). \tag{2}$$

The theoretical basis for symbol learning can be found in conceptual clustering approaches to learn shapes from examples. Wong and You [96] proposed a statistical-structural combined approach. They defined random graphs as a particular type of graphs which convey a probabilistic description of the data. A process is defined to infer a synthesized random graph minimizing an entropy measure. Segen [23] developed a graph-based learning method for non rigid planar objects able to learn a set of relations between primitives that characterize the shapes of the training set. Recently, Jiang et al. [13] proposed a genetic algorithm to compute the mean among a set of graphs. The algorithm was applied to graphs representing graphical symbols of electronic diagrams. The experimental results prove that the algorithm is able to obtain a representative symbol that smoothes the individual distortions in the noisy samples. Cordella et al. [97] describe the graph representing the symbols with a set of logic predicates. Then, they apply Inductive Logic Programming to modify these predicates from a set of examples. Another recent approach based on deformable models is described in [98], where for each test image, the deformation of the symbol that best fits the image is found. Then, the representative of the symbol is defined as the mean of all deformations. In conclusion, there is still room for improvements in symbol learning. Although there are plenty of contributions in the literature on the symbolic learning paradigm, a few of them experiment their proposed algorithms on graphical symbols frameworks, using real data with noise and distortion as source samples. On the other hand, learning from sets of complex symbols and improving the computational cost are still challenges on that issue.

4.5   Scalability

When the number of symbols to be recognized increases, the recognition performance can degrade because the uncertainty of the inter-class boundaries increases, and hence the probability of symbol confusion. From the point of view of computation time, if recognition requires sequential matching between the input symbol and each model symbol in the database, the time will clearly grow with the number of symbols. Very few symbol recognition methods take into account the scalability issue. Current methods work reasonably well with small databases but their performance seriously degrades when databases contain hundreds of prototypes. Since real applications use to have large databases of symbols, the scalability issue is mandatory to be addressed. It requires the development of new approaches able, firstly, to represent large databases of prototypes and, secondly, to recognize unknown symbols with invariance to the size of the database. One way to perform the search is to use an indexing procedure. The basic idea of indexing is to use a specific set of easily computable features of an input symbol in order to "rapidly" extract from the database those symbols containing this group of features. Some solutions have been proposed in restricted domains. The most outstanding contributions are based on graph representations [99, 100]. These approaches represent symbols by graphs and use variations of the graph adjacency matrix as indexing keys. The database is often organized in terms of a network in which symbols are hierarchically represented taking advantage of their common substructures. Although the matching cost is independent from the size of the database, these approaches are generally restricted to noise free environments and require an exponential time to compile the set of prototypes for the database. Another interesting approach, applied to logo recognition [60], is that of detecting some relevant features of the symbol and matching the input image only with those prototypes in the database sharing those features.

4.6   Performance Evaluation

Performance evaluation is an emerging interest of graphics recognition community. As we have seen in past sections, a lot of methods for symbol recognition have been designed. However, we have already noted the need for evaluation studies to estimate the accuracy, robustness and performance of different methods in some systematic and standard way. Usually, the algorithms are only evaluated by the subjective criteria of their own developers, and based on qualitative evaluation reported by human perception. In the last years, the Graphics Recognition community has reported interesting contributions on performance evaluation of vectorization systems, e.g. [101, 9] and several vectorization contests have been held in past editions of Workshop on Graphics Recognition [10, 11]. These contests have been directed towards the evaluation of raster-to-vector systems, designing metrics to measure the precision of vectorial representations. However, these metrics are not able to evaluate the impact of vectorization on higher-level processes such as symbol recognition. It could be interesting to extend them to be able to evaluate the influence of a vectorization method in symbol recognition.

Liu an Dori [102] distinguished three components to evaluate the performance of graphics recognition algorithms. First, the design of a sound ground truth covering a wide range of possible cases and degrees of distortion. Secondly, a matching method to compare the ground truth with the results of the algorithm. Finally, the formulation of a metric to measure the "goodness" of the algorithm. Following these criteria, a first contest in binary symbol recognition has been organized in the last *International Conference on Pattern Recognition* [103]. The dataset consisted of 25 electrical symbols with small rotation angles, scaled at different sizes and with three types of noise: quantization error, replacement noise and salt-and-pepper noise. Performance measures include misdetection, false alarm and precision of location and scale detection.

It is not an easy task to define a ground-truth and quantitative, domain independent evaluation measures for symbol recognition. The discussion is open, although we will outline some of the guidelines to develop general datasets and metrics. Datasets should be designed in order to include a high number of symbols from different classes and application domains to evaluate the generality and scalability of methods; they should also contain symbols with different kinds and degrees of distortion; non-segmented instances of the symbols are also required to test segmentation ability; finally, real drawings should also be provided to evaluate the performance in real applications. An important issue in designing a dataset is the proper organization and classification of the data. From the point of view of metrics, indices should be developed to measure issues such as recognition accuracy, computation time, degradation of performance with increasing degrees of distortion, degradation of performance with increasing number of symbols, generality, etc.

5   Conclusions and Perspectives

In this paper we have reviewed the state of the art on the symbol recognition problem. It is an update of previous reviews reported by Blostein [1], Chabbra [2] and Cordella [3] in the previous Graphics Recognition Workshops. In the first part of the paper, we have reviewed the literature from two points of view, namely the application domain and the techniques used for the recognition. Afterwards, we have discussed the challenges that the scientific community has in the following years in relation with symbol recognition and related issues.

Many symbol recognition techniques are available, but it is difficult to see a dominant one. The influence of the domain knowledge and the diagrammatic notation properties makes each family of applications to develop its own methods. The definition of a generic symbol recognition method is still a challenge. Several approaches have been proposed in terms of classical pattern recognition methods, either statistical or structural, but they tend to concentrate on a restricted range of requirements. There are other open issues beyond the matching itself. The effort should be made in the development of symbol recognition methods able to combine different classifiers that use different types of constituent features. Therefore, it is also desirable to find a representational model for symbols robust, manageable and general enough to uphold methods from different paradigms and also, to represent symbols from different domains. Symbol segmentation is feasible in those domains where the notation gives enough information about the differences between a symbol and the other graphical entities. However, when symbols appear embedded in the document segmentation and recognition are hardly separated. One solution is the definition of symbol signatures that allow to index into the drawing to locate areas where the symbol is likely to appear. Other issues have been outlined. Thus, the symbol prototypes should be learned, minimizing the intra-class distance and maximizing the inter-class distance. It is a straightforward task when symbols are represented by feature vectors, but becomes a non trivial issue when symbols are represented by structural models. Most symbol recognition methods are not scalable and are just tested with databases of a few number of prototypes, but real applications use to manage sets of hundreds of prototypes. Since the recognition performance uses to degrade with large databases, the robustness against scalability is strongly required to be studied by the research community. Finally, an open issue is the need for protocols to evaluate the performance of symbol recognition algorithms.

In addition to the technological and scientific challenges in the symbol recognition field, the perspectives should also be stated in terms of the potential applications that can take advantage of this technology. Classically, symbol recognition has been integrated in processes involving a paper-to-electronic format conversion. Although there is still room for improvements in the interpretation of paper-based diagrams, the background in symbol recognition can be perfectly used to solve new challenges associated with the evolution of new technologies. One of the promising applications where symbol recognition is a key procedure is document image retrieval by graphical content. There is a wide variety of applications: making queries in document databases to retrieve documents of a given company using its trademark, or indexing in a map database in terms of given symbols or those in the legend. But the same ideas can be applied in Web search engines. Up to now, only textual queries can be made, however extending it to search for graphical entities would be a great help in the navigability. Even in browsing large graphical documents, symbol recognition is a very interesting tool if the user was able to rapidly and interactively retrieve those graphical entities in the document similar with a selected one. The world of pen-based computers and, particularly, PDAs offers also great chances, in that case, taking advantage of the comprehensive work on on-line symbol recognition. Notice that we have stated new perspectives for symbol recognition but which can be supported on the classical techniques. Hence, in addition to investigate in the challenges discussed in section 4, the forthcoming activity within the graphics recognition community will also be concerned in exploring such new perspectives.

References

[1] Blostein, D.: General diagram-recognition methodologies. In Kasturi, R., Tombre, K., eds.: Graphics Recognition: Methods and Applications. Springer, Berlin (1996) 106–122 Vol. 1072 of LNCS. 105, 117

[2] Chhabra, A.: Graphic symbol recognition: An overview. In Tombre, K., Chhabra, A., eds.: Graphics Recognition: Algorithms and Systems. Springer, Berlin (1998) 68–79 Vol. 1389 of LNCS. 105, 117

[3] Cordella, L., Vento, M.: Symbol and shape recognition. In Chhabra, A., Dori, D., eds.: Graphics Recognition: Recent Advances. Springer-Verlag, Berlin (2000) 167–182 Vol. 1941 of LNCS. 105, 117

[4] Kasturi, R., Luo, H.: Research advances in graphics recognition: An update. In Murshed, N., Bortolozzi, F., eds.: Advances in Document Image Analysis, First Brazilian Symposium, BSDIA'97. Springer, Berlin (1997) 99–110 Vol. 1339 of LNCS. 105

[5] Chen, Y., Langrana, N., Das, A.: Perfecting vectorized mechanical drawings. Computer Vision and Image Understanding **63** (1996) 273–286. 105

[6] Dori, D., Liu, W.: Sparse pixel vectorization: An algorithm and its performance evaluation. IEEE Trans. on PAMI **21** (1999) 202–215. 105

[7] Nagasamy, V., Langrana, N.: Engineering drawing processing and vectorisation system. Computer Vision, Graphics and Image Processing **49** (1990) 379–397. 105

[8] Tombre, K., Ah-Soon, C., Dosch, P., Masini, G., Tabonne, S.: Stable and robust vectorization: How to make the right choices. In: Proceedings of Third IAPR Work. on Graphics Recognition. (1999) 3–16 Jaipur, India. 105

[9] Phillips, I., Chhabra, A.: Empirical performance evaluation of graphics recognition systems. IEEE Trans. on PAMI **21** (1999) 849–870. 105, 117

[10] Chhabra, A., Phillips, I.: The second international graphics recognition contest - raster to vector conversion: A report. In Tombre, K., Chhabra, A., eds.: Graphics Recognition: Algorithms and Systems. Springer, Berlin (1998) 390–410 Vol. 1389 of LNCS. 105, 117

[11] Chhabra, A., Philips, I.: Performance evaluation of line drawing recognition systems. In: Proceedings of 15th. Int. Conf. on Pattern Recognition. Volume 4. (2000) 864–869 Barcelona, Spain. 105, 117

[12] Groen, F., Sanderson, A., Schlag, F.: Symbol recognition in electrical diagrams using probabilistic graph matching. PRL **3** (1985) 343–350 . 106, 111, 126

[13] Jiang, X., Munger, A., Bunke, H.: Synthesis of representative graphical symbols by computing generalized median graph. In Chhabra, A., Dori, D., eds.: Graphics Recognition: Recent Advances. Springer-Verlag, Berlin (2000) 183–192 Vol. 1941 of LNCS. 106, 111, 115, 126

[14] Kuner, P., Ueberreiter, B.: Pattern recognition by graph matching. combinatorial versus continuous optimization. Int. Journal of Pattern Recognition and Artificial Intelligence **2** (1988) 527–542. 106, 111, 127

[15] Lee, S.: Recognizing hand-written electrical circuit symbols with attributed graph matching. In Baird, H., Bunke, H., Yamamoto, K., eds.: Structured Document Analysis. Springer Verlag, Berlin (1992) 340–358. 106, 111, 127

[16] Sato, T., Tojo, A.: Recognition and understanding of hand-drawn diagrams. In: Proceedings of 6th. Int. Conf. on Pattern Recognition. (1982) 674–677. 106, 127

[17] Messmer, B., Bunke, H.: Automatic learning and recognition of graphical symbols in engineering drawings. In Kasturi, R., Tombre, K., eds.: Graphics Recognition: Methods and Applications. Springer, Berlin (1996) 123–134 Vol. 1072 of LNCS. 106, 111, 127

[18] Muller, S., Rigoll, G.: Engineering drawing database retrieval using statistical pattern spotting techniques. In Chhabra, A., Dori, D., eds.: Graphics Recognition: Recent Advances. Springer-Verlag, Berlin (2000) 246–255 Vol. 1941 of LNCS. 106, 112, 127

[19] Lladós, J., Sánchez, G., Martí, E.: A string-based method to recognize symbols and structural textures in architectural plans. In Tombre, K., Chhabra, A., eds.: Graphics Recognition: Algorithms and Systems. Springer, Berlin (1998) 91–103 Vol. 1389 of LNCS. 106, 108, 127

[20] Valveny, E., Martí, E.: Hand-drawn symbol recognition in graphic documents using deformable template matching and a bayesian framework. In: Proceedings of 15th. Int. Conf. on Pattern Recognition. Volume 2. (2000) 239–242 Barcelona, Spain. 106, 108, 112, 127

[21] Chang, M., Chen, S.: Deformed trademark retrieval based on 2d pseudo-hidden markov model. PR **34** (2001) 953–967. 106, 108, 112, 126

[22] Cortelazzo, G., Mian, G., Vezzi, G., Zamperoni, P.: Trademark shapes descrition by string matching techniques. PR **27** (1994) 1005–1018. 106, 108, 126

[23] Segen, J.: From features to symbols: Learning relational models of shape. In Simon, J., ed.: From Pixels to Features. Elsevier Science Publishers B. V. (North-Holland) (1989) 237–248. 106, 111, 115, 127

[24] Bunke, H.: Attributed programmed graph grammars and their application to schematic diagram interpretation. IEEE Trans. on PAMI **4** (1982) 574–582. 106, 109, 111, 126

[25] Fahn, C., Wang, J., Lee, J.: A topology-based component extractor for understanding electronic circuit diagrams. IEEE Trans. on PAMI **2** (1989) 1140–1157. 111, 126

[26] Kiyko, V.: Recognition of objects in images of paper based line drawings. In: Proceedings of Third IAPR Int. Conf. on Document Analysis and Recognition, ICDAR'95. Volume 2., Montreal, Canada (1995) 970–973. 106, 111, 126

[27] Collin, S., Colnet, D.: Syntactic analysis of technical drawing dimensions. Int. Journal of Pattern Recognition and Artificial Intelligence **8** (1994) 1131–1148. 106, 107, 112, 126

[28] Dori, D.: A syntactic/geometric approach to recognition of dimensions in engineering machine drawings. Computer Vision, Graphics and Image Processing **47** (1989) 271–291. 107, 112, 126

[29] Joseph, S., Pridmore, T.: Knowledge-directed interpretation of mechanical engineering drawings. IEEE Trans. on PAMI **14** (1992) 928–940. 112, 126

[30] Min, W., Tang, Z., Tang, L.: Using web grammar to recognize dimensions in engineering drawings. PR **26** (1993) 1407–1916. 106, 107, 112, 127

[31] Fahmy, H., Blostein, D.: A survey of graph grammars: Theory and applications. In: Proceedings of 12th. Int. Conf. on Pattern Recognition (a). (1994) 294–298 Jerusalem, Israel. 106

[32] Sánchez, G., Lladós, J.: A graph grammar to recognize textured symbols. In: Proceedings of 6th Int. Conf. on Document Analysis and Recognition. (2001) Seattle, USA. 106, 112

[33] Kosmala, A., Lavirotte, S., Pottier, L., Rigoll, G.: On-Line Handwritten Formula Recognition using Hidden Markov Models and Context Dependent Graph Grammars. In: Proceedings of 5th Int. Conf. on Document Analysis and Recognition. (1999) 107 – 110 Bangalore, India. 106, 112, 126

[34] Lavirotte, S., Pottier, L.: Optical formula recognition. In: Proceedings of 4th Int. Conf. on Document Analysis and Recognition. (1997) 357 – 361 Ulm, Germany. 106, 109, 111, 127

[35] Bley, H.: Segmentation and preprocessing of electrical schematics using picture grafs. Computer Vision, Graphics and Image Processing **28** (1984) 271–288. 106, 112, 126

[36] Habacha, A.: Structural recognition of disturbed symbols using discrete relaxation. In: Proceedings of 1st. Int. Conf. on Document Analysis and Recognition. (1991) 170–178 Saint Malo, France. 106, 111, 126

[37] Vaxiviere, P., Tombe, K.: Celesstin: CAD conversion of mechanical drawings. Computer **25** (1992) 46–54. 106, 112, 128

[38] Myers, G., Mulgaonkar, P., Chen, C., DeCurtins, J., Chen, E.: Verification-based approach for automated text and feature extraction from raster-scanned maps. In Kasturi, R., Tombre, K., eds.: Graphics Recognition: Methods and Applications. Springer Verlag, Berlin (1996) 190–203. 106, 107, 112, 127

[39] Hartog, J., Kate, T., Gerbrands, J.: Knowledge-based segmentation for automatic map interpretation. In Kasturi, R., Tombre, K., eds.: Graphics Recognition: Methods and Applications. Springer, Berlin (1996) Vol. 1072 of LNCS. 106, 107, 112, 126

[40] Ah-Soon, C., Tombre, K.: Architectural symbol recognition using a network of constraints. PRL **22** (2001) 231–248. 106, 108, 112, 126

[41] Aoki, Y., Shio, A., Arai, H., Odaka, K.: A prototype system for interpreting hand-sketched floor plans. In: Proceedings of 13th. Int. Conf. on Pattern Recognition. (1996) 747–751 Vienna, Austria. 106, 108, 126

[42] Pasternak, B.: Processing imprecise and structural distroted line drawings by an adaptable drawing interpretation system. In Dengel, A., Spitz, L., eds.: Document Analysis Systems. World Scientific (1994) 349–365. 106, 112, 127

[43] Cheng, T., Khan, J., Liu, H., Yun, Y.: A symbol recognition system. In: Proceedings of Second IAPR Int. Conf. on Document Analysis and Recognition, ICDAR'93. (1993) 918–921 Tsukuba, Japan. 106, 110, 126

[44] Reiher, E., Li, Y., Donne, V., Lalonde, M., C.Hayne, Zhu, C.: A system for efficient and robust map symbol recognition. In: Proceedings of the 13th IAPR Int. Conf. on Pattern Recognition. Volume 3., Viena, Austria (1996) 783–787. 106, 107, 110, 127

[45] Anquetil, E., Coüasnon, B., Dambreville, F.: A symbol classifier able to reject wrong shapes for document recognition systems. In Chhabra, A., Dori, D., eds.: Graphics Recognition - Recent Advances. Springer, Berlin (2000) 209–218 Vol. 1941 of LNCS. 106, 107, 110, 126

[46] Miyao, H., Nakano, Y.: Note symbol extraction for printed piano scores using neural networks. IEICE Trans. Inf. and Syst. **E79-D** (1996) 548–553. 107, 110, 127

[47] Yadid-Pecht, O., Gerner, M., Dvir, L., Brutman, E., Shimony, U.: Recognition of handwritten musical notes by a modified neocognitron. Machine Vision and Applications **9** (1996) 65–72. 107, 110, 128

[48] Yang, D., Webster, J., Rendell, L., Garret, J., Shaw, D.: Management of graphical symbols in a cad environment: a neural network approach. In: Proceedings of Int. Conf. on Tools with AI. (1993) 272–279 Boston, Massachussets. 106, 110, 128

[49] Cesarini, F., Francesconi, E.; Gori, M., Marinai, S., Sheng, J., Soda, G.: A neural-based architecture for spot-noisy logo recognition. In: Proceedings of Fourth IAPR Int. Conf. on Document Analysis and Recognition, ICDAR'97. Volume 1. (1997) 175–179 Ulm, Germany. 106, 108, 110, 126

[50] Francesconi, E., Frasconi, P., Gori, M., Mariani, S., Sheng, J., Soda, G., Sperduti, A.: Logo recognition by recursive neural networks. In Tombre, K., Chhabra, A., eds.: Graphics Recognition - Algorithms and Systems. Springer, Berlin (1998) Vol. 1389 of LNCS. 106, 108, 110, 126

[51] Kim, S., Suh, J., Kim, J.: Recognition of logic diagrams by identifying loops and rectilinear polylines. In: Proceedings of Second IAPR Int. Conf. on Document Analysis and Recognition, ICDAR'93. (1993) 349–352 Tsukuba, Japan. 106, 110, 126

[52] Parker, J., Pivovarov, J., Royko, D.: Vector templates for symbol recognition. In: Proceedings of 15th. Int. Conf. on Pattern Recognition. Volume 2. (2000) 602–605 Barcelona, Spain. 106, 110, 127

[53] Adam, S., Ogier, J., Cariou, C., Gardes, J., Mullot, R., Lecourtier, Y.: Combination of invariant pattern recognition primitives on technical documents. In Chhabra, A., Dori, D., eds.: Graphics Recognition - Recent Advances. Springer, Berlin (2000) 238–245 Vol. 1941 of LNCS. 106, 107, 110, 126

[54] Arias, J., Lai, C., Surya, S., Kasturi, R., Chhabra, A.: Interpretation of telephone system manhole drawings. PRL **16** (1995) 355–369. 107, 110, 126

[55] Boatto, L., Consorti, V., Del Buono, M., Di Zenzo, S., Eramo, V., Espossito, A., Melcarne, F., Meucci, M., Morelli, A., Mosciatti, M., Scarci, S., Tucci, M.: An interpretation system for land register maps. Computer **25** (1992) 25–33. 106, 107, 112, 126

[56] Samet, H., Soffer, A.: A legend-driven geographic symbol recognition system. In: Proceedings of 12th. Int. Conf. on Pattern Recognition (b). (1994) 350–355 Jerusalem, Israel. 127

[57] Samet, H., Soffer, A.: Marco: Map retrieval by content. IEEE Trans. on PAMI **18** (1996) 783–797. 107, 110, 127

[58] De Stefano, C., Tortorella, F., Vento, M.: An entropy based method for extracting robust binary templates. Machine Vision and Applications **8** (1995) 173–178. 106, 107, 110, 127

[59] Armand, J.: Musical score recognition: a hierarchical and recursive approach. In: Proceedings of Second IAPR Int. Conf. on Document Analysis and Recognition, ICDAR'93. (1993) 906–909 Tsukuba, Japan. 106, 107, 110, 126

[60] Doermann, D., Rivlin, E., Weiss, I.: Applying algebraic and differential invariants for logo recognition. Machine Vision and Applications **9** (1996) 73–86. 106, 108, 110, 116, 126

[61] Soffer, A., Samet, H.: Using negative shape features for logo similarity matching. In: Proceedings of 14th. Int. Conf. on Pattern Recognition (1). (1998) 571–573. 108, 110, 127

[62] Suda, P., Bridoux, C., Kammerer, Maderlechner, G.: Logo and word matching using a general approach to signal registration. In: Proceedings of Fourth IAPR

Int. Conf. on Document Analysis and Recognition, ICDAR'97. Volume 1. (1997) 61–65 Ulm, Germany. 106, 110, 127

[63] Lee, H., Lee, M.: Understanding mathematical expressions using procedure-oriented transformation. PR **27** (1994) 447–457. 106, 109, 110, 127

[64] Yu, Y., Samal, A., Seth, C.: A system for recognizing a large class of engineering drawings. IEEE Trans. on PAMI **19** (1997) 868–890. 106, 109, 110, 128

[65] Okazaki, A., Kondo, T., Mori, K., Tsunekawa, S., Kawamoto, E.: An automatic circuit diagram reader with loop-structure-based symbol recognition. IEEE Trans. on PAMI **10** (1988) 331–341. 106, 110, 127

[66] Jorge, J., Fonseca, M.: A simple approach to recognise geometric shapes interactively. In Chhabra, A., Dori, D., eds.: Graphics Recognition - Recent Advances. Springer, Berlin (2000) 266–274 Vol. 1941 of LNCS. 106, 109, 110, 126

[67] Yu, B.: Automatic understanding of symbol-connected diagrams. In: Proceedings of Third IAPR Int. Conf. on Document Analysis and Recognition, ICDAR'95. (1995) 803–806 Montreal, Canada. 106, 110

[68] Tombre, K., Dori, D.: Interpretation of engineering drawings. In Bunke, H., Wang, P., eds.: Handbook of character recognition and document image analysis. World Scientific Publishing Company (1997) 457–484. 106, 112, 127

[69] Antoine, D., Collin, S., Tombre, K.: Analysis of technical documents: The RE-DRAW system. In Baird, H., Bunke, H., Yamamoto, K., eds.: Structured document image analysis. Springer Verlag (1992) 385–402. 106, 107, 112, 126

[70] Madej, D.: An intelligent map-to-CAD conversion system. In: Proceedings of 1st. Int. Conf. on Document Analysis and Recognition. (1991) 602–610 Saint Malo, France. 106, 107, 112, 127

[71] Randriamahefa, R., Cocquerez, J., Fluhr, C., Pépin, F., Philipp, S.: Printed music recognition. In: Proceedings of Second IAPR Int. Conf. on Document Analysis and Recognition, ICDAR'93. (1993) 898–901 Tsukuba, Japan. 106, 107, 112, 127

[72] Ramel, J., Boissier, G., Emptoz, H.: A structural representation adapted to handwritten symbol recognition. In Chhabra, A., Dori, D., eds.: Graphics Recognition: Recent Advances. Springer-Verlag, Berlin (2000) 228–237 Vol. 1941 of LNCS. 106, 109, 112, 127

[73] Kasturi, R., Bow, S., El-Masri, W., Shah, J., Gattiker, J., Mokate, U.: A system for interpretation of line drawings. IEEE Trans. on PAMI **12** (1990) 978–992. 106, 109, 112, 126

[74] Ventura, A., Schettini, R.: Graphic symbol recognition using a signature technique. In: Proceedings of 12th. Int. Conf. on Pattern Recognition (b). (1994) 533–535 Jerusalem, Israel. 106, 128

[75] Wilfong, G., Sinden, F., Ruedisueli, L.: On-line recognition of handwritten symbols. IEEE Trans. on PAMI **18** (1996) 935–940. 106, 109, 128

[76] Ablameyko, S.: An Introduction to Interpretation of Graphic Images. SPIE Optical Engineering Press (1997) 106

[77] Dosch, P., Masini, G., Tombre, K.: Improving arc detection in graphics recognition. In: Proceedings of 15th. Int. Conf. on Pattern Recognition. Volume 2. (2000) 243–246 Barcelona, Spain. 106

[78] Ablameyko, S., Bereishik, V., Frantskevich, O., Homenko, M., Paramonova, N.: Knowledge-based recognition of crosshatched areas in engineering drawings. In Amin, A., Dori, D., Pudil, P., Freeman, H., eds.: Advances in Pattern Recognition. Vol. 1451 of LNCS (1998) 460–467. 106

[79] Lladós, J., Martí, E., López-Krahe, J.: A Hough-based method for hatched pattern detection in maps and diagrams. In: Proceedings of 5th Int. Conf. on Document Analysis and Recognition. (1999) 479–482 Bangalore, India. 107

[80] Fahmy, H., Blonstein, D.: A graph grammar programming style for recognition of music notation. Machine Vision and Applications **6** (1993) 83–99. 107, 111, 126

[81] Landay, J., Myers, B.: Sketching interfaces: Toward more human interface design. IEEE Computer **34** (2001) 56–64. 109, 127

[82] Wenyin, L., Jin, X., Qian, W., Sun, Z.: Inputing composite graphic objects by sketching a few constituent simple shapes. In: Proceedings of Fourth IAPR Work. on Graphics Recognition. (2001) 73–84 Kingston, Canada. 109

[83] Blostein, D., et al.: User interfaces for on-line diagram recognition. In: Proceedings of Fourth IAPR Work. on Graphics Recognition. (2001) 95–106 Kingston, Canada. 109

[84] Paek, S., Smith, J.: Detecting image purpose in world-wide documents. In Lopresti, D., Zhou, J., eds.: Document Recognition V. SPIE, Bellingham, Whashington, USA (1998) 151–158 Vol. 3305 of Proceedings of SPIE. 109

[85] Duda, R., Hart, P., Stork, D.: Pattern Classification and Scene Analysis. John Wiley and Sons, New York (2000) 109

[86] Fu, K.: Syntactic Pattern Recognition and Applications. Prentice-Hall, Englewood Cliffs, N. J. (1982) 109

[87] Jain, A., Duin, R., Mao, J.: Statistical pattern recognition: a review. IEEE Trans. on PAMI **22** (2000) 4–37. 109

[88] Pavlidis, T.: Structural Pattern Recognition. Springer-Verlag, New York (1977) 109

[89] Trier, O., Jain, A., Taxt, T.: Feature extraction methods for character recognition - a survey. PR **29** (1996) 641–662. 110

[90] Furuta, M., Kase, N., Emori, S.: Segmentation and recognition of symbols for handwritten piping and instrument diagram. In: Proceedings of the 7th IAPR Int. Conf. on Pattern Recognition. (1984) 626–629. 110

[91] Lin, X., Shimotsuji, S., Minoh, M., Sakai, T.: Efficient diagram understanding with characteristic pattern detection. Computer Vision, Graphics and Image Processing **30** (1985) 84–106. 110

[92] Lladós, J.: Combining Graph Matching and Hough Transform for Hand-Drawn Graphical Document Analysis. Application to Architectural Drawings. PhD thesis, Universitat Autònoma de Barcelona and Université de Paris 8 (1997) 111

[93] Burr, D.: Elastic matching of line drawings. IEEE Trans. on PAMI **3** (1981) 708–713. 112

[94] Doermann, D.: The indexing and retrieval of document images: A survey. Technical report, University of Maryland (1998) Technical Report CS-TR-3876. 113

[95] Casacuberta, F., de Antonio, M.: A greedy algorithm for computing approximate median strings. In: VII Spanish Simposium on Pattern Recognition and Image Analysis. (1997) 193–198 Barcelona. 115

[96] Wong, A., You, M.: Entropy and distance of random graphs with application to structural pattern recognition. IEEE Trans. on PAMI **7** (1985) 599–609. 115

[97] Cordella, L., Foggia, P., Genna, R., Vento, M.: Prototyping structural descriptions: an inductive learning approach. In Amin, A., Dori, D., Pudil, P., Freeman, H., eds.: Advances in Pattern Recognition. Springer Verlag, Berlin (1998) 339–348. 116

[98] Valveny, E., Martí, E.: Learning structural descriptions of graphic symbols using deformable template matching. In: Proceedings of 6th Int. Conf. on Document Analysis and Recognition. (2001) Seattle, USA. 116

[99] Messmer, B.: Efficient Graph Matching Algorithms for Preprocessed Model Graphs. PhD thesis, University of Bern (1995) 116

[100] Sossa, H., Horaud, R.: Model indexing: The graph-hashing approach. In: Proceedings of IEEE Conf. on Computer Vision and Pattern Recognition. (1992) 811–814 Champaign, Illinois. 116

[101] Liu, W., Dori, D.: A protocol for performance evaluation of line detection algorithms. Machine Vision and Applications **9** (1997) 240–250. 117

[102] Liu, W., Dori, D.: A proposed scheme for performance evaluation of graphics/text separation algorithms. In Tombre, K., Chhabra, A., eds.: Graphics Recognition: Algorithms and Systems. Springer, Berlin (1998) 359–371 Vol. 1389 of LNCS. 117

[103] Aksoy, S., Ye, M., Schauf, M., Song, M., Wang, Y., Haralick, R., Parker, J., Pivovarov, J., Royko, D., Sun, C., Farneboock, G.: Algorithm performance contest. In: Proceedings of 15th. Int. Conf. on Pattern Recognition. Volume 4. (2000) 870–876 Barcelona, Spain. 117

[104] Yamada, H., Yamamoto, K., Hosokawa, K.: Directional mathematical morphology and reformalized Hough transformation for the analysis of topographic maps. IEEE Trans. on PAMI **15** (1993) 380–387. 128

A Bibliography Summary

Table 3. Bibliography summary

Reference	Application	Segmentation	Primitives	Recognition method	Notes
Adam:00 [ ]	Utility maps	Connected components	Fourier-Melin transform, Zernike moments, circular primitives	Learning vector quantization, 1NN	
Ah-Soon:01 [ ]	Architecture	Integrated with recognition	Vectors	Network of constraints	
Anquetil:00 [ ]	Musical scores	Connected components	Pixels	Neural network	
Antoine:92 [ ]	Cadastral maps	Integrated with recognition	Vectors	Specific for each type of symbol	
Aoki:96 [ ]	Architecture	Low level: segmentation of lines and geometric primitives	Low level geometric entities (squares, circles, lines, etc.)	Pattern matching (low-level primitives) and rules (symbols)	
Arias:95 [ ]	Telephone maps	Integrated with recognition	Vectors	Pattern matching	
Armand:93 [ ]	Musical scores	Specific to isolate notes and staff lines	Geometric feature vector	Nearest Neighbour	
Bley:84 [ ]	Logic diagrams	Based on run-length encoding	Graph	Production system	
Boatto:92 [ ]	Cadastral maps	Based on run-length encoding	Vectors and geometric features	Similarity in terms of geometric and topologic features	
Bunke:82 [ ]	Logic diagrams and flowcharts	Integrated with recognition	Graph	Graph grammar	Distortion model that allows hand drawn diagrams
Cesarini:97 [ ]	Logos	Connected components	Image normalized to a fixed size	Autoassociative neural networks	
Chang:01 [ ]	Logos	Assumes presegmented symbols	Log-polar space	2D HMM	
Cheng:93 [ ]	Logic diagrams	User driven	Invariant moments	Hierarchic neural network	
Collin:94 [ ]	Engineering drawings	Assumes presegmented symbols	Lines, arrowheads, etc.	Plex-grammars	
Cortelazzo: 94 [22]	Logos	Assumes presegmented symbols	Strings	String matching	
Doermann: 96 [ ]	Logos	Text-graphics separation	Contours	Comparison of algebraic invariants from boundary	
Dori:89 [ ]	Engineering drawings	Assumes presegmented symbols	Graph	Web grammars (dimensions) and lookup table (low level primitives)	
Fahmy:93 [ ]	Musical scores	Assumes presegmented symbols	Graph	Graph grammar	
Fahn:89 [ ]	Logic diagrams	Integrated with recognition	Vectors	Non contextual grammar	
Francesconi: 98 [ ]	Logos	Assumes presegmented symbols	Tree of boundary segments with geometric attributes	Recursive neural network	
Groen:85 [ ]	Logic diagrams	Based on loops	Graph	Probabilistic graph matching	
Habacha:91 [ ]	Logic diagrams	Integrated with recognition	Graph	Discrete relaxation	
Hartog:96 [ ]	Utility maps	Integrated with recognition	Pixels	Rule-based system	
Jiang:00 [ ]	Logic diagrams	Assumes presegmented symbols	Graph	Graph Matching and genetic algorithm	Symbol learning by mean graph computation
Jorge:00 [ ]	On-line diagrams	Based on the sketches and time-outs	Similarity between geometric features	Decision trees with fuzzy logics	
Joseph:92 [ ]	Engineering drawings	Integrated with recognition	Vectors	Recognition driven by a grammar	The grammatical rules are the control mechanism that guides the recognition strategy
Kasturi:90 [ ]	Line drawings	Based on minimum redundancy loops	Graph	Specific for each type of symbol	
Kim:93 [ ]	Logic diagrams	Based on loops	Fourier descriptors and moments	Nearest Neighbour	
Kiyko:95 [ ]	Logic diagrams	Integrated with recognition	Skeleton	Grammar	Also applied to hand drawn sketches
Kosmala:99 [ ]	Mathematical formulas	Connected components	Pixels	HMM and graph grammar	

Table 3. (continued)

Reference	Application	Segmentation	Primitives	Recognition method	Notes
Kuner:88 [ ]	Logic diagrams	The input image is divided in windows	Skeleton graph	graph matching and rule-based system to solve ambiguities	
Landay:01 [ ]	User Interfaces	Assumes presegmented symbols	Sequence of coordinates	Statistical PR	On-line recognition
Lavirotte:97 [ ]	Mathematical formulas	Connected components	Pixels	graph grammar	
Lee:92 [ ]	Logic diagrams	Assumes presegmented symbols	Graph	graph matching	
Lee:94 [ ]	Mathematical formulas	Connected components	Feature vectors	Nearest Neighbour and dynamic programming to modelize distortion	
Llados:98 [ ]	Architecture	Integrated with recognition	Graph	Error-tolerant graph matching	Applied to hand drawn diagrams
Madej:91 [ ]	Cadastral maps	Based on symbol-dependent heuristics	Hierarchical graph	Heuristic	
Messmer:96 [ ]	Engineering drawings	Integrated with recognition	Vectors	Graph Matching	Symbol models are previously compiled in a network to reduce the computational cost of recognition
Min:93 [ ]	Engineering drawings	Assumes presegmented symbols	Arrowheads, lines, etc.	Web grammar	
Miyao:96 [ ]	Musical scores	Combines projections, runs and meshes	Feature vectors	Neural network	
Muller:00 [ ]	Engineering drawings	Integrated with recognition	Cosine transform	2D Hidden Markov Model	Database retrieval by sketching
Myers:96 [ ]	Geographic maps	Hypotheses generation concerning to image locations likely to contain symbols	Pixels	hypothesis-and-test	
Okazaki:88 [ ]	Logic diagrams	Based on lines and connections	Primitives found by pattern matching	Decision tree	
Parker:00 [ ]	Logic diagrams	Assumes presegmented symbols	Vectors	Template matching	Also applied to hand drawn textual symbols
Pasternak:94 [ ]	General	Integrated with recognition	Arcs and lines	Driven by a language of descriptions	
Ramel:00 [ ]	Formula	Integrated with recognition	Graph	Heuristic	
Randriamahefa: 93 [ ]	Musical scores	Based on horizontal lines and connected components	Attributed graph	Rule-based system	
Reiher:96 [ ]	Geographic maps	Integrated with recognition	Pixels	Hausdorff distance and neural network	
Samet:94 [ ]	Geographic maps	Based on color information	Pixels	k-nearest neighbours	The recognition is driven by the information found in the legend
Samet:96 [ ]	Geographic maps	Connected components	Globals (Moments, circularity, eccentricity) and locals (crossings, gaps)	Voting by weighted bounded several-nearest neighbor classifier	
Sato:82 [ ]	Electronic diagrams	"Image lines are followed using a window to separate long lines and "complex regions"	Lines and regions	Relational matching	
Segen:89 [ ]	Shapes	Assumes presegmented symbols	Boundary	Bondary saliency points comparison	
Soffer:98 [ ]	Logos	Assumes presegmented symbols	Connected components	Comparison of connected component attributes	Indexing in document databases
Stefano:95 [ ]	Geographic maps	Integrated with recognition	Pixels	Template matching	
Suda:97 [ ]	Logos	Assumes presegmented symbols	Image sampling using a fixed grid	Distance between feature vectors (zoning)	
Tombre:97 [ ]	Engineering drawings	Integrated with recognition	Vectors	Specific for each type of symbol	
Valveny:00 [ ]	Architecture	Assumes presegmented symbols	Pixels and vectors	Deformable models	Also applied to hand drawn symbols

Table 3. (continued)

Reference	Application	Segmentation	Primitives	Recognition method	Notes
Vaxiviere:92 [ ]	Engineering drawings	Integrated with recognition	Vectors	Rule-based system	
Ventura:94 [ ]	CAD	Integrated with recognition	Vectors and background areas	Signature indexing	
Wilfong:96 [ ]	Characters	Assumes presegmented symbols	Sequence of coordinates	Curvature distance	On-line recognition
Yadid-Pecht:96 [ ]	Musical scores	Specific to isolate notes and staff lines	Pixels	"Neural networks; Neocognitron"	
Yamada:93 [ ]	Topographic maps	Integrated with recognition	Pixels	Hough Transform approach combined with directional morphologic operators	The recognition is driven by the information found in the legend
Yang:93 [ ]	Architecture	The drawing is analyzed dynamically	Pixels	Neural network	
Yu:97 [ ]	Line drawings	Using rules on the lines properties to separate them from connections	Loops	Hierarchical matching	

An Error-Correction Graph Grammar to Recognize Texture Symbols

Gemma Sánchez[1,2], Josep Lladós[1], and Karl Tombre[2]

[1] Computer Vision Center, Dept. Informàtica, Univ. Autònoma de Barcelona, Spain
{gemma,josep}@cvc.uab.es
http://www.cvc.uab.es
[2] Loria, Vandœuvre-lès-Nancy CEDEX, France
{Karl.Tombre}@loria.fr
http://www.loria.fr

Abstract. This paper presents an algorithm for recognizing symbols with textured elements in a graphical document. A region adjacency graph represents the document. The texture symbols are modeled by a graph grammar. An inference algorithm is applied to learn such grammar from an instance of the texture. For recognition, a parsing process is applied. Since documents present distortions, error-correcting rules are added to the grammar.

1 Introduction

The detection and recognition of symbols is an important area within graphics recognition. There are typically two kinds of structures that can be considered to form a symbol: those consisting of fixed primitives, i.e. parts that can be represented by a prototype pattern, and structures consisting of a repetitive pattern, such as crosshatched areas or tiled patterns. Each type of symbol induces its own recognition strategy. Prototype-based symbols can be recognized by many kinds of pattern matching approaches, whereas repetitive structures need a strategy able to iteratively capture the repetitions. In a previous work, we have studied prototype-based symbol recognition and a preliminary approach of texture symbol recognition using separate strategies [10]. In this work a syntactic model for detecting and recognizing texture-based components is proposed.

A number of contributions address the problem of detecting and recognizing structured textured areas having a particular meaning in the domain where they appear. However, these approaches lack of generality, i.e. they propose *ad hoc* methods for certain structures. For example, several authors have proposed solutions for detecting hatched patterns in cadastral maps and engineering drawings [2, 3], or for linear textures such as isolines in geographic maps [1, 8]. There is still room for progress in the development of general structured texture models in graphic documents. From a more general perspective, some other Computer Vision areas bring us interesting approaches, based on some kind of relational structures, that can serve as basis for our purpose. Fu [12] when established the

D. Blostein and Y.-B. Kwon (Eds.): GREC 2002, LNCS 2390, pp. 128–138, 2002.

basis of syntactic pattern recognition, proposed grammatical models for texture recognition in images. Hamey [7] modelized and detected regular repetitive textures in real-world scenes comparing structural relationships among prominent features in the image. Matsuyama [13] formulated a model in which structured textures are composed by primitives, defined in terms of salient features of the image, and relative positional vectors between these primitive elements. Notice that these models inherently involve the definition of a structured texture: a double regular repetition, namely the primitive element and the spatial relationship. In this paper, we propose a syntactic model to represent, discriminate and recognize a wide class of structured texture patterns in graphical documents. For that purpose, we adapt the ideas given in the area of structural texture analysis to the particularities of structures in graphical documents.

Our work is applied to the recognition of symbols in architectural drawings. Structured textured areas characterize symbols such as stairs, walls, roof, tiles, etc. Some of them are two-dimensional (see Figs. 1(a)(b)). Others are one-dimensional (see Figs. 1(c)(d)). Some are formed by only one kind of shape, or primitive (see Figs. 1(a)(c)(d)), and others with two (see Fig. 1(b)). Any of them is characterized by one or more primitives placed following a rule. As stated above, the recognition of texture symbols is combined with prototype-based symbols. Actually, some symbols can contain both structures. To recognize them, it is necessary to combine the two respective recognition procedures. We define a graph model to represent documents after the vectorization step. Prototype-based structures are recognized by a graph matching approach. See [11] for more details on that algorithm. Texture structures are represented by a graph grammar. To decide whether a subgraph consisting of repetitive pattern represents a *word* accepted by such grammar, a parsing process is required.

We propose an attributed *Region Adjacency Graph* (RAG) $H(V, E, LV, LE, FV, FE)$ where the set of nodes V represent closed loops or regions of the input document and E is the set of edges representing region adjacencies. FV and FE are functions that give attributes to nodes and edges respectively. Thus, every graph vertex, $v \in V$, is given as attribute a cyclic string that represents the sequence of lines forming the shape contour of the region, and every graph edge (v, v') is given as attribute the vector connecting the centers of the shapes v and v'. LV and LE are the functions that give labels to the vertex and the edges

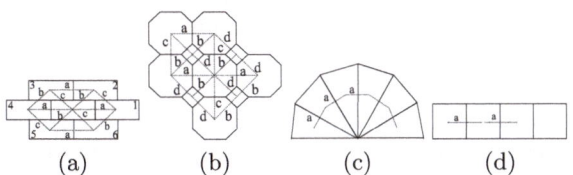

(a)   (b)   (c)   (d)

Fig. 1. (a) 1-primitive 2D texture symbol. (b) 2-primitives 2D texture symbol. (c) 1-primitive 1D texture symbol. (d) 1-primitive 1D texture symbol

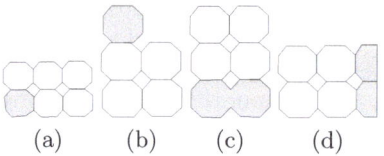

(a)   (b)   (c)   (d)

Fig. 2. (a) Shape distortions. (b) Placement rules distortions. (c) Merging of the shapes. (d) Occlusion of the shapes

respectively. These labels are used by the parser and identify types of regions and relations.

Taking a RAG as a core representation, this work defines a graph grammar to modelize texture symbols. Graph grammars have been used in syntactic pattern recognition to describe and classify n-dimensional repetitive patterns in terms of the space relations among their parts. Graph grammars have been used in diagram recognition [4], music notation [6], visual languages [15], drawing dimensions recognition [5, 14] and mathematical formula recognition [9]. In our work, given an input document, the symbol recognition process works as follows. First, subgraph isomorphisms are searched between prototype-based models and the input graph. For repetitive structures, a texture detection algorithm is applied based on regularity of graph nodes and edges. Afterwards, in subgraphs where a texture is found, a parsing process is applied driven by graph grammars representing known texture symbols. While the definition of graphs representing models of prototype-based symbols can be done just by using an instance of the symbol as the prototype, the definition of the graph grammar characterizing texture symbols requires a learning process. We have developed a grammatical inference algorithm to learn new texture symbols from instances of them. Additionally, since there is no standardized notation in architectural drawings, some unknown repetitive structures can appear in the drawing. The texture recognition algorithm detects that structure and the inference process computes the graph grammar representing that texture. This inferred graph grammar is added to the model database to recognize other symbols having the same structure.

In the RAG obtained from the input document some distortions may appear, due to the scanning process, to the vectorization, and to the finite nature of the texture. Particularly, four kinds of distortions may appear in the symbol (see Fig. 2): (i) Distortions in the shapes forming the texture, see Fig. 2(a), (ii) Distortions in their placement rules, see Fig. 2(b), (iii) Fusion of two or more texture elements, see Fig. 2(c), (iv) Partial occlusion of the texture elements, due to the finite nature of the textured area, see Fig. 2(d). In order to tolerate these distortions, the graph grammar is augmented with error-correction rules.

The paper is structured as follows: section 2 presents the grammar to model textured components and its inference process. Section 3 explains the error-correcting parser for recognition of texture symbols. Section 4 is devoted to conclusions and future work.

2   Graph Grammar Inference

In this section we present the definition of our grammar and the inference process from a given instance. Given a texture symbol X, we infer a context-sensitive error-correction graph grammar to recognize it. A graph grammar is a five tuple $G = (\Sigma, \Delta, \Omega, P, S)$ where Σ is the alphabet of non-terminal node labels, Δ is the alphabet of terminal node labels, Ω is the alphabet of terminal edge labels, P is the finite set of graph productions or rewriting rules, and S the set of initial graphs, usually consisting of one node with a non-terminal label. A graph production P is a four tuple $P = (h_l, h_r, T, F)$, where h_l is the left hand graph, h_r is the right hand graph, T is the embedding transformation $T = \{(n, n') | n \in V_{h_l}, n' \in V_{h_r}\}$ and F is the set of attribute transferring functions, for edges and nodes in h_r. The direct derivation of a graph H' from a host graph H by applying a production $P = (h_l, h_r, T, F)$, $H \xrightarrow{P} H'$ is defined by locating h_l^{host}, a subgraph of H isomorphic to h_l, and replacing h_l^{host} by h_r^{host}, a subgraph isomorphic to h_r. Let $H^{-h_l^{host}}$ be the graph remaining after deleting h_l^{host} from H, the edges between the subgraph h_r^{host} and $H^{-h_l^{host}}$ are *the embedding of h_r^{host} in $H^{-h_l^{host}}$*, and defined by T. In the present work we use induced isomorphism. An induced subgraph of a graph must include all local edges of the host graph, i.e. all edges that connect two nodes in the subgraph. For a more comprehensive explanation of graph grammars, see [16].

A texture symbol can be formed by different repetitive structures, each of them characterized by a primitive and a relational placement rule. In our work we consider as primitives any kind of polygonal or line segments. Each kind of placement rule is defined by a vector connecting the centroids of two adjacent primitives. Given a texture symbol X consisting of regular repetitions of n different primitives, $N_X = \{N_1, \ldots, N_n\}$, and m different kinds of placement rules $R_X = \{R_1, \ldots, R_m\}$, the inference process needs the following definitions:

Node Labels: For each primitive N_i in X, we define a hierarchy of non-terminal node labels L_i, and three terminal node labels l_i, cl_i, dl_i, representing the primitive, the primitive cut on a border, and more than one primitive merged into one shape respectively. Thus, $\Delta = \{L_1, \ldots, L_n, S'\}$ and $\Sigma = \{l_1, cl_1, dl_1 \ldots, l_n, cl_n, dl_n\}$. Where S' is the start node label. For each hierarchy L_i we define two derived labels, IL_i and RL_i, representing an inserted node and a real one, respectively. An inserted node represents a missing primitive in the input graph. Usually, these nodes represent gaps or borders of the textured area. A real node represents a correctly detected primitive.

Edge Labels: One terminal edge label e_i is defined for each kind of neighborhood (placement rule) R_i, having $\Omega = \{e_1, \ldots, e_m\}$. For each primitive $N_i \in N_X$ we denote as $NR_i = \{R_1^i, \ldots, R_{m_i}^i\}$, $\forall j = 1 \ldots m_i, R_j^i \in R_X$, the set of different kinds of neighborhoods that N_i has in counter-clockwise order, for example for the symbol in Fig. 1(a), N_i being the rectangle in the middle, $NR_i = \{a, b, c\}$. For each primitive $N_i \in N_X$ we denote as $NN_i = \{N_1^i, \ldots, N_{n_i}^i\}$, $\forall j = 1 \ldots n_i, N_j^i \in N_X$, the set of neighbors that N_i has in counter-clockwise order, being R_1^i the label of the edge (N_i, N_1^i). In the symbol shown in Fig. 1(a), denoting each node neighbor with the natural number the rectangle has in its corner, $N_1, \ldots, N_6$, $NN_i = \{N_1, \ldots, N_6\}$.

Assuming the previous definitions, the productions are inferred as follows:

1. For each $N_i \in N_X$ we define a production with the left hand side being the start node S', the right hand side a subgraph with a node n labeled as l_i, with all the closed loops starting on N_i, labeling the neighboring nodes of n with the following non-terminal node labels $\{L_1^i, \ldots, L_{m_i}^i\}$ in the counter-clockwise order, and the edges among all of them with their corresponding terminal label. In the grammar of Fig. 3, the productions number 1 and 10 correspond to the productions that are generated in this step.

2. Each non terminal node can have one or more terminal nodes as neighbors. Each terminal node has all its neighbors at least labeled as a non terminal node. Then, for each shape N_i, we generate one set of productions for each number of terminal nodes that N_i can have, i.e. we generate one set of productions when N_i has one terminal node as a neighbor, then when it has two, and so on until n_i. Each production in each of this set of productions is defined taking the first terminal neighbor, $N_j^i \in NN_i$, following one possible kind of neighboring, $R_k^i \in NR_i$. In the grammar of Fig. 3, for the shape that it is a square, four set of productions are generated using this process, corresponding to the case where N_i has as neighbours one, two, three and four terminal nodes respectively. Those sets consist of rules 2 and 3, 4 and 5, 6 and 7, and 8 respectively.

3. Inserted nodes allow error-correction of distorted, fragmented or merged shapes. For these inserted nodes, one rule is added that substitutes them by λ, the cut terminal label, cl_i, or the joined terminal label, dl_i. The selection of the corresponding rule is done during the parsing process. Each inserted node can finally match different terminal nodes of the input graph, depending of the type of distortion. Graph productions are considered to allow to delete an inserted node, when the input shape is completely different of the expected one, or to substitute them by the cut terminal label, if the shape corresponds to a part of the expected shape, or by the joined terminal label if the shape in the original graph matches partially with the expected shape. In the grammar of Fig. 3 rules number 9 and 40 correspond to error-correcting rules for the square and the octogon respectively.

Figure 3 shows the example of the graph grammar inferred from the texture in Fig. 1(b). There are two kinds of primitives forming the texture: the square and the octogon. Three terminal labels are defined for each primitive: o, co, do, for the octagon, and r, cr, dr, for the square, representing the normal primitive, the cut one, and the merging of different primitives into one shape, respectively. In addition, a hierarchy of non-terminal labels is defined for each primitive: O, with derived labels Ro and Io, for the octagon, and R, with derived labels Rr and Ir,

Fig. 3. Graph grammar representing the texture symbol of Fig. 1(b)

for the square. Derived labels Ro and Rr represent an existing node while Io and Ir represent an inserted one. For each production P, h_l and h_r are graphs whose nodes have a number or a number with $'$, this number represents the embedding rule, i.e. the node on the left side with number i, or the corresponding one on the right side with number i'. Rules 1 and 10 are the starting rules for the square and the octagon respectively, while rules 9 and 40 are the end rules and error-correction rules for the square and the octagon, respectively. Notice that each time a label O or R appears on the production, they can represent one of their both derived labels Io or Ro, and Ir or Rr respectively. However, we can only rewrite one real non-terminal node by a terminal one, the only exceptions being the rules to correct errors. We should also notice that each rule is representing itself and its symmetric rule, and the rules with labels with parameters $e1$ and $e2$ represent four kinds of rules: those without the nodes and edges parameterized with them, those with the nodes parameterized with $e1$, those with the nodes parameterized with $e2$, and those with all of them.

We report an example in the domain of architectural plan interpretation. In Fig. 5(a) we show the result of detecting repetitive structures on the input document. Each texture is characterized by the primitive element and the placement rule. Such information is the basis for the inference of the grammar. Notice that some textures in the document are instances of the same class. The inference process, detects texture primitives and relational rules belonging to the same class

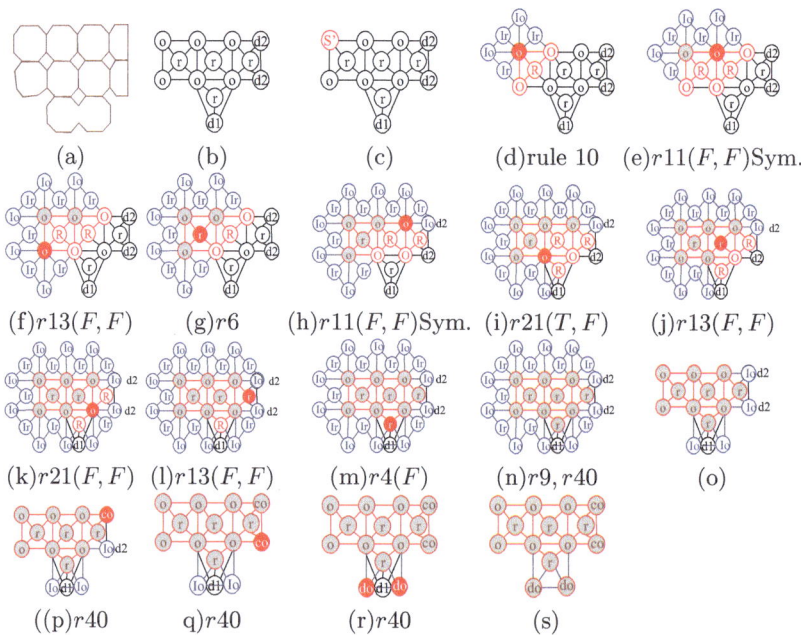

Fig. 4. Parsing process of a graph

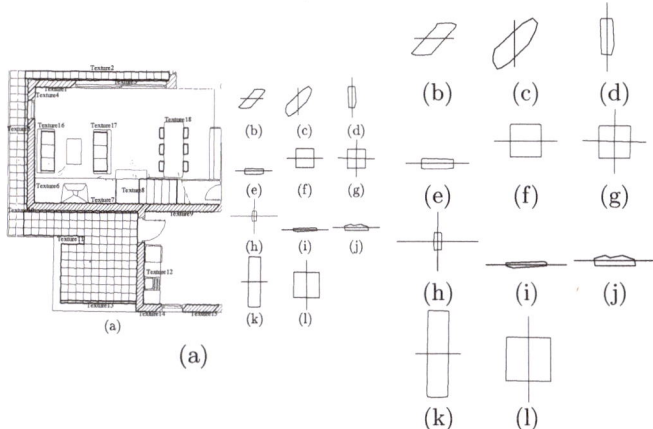

Fig. 5. (a) Texture detection. (b) Textures: 1,7,9,14,15. (c) T: 4, 6, 12. (d) T. 5. (e) T. 13. (f) T. 2. (g) T. 11. (h) T. 18. (i) T. 3. (j) T. 10. (k) T. 8. (l) T: 16,17

and one representative grammar is inferred from them. Figures 5(b)-(l) show the real representative primitives after different instances of the same textures have been detected. The representative primitive of each grammar is computed by the mean string algorithm described in [17]. Afterwards, a grammar has been inferred from each texture following the steps described above.

3   Recognition Process: Parsing

Given a host graph H representing an architectural drawing, and a graph grammar G representing the texture symbol, the parsing process on H, to recognize the texture symbol, is done in the following way: First, nodes in H are taken onc by onc until a start rule can be applied to one of them. Let us denote by n this selected node. This means that there is a subgraph isomorphism between a subgraph around n and the graph h_r of this starting rule. The similarity between the shapes is computed by the string edit distance explained in [10]. In this process, n is marked as an element forming the texture symbol, and the set of nodes and edges of H forming the subgraph isomorphic to h_r are labeled as the rule points, being the nodes labeled as non-terminal inserted into a list, Lnt, to be analyzed by the parser, while the nodes and edges not found are labeled as inserted nodes and inserted into a separate list, Li, to be analyzed during a post-processing phase, to find possible errors and to finalize the texture symbols. These inserted nodes are pointing to existing nodes in H, which are in a relative position with respect to n similar to the expected corresponding nodes in the h_r of the grammar rule. Then, for each non-terminal node n_i in Lnt, the rule to be applied each time is directly selected by counting the number of terminal neighbor nodes it has and in which position they start. Then the rule is applied as in the previous step, using a subgraph isomorphism and inserting into Lnt and Li the non-terminal and inserted nodes to be analyzed, respectively. Node n_i is marked as an element forming the texture symbol. Once all the non-terminal nodes in Lnt are analyzed, a set of nodes in H have been labeled as texture-belonging, and other surrounding nodes, as inserted. Then, for each inserted node n in Li, we apply the error-correcting or ending rules. Ending rules delete the inserted nodes while error-correcting rules test if the inserted node represents a cut shape or a split shape. Error-correcting rules have an associated cost that quantifies the distortion of the texture symbol regarding to the model. Once this cost reaches a given threshold, no more error-corrections are possible, and the remaining inserted nodes are deleted.

Fig. 4(a) shows the input structure and Fig. 4(b) the corresponding graph. An starting node S' is selected (Fig. 4(c)). Since an isomorphism between the subgraph centered at S' and the right hand side of the rule 10 exists, this rule is applied (Fig. 4(d)). Ir and Io represent inserted ones corresponding to each type of primitive, o and r represent terminal nodes, and R and O represent non-terminal nodes. A non-terminal node is chosen and rule 11 is applied in its symmetric form with parameters $e1$ and $e2$ as false (Fig. 4(e)). The nodes labeled in grey are nodes already parsed as being part of the symbol. Iteratively, other productions are applied (Figs. 4(e)–(m)). In these figures, T and F denote the existence of the nodes labeled with $e1$ ad $e2$, and Sym indicates that the rule is applied in its symmetric form. In Fig. 4(f) rule $13(F, F)$ is applied. Notice that although this shape is distorted, it can be matched with the original primitive because of the error model involved in the parser. In Fig. 4(n) all the non-terminal nodes have been analyzed. Nodes displayed in grey are those recognized as belonging to the texture symbol. For the inserted nodes appearing around, rule 40 allows to correct the involved errors by replacing them by the terminal nodes oc (cut shape) and os (merged shapes).

4   Conclusions

In this paper, a model to represent texture symbols is presented. The model is based on a graph grammar defined in terms of region adjacency graphs. The grammar also contain error-correcting rules to tolerate four kind of distortions in the texture: Distortions in the shapes forming the texture, distortions in their neighborhood, shapes which are split, and shapes which appear only partially. This grammatical model allows to represent structured textures formed by regular repetitions of polygons or segments. The error-correcting model involved in the grammar is a preliminary approach to tolerate distortions. Next step is to define an edit cost associated to error-correction graph productions. Two tasks are required for a complete texture interpretation system. Firstly, grammars representing texture symbols have to be learned from instances. With that purpose, an inference algorithm has been proposed. This algorithm is also used when an unknown texture is detected in the document to define it as a new symbol. Secondly, the recognition of texture symbols in the document requires a parsing algorithm. This process is combined with a graph matching procedure that looks for other symbols represented by prototypes. The grammatical model for texture symbol detection, recognition and inference has been proved in the domain of architectural floor plan analysis. This model represents a generalization in the field of graphics recognition regarding previous works that focused only on particular textures such as crosshatched or linear ones.

References

[1] S. Ablameyko. *An introduction to interpretation of graphic images*, volume TT27 of *Tutorial Texts in Optical Engineering*, chapter Recognition of cartographic objects, pp. 92–123. SPIE The international society for Optical Engineering, 1997. 128

[2] D. Antoine, S. Collin, K. Tombre. Analysis of technical documents: The REDRAW system. In H. Baird, H. Bunke, K. Yamamoto, eds.: Structured document image analysis. Springer Verlag (1992) 385–402. 128

[3] L. Boatto, et al. An interpretation system for land register maps. *Computer*, 25:25–33, 1992. 128

[4] H. Bunke. Attributed Programmed Graph Grammars and Their Application to Schematic Diagram Interpretation. *IEEE Transactions on PAMI*, 4(6):574–582, November 1992. 130

[5] D. Dori. A syntactic/geometric approach to recognition of dimensions in engineering machine drawings. *Computer Vision, Graphics and Image Processing*, 1989. 130

[6] H. Fahmy and D. Blostein. A graph grammar programing style for recognition of music notation. *Machine Vision and Aplications* 6:83–99, 1993. 130

[7] L. G. C. Hamey. *Computer Perception of Repetitive Textures*. PhD thesis, Computer Science Department, Carnegie Mellon University, Pittsburg, 1988. 129

[8] T. Kasvand. Linear textures in line drawings. In *Proc. in the 8th Int. Conf. on Pattern Recognition*, 1986. 128

[9] A. Kosmala, S. Lavirotte, L. Pottier, and G. Rigoll. On-Line Handwritten Formula Recognition using Hidden Markov Models and Context Dependent Graph Grammars. In *Proc. of 5th Int. Conf. on Document Analysis and Recognition*, 1999. 130

[10] J. Lladós, G. Sánchez and E. Martí. A string-based method to recognize symbols and structural textures in architectural plans. *Graphics Recognition, Algorithms and Systems. K. Tombre and A. K. Chhabra (eds). Lecture Notes in Computer Science, Springer-Verlag*, 1389:91–103, 1998. 128, 135

[11] J. Lladós, E. Martí and J. J. Villanueva. Symbol Recognition by Error-Tolerant Subgraph Matching between Region Adjacency Graphs. *IEEE Transactions on PAMI*. 23(10):1137–1143, October, 2001. 129

[12] S. Y. Lu and K. S. Fu. A syntactic approach to texture analysis. *Computer Graphics and Image Processing*, 7:303–330, 1978. 128

[13] T. Matsuyama, K. Saburi, and M. Nagao. A structural analyzer for regularly arranged textures. *Computer Graphics and Image Processing*, 18, 1982. 129

[14] W. Min, Z. Tang, and L. Tang. Using web grammar to recognize dimensions in engineering drawings. *Pattern Recognition*, 26(9):1407–1916, 1993. 130

[15] J. Rekers and A. Schürr. Defining and Parsing Visual Languages with Layered Graph Grammars. *Journal of Visual Languages and Computing*, 8(1):27–55, London: Academic Press (1997). 130

[16] G. Rozenberg. Handbook of Graph Grammars and Computing by Graph Transformation. Vol. I, Foundations, World Scientific, 1997. 131

[17] G. Sánchez, J. Lladós and K. Tombre. A Mean String Algorithm to Compute the Average Among a Set of 2D Shapes. *P*attern Recognition Letters. 23(1-3):203–213, January 2002. 135

Perceptual Organization as a Foundation for Graphics Recognition

Eric Saund, James Mahoney, David Fleet, and Daniel Larner

Xerox Palo Alto Research Center, Palo Alto, CA, USA
{saund,jvmahoney,fleet,larner}@parc.xerox.com
http://www.parc.xerox.com/spl/groups/pda/

Abstract. This paper motivates an approach to graphics recognition grounded in a framework for human and machine vision known as *Perceptual Organization*. We review some of the characteristics of this approach that distinguish it from traditional engineering of document recognition systems, and we suggest why and how the techniques and philosophy of Perceptual Organization might lead to advances in the very practical matters of interpreting diagrams, drawings, and sketches.

1 What is Perceptual Organization?

Perceptual Organization (PO) refers to a cross-disciplinary tradition in the study of the *intermediate* stages of vision. Perception lies above the level of sensory processes such as image enhancement, contrast normalization, and feature measurement, but below the levels of end tasks such as object recognition, data indexing, and scene modeling. Prior to interpreting semantically meaningful objects, PO builds representations for visual structure, including geometric configurations, texture and color properties, and motion: What are the readily apparent regions of an image? What grabs an observer's attention? What features tend to group together to form larger coherent units? What emergent patterns become apparent? Perceptual Organization is as far as your visual system gets when you view abstract art: you can make sense out of parts of the image locally and perhaps assemble some limited gist of the whole, but you don't necessarily know what it is you're supposed to be looking at. That abstract art is in fact interesting to look at says something important about how the most powerful extant vision systems work [1].

Classically, Perceptual Organization has been concerned with the kinds of visual structure identified by the school of Gestalt Psychology, dating from the early 20th century [7,11,15,28]. The "Gestalt Laws" of *smooth continuation, common fate, symmetry, similarity, proximity, closure,* etc. are taught in some form in almost every introductory Psychology class. The Gestalt phenomena were traditionally studied by psychophysicists who devised wickedly clever visual displays attempting to expose broad principles governing human vision.

More recently, computer vision researchers began attempting to model these processes computationally, while beginning to focus on general issues associated with intermediate stages of visual analysis [25,16,29,31]. As expressed by

D. Blostein and Y.-B. Kwon (Eds.): GREC 2002, LNCS 2390, pp. 139-147, 2002.

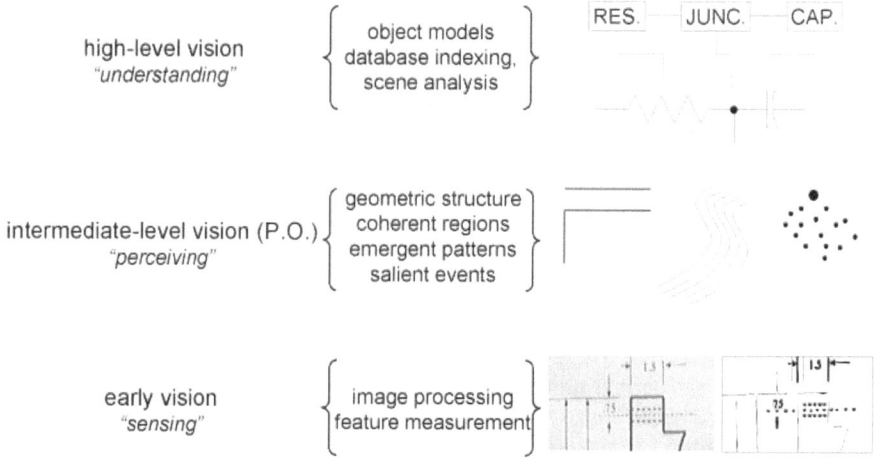

Fig. 1. Perceptual Organization operates at the stage of intermediate level vision.

Witkin and Tenenbaum, at the very heart of visual analysis is our "ability to impose organization on sensory data - to discover regularity, coherence, continuity, etc., on many levels." By now hundreds of papers have been published on computational models of intermediate level visual processing, and a regular workshop is held entitled, "Perceptual Organization in Computer Vision" [2, 3, 10]. We believe that the philosophy, methodology, and techniques of Perceptual Organization are now ready to contribute to areas in document image analysis. Graphic images, in particular, present an ideal context in which to carry out this cross-fertilization of disciplines.

2   What Would it Mean to Apply PO to Graphic Documents?

Let us consider three examples of how general processes for detecting intermediate level visual structure might be applied to graphics recognition, in contrast to a straw-man "engineering" approach.

Example 1: text/line art segmentation.   Text and graphical material are often separated early in graphics recognition programs, so that text recognition and spatial analysis procedures can be applied respectively to appropriate input data [13]. Traditional methodology is to focus on the properties that specifically distinguish text from line art, then build a classifier to exploit such properties of ideal text as connected component dimension, connected component spacing, and linear alignment of components' edges along parallel toplines and baselines.

Fig. 2. Demonstrations of Perceptual Organization. a. *smooth continuation:* which decomposition into parts seems more likely? b. *closure:* which dot appears to be on an object?

A PO approach might approach this problem somewhat differently. Instead of focusing on the text/line art distinction specifically, PO would call for a more general process of identifying compact regions having consistent texture properties (e.g. [12,18]). These would call out not only ideally printed text, but also graphic elements that resemble text in appearance, certain kinds of smudges and noise, and scrawled handwritten notes that violate the nominal constraints that most text obeys. These objects would then serve as a starting point for more specialized text classification and recognition.

Example 2: hash region detection. Mechanical drawings depict cutaway views by parallel hash lines. These are typically detected by specialized procedures for finding repeating parallel-lines, parameterized by spacing, line weight, and orientation. These procedures are launched when certain cues are encountered such as a line found inside a contour classified as "object boundary".

A PO approach might instead perform an initial global region segmentation step designed to identify regions distinguished by uniform textural properties such as average lightness, distribution of characteristic grain size, spatial frequency or image orientation, and distribution of texture primitives like curve ends, corners, and bars. This process would not only detect hash line regions, but also graphic representations for dirt, concrete, wood grain, etc., as well as blocks of text. The intermediate stage texture regions would serve as candidates for more specialized classification. This processing might also help in deciding which contours comprise an object boundary by virtue of enclosing a uniform texture region.

Example 3: detection of enclosing boxes and ovals. Charts and diagrams frequently represent conceptual objects as text enclosed by rectangular boxes, or ovals. A standard approach to detecting these is to exploit their geometric regularities. Boxes consist of two sets of parallel lines at right angles; Hough techniques prove useful here. Ovals can be fit by a piecing together of the outputs of circular arc finding routines.

A PO approach might instead begin by detecting compact objects formed by closed paths [22]. This would turn up not only rectangles and ovals, but other enclosing graphic shapes like octagon, and triangles, as well as salient graphical structure involving closed paths. As in the first two examples, the intermediate level structure provides building blocks for more specialized recognition routines.

Perceptual Organization, as such, does not "solve" any of these problems; much research remains to be done to flesh out the algorithms and architectures. Rather, our point is that PO approaches the problem in a different way, placing emphasis on finding intermediate level structure, prior to turning to knowledge about a particular task or image domain.

3   Characteristics of a Perceptual Organization Approach

These examples illustrate several characteristics of the Perceptual Organization approach. Some of these characteristics may not be held as standard in the P.O. community, but they reflect our conception of what is most relevant and powerful in the approach.

- *weak prior models:* It is well understood that knowledge about an image source can be exploited to direct search and to select appropriate representations for inferences to match task requirements. Classically, document image analysis has pursued *strong prior models* incorporating as much prior knowledge as possible. This approach can become brittle as actual data departs from the assumptions of prior models. By contrast, PO suggests the use of *weak models*, to identify salient visual structure common to broad ranges of image domains. As a result, knowledge is kept modular, postponing the use of domain-specific models.
- *rich and redundant representations:* Perceptual Organization seeks to make explicit many types visual structure in ways that may amount to an overcomplete or redundant description of an image. For example, a rectangle can be described economically with five numbers, while its perceptually significant aspects include the presence of a blob with a rough size, orientation, and elongation; straight lines; parallels; corners; a closed region; and more (Figure 3). By articulating these, a P.O. approach makes a wealth of potentially relevant information available for use in later tasks.
- *incremental description:* In PO, typically each assertion is limited in scope. Images are described in terms of large numbers of simple statements instead of all-encompassing models with a multitude of parameters.
- *partial, conflicting, overlapping, and ambiguous assertions:* Perceptual Organization is content to deliver partial interpretations of data, and to have some of these interpretations overlap or conflict with one another. For example, two squares sharing a common edge may both give rise to assertions that interpret the edge as "belonging" to one square or the other, or both.

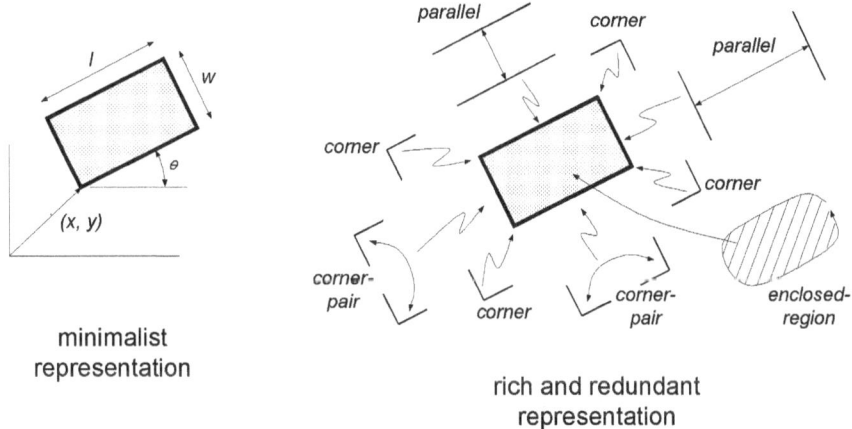

Fig. 3. In a minimalist representation, a rectangle can be described with five parameters. Perceptual Organization constructs rich and redundant representations making explicit many properties and relations.

4   Why Take a PO Approach?

There are a variety of reasons to believe that Perceptual Organization, the intermediate level representation of generic visual structure, offers several benefits for document image analysis.

Documents are human artifacts designed to be viewed and interpreted by human visual systems which employ processes of Perceptual Organization. It is no accident that properties of proximity, symmetry, closure, and smooth continuation are integral to the design of graphic documents. Therefore, it is sensible to believe that modeling of some of these processes in machines will support processes that mirror the power of human strategies for interpreting and understanding them.

Perceptual Organization will lead to more natural and intuitive user interfaces for document image analysis since the elements of visual structure identified by PO provide an impedance match to users. If a graphic editor program could let users select graphic elements on the basis of the visual objects they comprise instead of on an element-by-element basis, for example the entire set of markings comprising a texture fill of a region, users could make modifications with much greater facility [6, 23, 24, 27].

Perceptual Organization is sound engineering. It attempts to introduce knowledge in a modular fashion. Techniques that don't rely on detailed knowledge of any particular image document type hold the promise of generality and stability across applications and image domains. This will include images of structured, formal documents, as well as informal documents such as hand-drawn sketches and annotations.

Perceptual Organization promises a broad foundation for a variety of applications. Intermediate level visual events become building blocks for later processes. In this sense, Perceptual Organization encompasses classical feature detection, where the features can be more complex and more abstractly construed than simple local measurements. Indeed, some work in graphics recognition can be viewed in this framework, including advances in region segmentation, adaptive thresholding and preclassification, and grouping approaches to assembling larger structures from smaller partial collections [4, 14, 17, 19–21, 26, 30].

5   Sophistication and Robustness

The benefit of PO will be the ability to support machine conversion and interactive tasks with graphic documents that current systems either cannot cope with or else require specialized programming and training to handle. For example, hand-drawn sketches are, for people, not much more difficult to read than cleanly drafted drawings, yet the subtle details of whether lines, curves, junctions, regions, and text obey the constraints of drafting technique lead to failure of classical systems.

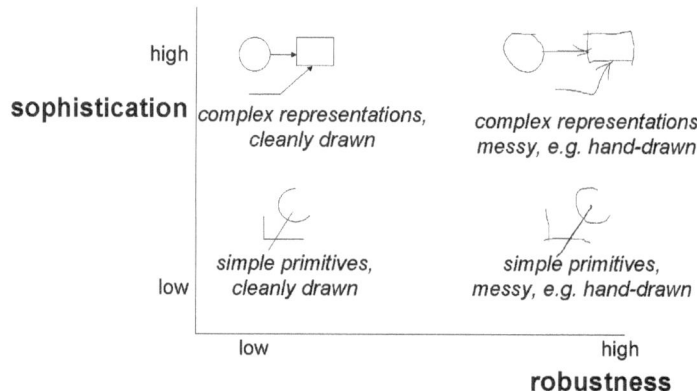

Fig. 4. Classical document recognition approaches pursue sophisticated models incorporating greater domain knowledge. Perceptual Organization seeks, first, robust descriptions of intermediate level visual structure that will apply across document genres.

Accordingly, two key objectives in improving graphics recognition systems are to achieve greater sophistication and to achieve greater robustness (see Fig. 4). *Sophistication* refers to the knowledge complexity of the interpretive models employed. A system that knows only about line, rectangle, and ellipse elements is less sophisticated than one that knows about the syntax of how these combine to form organization charts and process flow diagrams. *Robustness* refers to the range of variability the recognition system can tolerate. A system that can only recognize printed CAD drawings is less robust than one that can also recognize the equivalent rendered as a hand-drawn sketch.

Most attention in classical document image analysis has been devoted to the vertical axis, organizing knowledge so that domain-dependent constraints can interpret graphics in terms of complex models while applying top-down methods to cope with noisy data. Perceptual Organization is concerned with achieving robustness first. This is not simply a matter of "modeling the noise" within a knowledge-intensive system. Instead, according to a PO approach, graphically different figures that are perceived equivalently by human observers are not considered a matter of noise, but of truly equivalent intermediate level visual structure.

6   Conclusion

People perceive figures like Figure 5 in the same way they perceive cubist art. This image clearly contains visual structure; it kind of looks like something meaningful, in local regions at least, but it doesn't quite make sense in terms of any global semantic model. The intermediate level representations our visual systems build give us a language for discussing and evaluating the piece intelligently, and, they support our cognitive explorations of mappings to objects and domain elements. What is the material represented by the texture inside the circle? Could the arrow be a dial? Are there repeated elements among what appear to be symbols? Is that a funnel connected to a coil over there?

In our view, the construction of comparable intermediate level representations in machine vision systems stands as a worthy goal. Computers may or may not ever appreciate geometry and form on an aesthetic level—as art. But the processes and representations of Perceptual Organization that give humans the means to discern visual structure at a pre-semantic level are becoming more widely appreciated as they are becoming better understood and better modeled. These are arguably the foundation of humans' abilities to comprehend the world through vision. Similarly, we suggest that Perceptual Organization be pursued as a foundation for the field of Graphics Recognition.

References

1. Arnheim, R., *Art and Visual Perception*, Univ. of California Press, Berkeley (1954, 1974)
2. Boyer, K., and Sarkar, S., eds., *Computer Vision and Image Understanding* Special Issue on Perceptual Organization, V. 76, No. 1 (1999)
3. Boyer, K., and Sarkar, S., eds., *Perceptual Organization for Artificial Vision Systems*, Kluwer, Boston, (2000)
4. Canham, R.O., Smith, S.L., Tyrrell, A.M., "Recognition and Grading of Severely Distorted Complex Geometric Shapes from within a Complex Figure", *Pattern Analysis and Applications*, V.3, No. 4 (2000) 335-347

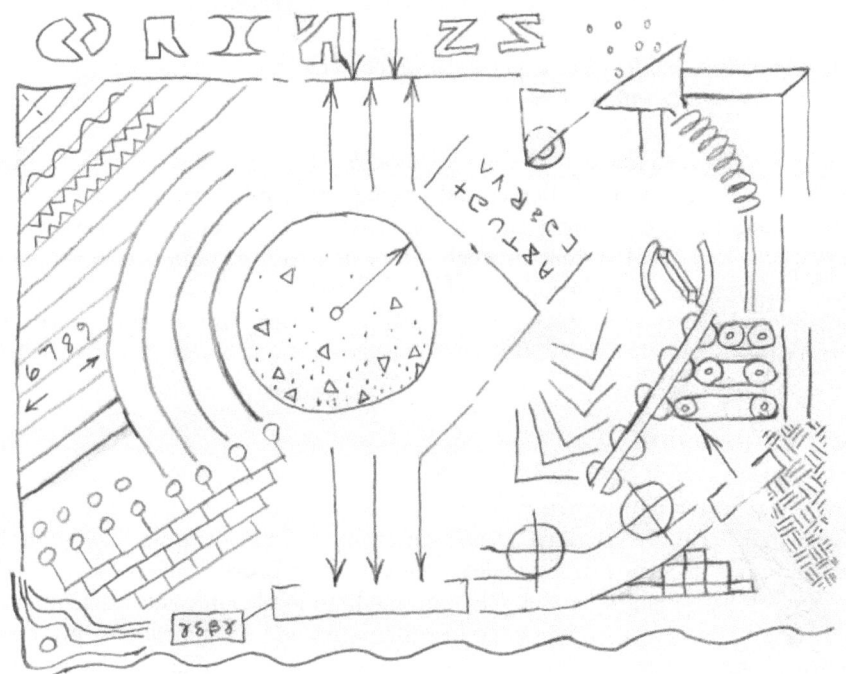

Fig. 5. A graphical diagram in the "cubist" motif.

5. Freeman, W., and Perona, P., "A factorization approach to grouping", European Conference on Computer Vision (1998)

6. Galindo, D., and Faure, C., "Perceptually-Based Representation of Network Diagrams", *Proc. 4th ICDAR* (1997) 352-356

7. Green, C., "Introduction to: 'Perception: An introduction to the Gestalt-theorie' by Kurt Kaffka (1922)", "http://psychclassics.yorku.ca/Koffka/Perception/intro.htm" (2000)

8. Havaldar, P., Medioni, G, and Stein, F., "Perceptual Grouping for Generic Recognition", *Int. Journal of Computer Vision*, V. 12 (1996) 59-80

9. Ip, H.H.S., and Wong, W.H., "Detecting Perceptually Parallel Curves: Criteria and Force-Driven Optimization", *Computer Vision and Image Understanding*, V. 68. No. 2 (1997) 190-208

10. Jacobs, D., and Lindenbaum, M., eds., *POCV 2001: The Third Workshop on Perceptual Organization in Computer Vision*, CIS Report #CIS-2001-05, Center for Intelligent Systems, Technion, Israel (2001)

11. Kanizsa, G., *Organization in Vision: Essays on Gestalt Perception*, Praeger, New York (1979)

12. Kass, M, and Witkin, A., "Analyzing Oriented Patterns", *Computer Vision Graphics and Image Processing* V. 37 (1997) 362-385

13. Kasturi, R., Raman, R., Chennubhotla, C., and O'Gorman, L., "An Overview of Techniques for Graphics Recognition", in Baird, H.S., Bunke, H., and Yamamoto, K., eds., *Structured Document Image Analysis*, Springer-Verlag, Berlin (1992)

14. Kise, K., Sato, A., and Iwata, M., "Segmentation of Page Images Using the Area Voronoi Diagram", *Computer Vision and Image Understanding* V. 70, No. 3 (1998) 370-382.

15. Koffka, K., "Perception: An introduction to Gestalt-theorie", *Psychological Bulletin*, V. 19 (1922) 531-585

16. Lowe, D., and Binford, T., "Perceptual Organization as a Basis for Visual Recognition", *Proc. AAAI-83* (1983)

17. Mahoney, J., and Fromherz, M., "Interpreting sloppy stick figures by graph rectification adn constraint-based matching," submitted to LNCS volume arising from GREC 2001 (2001)

18. Malik, J., Belongie, S., Shi, J., and Leung, T., "Textons, Contours and Regions: Cue Integration in Image Segmentation", *Proc. Seventh Int. Conf. on Computer Vision (ICCV '99)*, Corfu, Greece (1999)

19. Maurizio, P., "Deskewing Perspectively Distorted Documents: An Approach Based on Perceptual Organization", HPL-2001-100, HP Labs Technical Report (2001)

20. Ramel, J-Y., GBoissier, G., and Emptoz, H., "A structural representation to handwritten symbol recognition", *3rd IAPR Int. Workshop on Graphics Recognition, GREC '99* (1999) 259-266

21. Saund, E., "Labeling of Curvilinear Structure Across Scales by Token Grouping", *Proc. IEEE Conf. Computer Vision and Pattern Recognition* (1992) 257-263

22. Saund, E., "Finding Perceptually Closed Paths in Sketches and Drawings," *POCV 2001: The Third Workshop on Perceptual Organization in Computer Vision*, CIS Report #CIS-2001-05, Center for Intelligent Systems, Technion, Israel (2001)

23. Saund, E., and Moran, T., "A Perceptually-Supported Sketch Editor", *Proc. ACM Symposium on User Interface Software and Technology (UIST '94)*, (1994) 175-184

24. Saund, E., and Moran, T., "Perceptual Organization in an Interactive Sketch Editing Application", *Proc. 5th Int. Conf. on Computer Vision*, (1995) 597-604.

25. Stevens, K., "Computation of locally parallel structure," *Biological Cybernetics*, Vol. 29 (1978) 19-26

26. Syeda-Mahmood, T., "Indexing of Technical Line Drawing Databases", *IEEE TPAMI*, V. 21, No. 8 (1999) 737-751

27. Thorisson, K., "Simulated Perceptual Grouping: An Application to Human-Computer Interaction", *Proc. Sixteenth Annual Conf. of the Cognitive Science Society*, (1994) 876-881

28. Wertheimer, M., "Laws of Organization in Perceptual Forms", in Ellis, W., ed, *A source book of Gestalt psychology*, Routledge & Kegan Paul, London, 1938 (1923)

29. Witkin, A.P., and Tenenbaum, J.M., "On the role of structure in vision", in Beck, J., Hope, B., and Rosenfeld, A. (eds.), *Human and Machine Vision*, Academic Press (1983) 481-543

30. K. Yip and F. Zhao, "Spatial Aggregation: Theory and Applications." *J. of Artificial Intelligence Research*, V. 5 (1996) 1-26

31. Zucker, S., "Computational and Psychophysical Experiments in Grouping: Early Orientation Selection," in Beck, J., Hope, B., and Rosenfeld, A. (eds.), *Human and Machine Vision*, Academic Press (1983) 545-567

Exploiting Perceptual Grouping for Map Analysis, Understanding and Generalization: The Case of Road and River Networks

Robert C. Thomson[1] and Rupert Brooks[2]

[1]Robert Gordon University, School of Computing
St Andrew St., Aberdeen, UK
r.c.thomson@rgu.ac.uk
[2]National Atlas of Canada , Canada Centre for Remote Sensing
615 Booth Street, Ottawa, Ontario K1A 0E9, Canada
brooks@nrcan.gc.ca
http://www.cyberus.ca/~rbrooks

Abstract. A successful project in automated map generalization undertaken at the National Atlas of Canada made extensive use of the implicit perceptual information present in road and river networks as a means of analysing and understanding their basic structure. Using the perceptual grouping principle of good continuation', a network is decomposed into chains of network arcs, termed strokes'. The network strokes are then automatically ranked according to derived measures. Deleting strokes from the network following this ranking sequence provides a simple but very effective means of generalizing (attenuating) the network. This technique has practical advantages over previous methods. It has been employed in road network generalization, and applied in the selection of hydrologic data for a map covering Canada's northern territories. The method may find further application in the interpretation of other forms of documents, such as diagrams or handwriting.

1   Introduction

A successful project in map generalization has been undertaken at the National Atlas of Canada. This work has exploited perceptual grouping principles to gain a level of map understanding that can support useful and efficient automated generalization.

Perceptual grouping principles present a fundamental set of tools for the understanding of images. They identify basic, natural' elements in images, particularly images designed for human perception, such as maps and diagrams. This intermediate level of structure more closely corresponds to real world elements and permits the formation of object hypotheses. This process is essentially domain-independent [1].

The network generalization technique presented below makes particular use of the perceptual grouping principle of good continuation'. This allows a network to be

D. Blostein and Y.-B. Kwon (Eds.): GREC 2002, LNCS 2390, pp. 148-157, 2002.

partitioned into a set of elements called 'strokes', which are chains of network arcs. The strokes' relative importance can be estimated from more intuitive properties than those of the basic graph elements. The deletion of the strokes according to the derived sequence then provides a simple but very effective method of generalizing the network.

An implementation of this approach has been extensively tested and found to be applicable to both road and river networks. The technique has, for example, been used in practice with good results to select the rivers from the 1:1M scale GeoBase Level 0 dataset shown on a printed map of Canada's Northern Territories published in 2000 [2].

2   The Generalization of Geographic Networks

2.1   Generalization

This paper considers the generalization of geographical networks, primarily road and river networks. 'Generalization' can encompass a range of procedures in spatial database handling and the cartographic representation of data. In the cartographic domain, it may be defined as the "selection and simplified representation of detail appropriate to the scale and/or purpose of the map" [7]. In the digital context, notably in geographic information systems (GIS), generalization is recognized as a vital computational procedure in presenting and integrating spatial data [20]. Generalization may be understood as a process of controlled distribution refinement affecting both thematic and geometric domains [5]: attenuating a spatial pattern while retaining the most important elements for a given context and, moreover, preserving the character of the pattern.

Map generalization is most clearly entailed in scale reduction. Although it is relatively easy to judge how much detail should be eliminated for a given scale change, the question of which particular elements to eliminate or retain is entirely different. To address this question the cartographer must understand the relative contribution of map elements to the message of the map.

2.2   Generalizing Hydrologic and Road Networks

Network generalization can, ultimately, be viewed as a sorting process. Network elements must be ranked in order of importance according to context (the target scale and map use). For a reduced level of detail only a certain percentage of the elements will be used. The problem of hydrologic network generalization has thus been seen historically as that of deriving weighting values for each river segment. By thresholding the data using these values the network can then be attenuated to the desired degree. Established stream ordering methods [6], taken together with segment length, give a relatively simple but effective ordering of the segments. Stream order calculations require, of course, knowledge of the stream directions. Complications are introduced into the ordering scheme by stream braiding and delta structures; further complications arise when lakes within the network are also subject to removal by generalization. However, these difficulties can be overcome and a reliable generalization scheme for hydrologic networks established on this foundation [13].

Road networks, unlike river networks, contain loop structures (foiling attempts to derive equivalents to stream orders), have no equivalent to a flow direction, and road segments may have associated semantic data to take into account, such as road class or surface. If the context for the generalization includes a set of locations of special importance on the network, it is possible to derive estimates of the relative importance of the road segments, for example by way of considering optimal paths between the designated locations [14] [19], and so proceed to generalize the network.

A consideration of examples makes it clear that, even without such a site selection process to provide contextual information, road networks can usually still be generalized (by a human cartographer) purely on the basis of their geometric, topological, and thematic properties (*i.e.* the road classification information). Indeed, it is usually possible to infer the relative importance of road segments in a network in the absence of all thematic information.

We suggest that this generalization capability follows from how the human visual system perceives an image as a collection of perceptually salient, 'natural', elements. When the abstraction process needed to represent a network as a map preserves enough descriptive geometric detail, the 'natural', perceptual units make sense as 'natural' road or river segments. Further, the relative perceptual significance of these perceived elements corresponds closely to their relative functional importance in the network. This property is especially pronounced in a map, which has been carefully designed to convey information to the human visual system. (In situations where the correspondence between functional importance and perceptual salience is weakened, appropriate adjustments can be made.)

These propositions will be expanded upon below, and a description given of how this translates into a practical method for automatic network generalization, applicable to many types of geographical networks, including both roads and rivers. The basis for this development is perceptual grouping, which here, as in other vision-related situations, identifies important features of an image and gives clues as to the relationships in the real world of the entities represented by those image features.

3   Perceptual Grouping

Perceptual grouping is a fundamental component of perceptual organization – the ability of a visual system to spontaneously organize detected features in images, even in the absence of high level or semantic knowledge [11] [22]. Perceptual organization principles are widely recognized as the basis for parsing the visual world into surfaces and objects according to relatively simple visual characteristics, in a process that operates independently of the domain being represented. Many perceptual grouping (Gestalt) principles have been identified, such as grouping by proximity, similarity, symmetry, closure, parallelism, collinearity, and continuity [21]. These principles play a vital role in understanding 2D images of 3D scenes [9] [22]. These principles have also been recognized as crucial in the interpretation of other forms of images, such as cross-sectional images [17] and maps [10].

Direct analogies have long been noted between what is required for successful map generalization and the principles of perceptual grouping [5].   Perceptual grouping supports an intermediate level of map understanding that is domain independent yet sufficiently powerful to identify features (segment) and infer their relative importance, so supporting generalization, and other analyses. We also suggest that the character of a map is a function of the perceptual groups it contains, and so the preservation of salient perceptual groups, albeit in some attenuated form, should help retain that character.

4   Network Generalization via Perceptual Grouping

The perceptual grouping principle that most clearly contributes to the generalization of networks is the principle of good continuation: "that elements that appear to follow in the same direction .tend to be grouped together" [4]. Thus a series of relatively smoothly connected elements will be naturally "grouped" and perceived as a single object, perhaps intersected by others. We use the term "strokes" for these elements, prompted by the idea of a curve drawn in one smooth movement.

The arc-node topological data model is commonly used in GIS systems. In this model, a dataset of linear features is maintained as a planar graph. A network in such a system has two types of object: the linear arcs and their terminal nodes. In our data model, a stroke comprises a set of arcs and nodes. This set represents a path through the network from one terminal node to another (possibly the same) which will utilize all the arcs in the set, without repetition – although the path may self-intersect and so nodes may repeat. In the traditional Arc-Node topology, for example, the network shown in figure 1 would have eight arcs and nine nodes. These elements form four intersecting strokes.

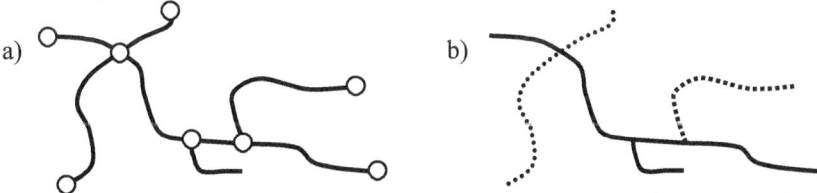

Fig.1. A simple network with 8 arcs and 9 nodes (a) resolves into 4 strokes (b)

4.1   Stroke Building

The stroke building process assumes the creation of a proper planar graph data structure for the given dataset. At each node, a decision is made as to which incident arc pairs (if any) should be connected together. The simplest criterion to use is the angle of deflection that the join would imply. Additional criteria that have proved useful in practice are the class (for road networks), direction of flow (for river networks), and the name of the feature (for roads and rivers). Once decisions have been made at each node the strokes are assembled as sets of arcs that connect.

In the implementation used at the National Atlas of Canada, stroke building is a local process, considering each node in turn, and dependent completely on the network properties at that node neighbourhood. This implies that strokes can be locally adjusted, with the update requiring only small processing time.

It may be noted that the process is guaranteed to give the same resulting set of strokes regardless of the order of consideration of the nodes or the arcs around each node. The stroke construction process is also robust in that no degenerate cases cause the algorithm to fail. For meaningful results, however, it is essential that the dataset have correct connectivity and, in the case of hydrologic networks, correct flow direction. In both cases, better results can be achieved by using additional attribute information. For example, good results are obtained by constraining the process so that arcs with the same name attribute (e.g. Peace River, Bank Street) are connected into a stroke. Where such information is incomplete, the strokes are built according to purely geometric considerations.

Once the strokes are constructed the network must still be generalized. First all the strokes must be ordered by importance. Then some suitable proportion of them can be removed to reduce the data density. It may be necessary, or at least useful, to finetune the stroke-based determination of generalization order by additional processing. For transportation networks this took the form of procedures previously developed to prevent unnecessary dead-ends being introduced during network attenuation, with spanning trees used to maintain certain connections [19].

4.2   Road Networks

To devise an ordering scheme for a road network, three simple principles were applied:

- All else being equal, a longer stroke is more important than a shorter one.
- All else being equal, a stroke made up of roads of higher quality is more important than one made up of roads of lower quality (e.g. highways *vs* tracks).
- No connected components of the original network would be split into two disconnected parts.

It is relatively straightforward to define a salience value for each stroke based on length and weighted by a factor derived from the road quality. The ordered list based on the raw salience may then be adjusted to avoid disconnecting the network. This is achieved by a relatively simple procedure [20] that creates an ordered list from which strokes can be removed in sequence without disconnecting the network.

Figures 2 and 3 show the effect of this generalization on the road network data for the Ottawa area. Figure 2 shows the complexity of the original dataset. The darker lines in figure 3 show the resulting selection, after generalization using the method described above. No postprocessing has been applied to this result so that the effect of the algorithm is not obscured. The dataset contains a total of 28 841 arc features in 8497 strokes with a total length of 7095 km. After generalization, 254 of these strokes have been kept, comprising 5500 arcs with a total length of 2546 km. The importance of length in computing the salience of a stroke shows clearly in these figures.

Fig.2. Raw road data for Ottawa and surrounding area

Fig.3. Generalized road network for Ottawa and surrounding area

4.3 Hydrologic Networks

For river networks, it has been found [15] that the Horton stream ordering method closely approximates the generalization decisions made by a human cartographer.

Horton ordering works in two stages. First, fingertip streams are numbered as one and then removed from consideration; this process is then repeated, incrementing the assigned number each time, until the entire network has been ordered (this is also known as Strahler ordering). Secondly, a main stream is selected and assigned the highest order occurring on its length. Of the remaining tributaries, a main stream is selected in each and assigned the maximum order which occurs on its length. This process continues until all components have been assigned an order [6].

By careful selection of building rules, the strokes can be constructed so that they can be ordered using Horton's method. The strokes are used to define the main paths, needed for the second processing stage. At each node, a choice of the main path is made which is then connected to form a stroke. The maximum first stage (Strahler) order occurring on any element of a stroke is assigned as its Horton order. The definition of what constitutes the main stream"is a key component of this process. Horton's original criteria were that the longest and straightest path should be used to define the main stream. Other determinations of the main stream have included the straightest path [13] or the largest drained area [12]. The use of the stroke data model does not enforce any particular choice of main stream; the rules for stroke-building can use these or more sophisticated approaches for selecting the main stream. Braided streams add an additional level of complexity to the ordering process, but a simple extension of the construction process [16] can prevent disconnections.

Fig. 4. Generalized river network (Attawapiskat River)

Figure 4 shows the generalization process using strokes as applied to a river network. Initially no river carried any thematic information; those not selected by the generalization process are shown faded. . This result was derived from the final released dataset, and some slight post processing has been applied to the resulting selection.

5   Practical Experience at the National Atlas of Canada

An automated generalization system was built for production use at the National Atlas of Canada. This was heavily based on previous work [13] [14] [18] [19]. The system has been used to generalize the hydrology for a 1:4M scale map of Canada's three northern territories from source material in the GeoBase Level 0 hydrology dataset. As part of an ongoing program [3], the source data had been cleaned and attributed, connectivity corrected and directionality computed.

The stroke-building algorithm itself is very efficient. Measured computational cost in practice appears approximately linear in the number of nodes. The algorithm will operate on erroneous datasets. Although the results will be progressively degraded as source data degrades, this robustness allows operation in a production environment on large datasets, where detecting and correcting all errors in a dataset may not be an option. For production generalization, a number of constraints were applied to the stroke building process to increase the cartographic accuracy.

- Strokes could never connect two arcs with incompatible direction of flow.
- Strokes would always follow a named path, if one existed.
- Strokes would never cross from a braid to a non-braid stream.
- Strokes could never jump more than two Strahler stream orders
- The user could enter constraints to force the stroke to be built along a certain path at a node. These constraints overruled all other considerations.

With these constraints satisfied, the stroke building algorithm would find the "longest and straightest" paths. The final selection that resulted from the generalization technique described here was processed using cartographic generalization techniques, including line simplification, displacement and exaggeration.

Once built, the stroke model was also used for non-generalization applications. For example, it assisted in matching name attributes to river tributaries, greatly increasing the efficiency of that process. The Canadian GeoNames Database (CGNDB) records all officially recognised place and physical feature names in Canada. Each name is represented by a single latitude / longitude point in the database. For network features this is not a particularly useful representation: the aim was to attach the name attributes directly to the river features in the database, defining their entire extent.

CGNDB points were usually present at river mouths only. Connecting names to rivers thus required finding which river mouth was referred to, and following that river upstream. The perception of main stream inherent in the strokes helped to partially automate this process. The longest stroke ending near each CGNDB point was found and traced upstream, with the name information attached to all features on this path. Results were verified by an operator and adjusted as needed.

The National Atlas of Canada is also currently working with Environment Canada to define upstream drainage basins for the latter's gauging station holdings. This information is implicitly contained in the stroke model and the defined ordering, since upstream arcs must be removed before downstream ones. To make the upstream relationships explicit, a hierarchical coding structure was defined on the strokes themselves. Selection of upstream and downstream components could then be reduced to a simple SQ query. The stroke structure was particularly useful here, as it is robust in the presence of braids and areas that drain in multiple directions

6   Conclusions

The perceptual strokes technique allows the delineation of meaningful units of a network in a manner that echoes the human perception. Relatively simple properties of these units may then be used to determine an order of importance.

This method is robust enough to use on faulty data with reasonable results. Furthermore, local constraints could be added to the stroke building method without affecting its overall performance. This allowed additional information, such as names, where available, to be used to adjust the system's performance. Tests suggest that the stroke-based generalization method will apply to other network types, including pipeline and electrical networks. The source datasets for this work are cartographic databases, which provide the basic arc-node topology of the network, geometric (polyline) descriptions of the arcs, and some thematic attributes. In principle all this information could also be recovered from a scanned map image.

We venture that this perceptually-motivated stroke-building approach may find wider application in the automatic analysis or interpretation of other forms of document, such as line diagrams or handwriting. In particular, the possibility that the method may support the recovery of drawing order from single-stroke handwriting images [8] invites investigation.

Acknowledgments

This work builds on methodologies and techniques developed at CCRS by, and under the direction of, Dr Dianne Richardson. Statistics Canada generously provided road network data from their database which was used for the road network experiments.

References

1. Boyer, K. L., Sarkar, S.: Perceptual Organization in Computer Vision: Status, Challenges and Potential. Computer Vision and Image Understanding, 76 (1999) 1-5
2. Brooks, R.: National Atlas of Canada Producing First Map Using Automated Generalisation of Framework Data. Cartouche 39 (2000)
3. Brooks, R.: The New and Improved Base Framework for National Atlas Data. Proc. 19th Int. Cartographic Conference (1999)
4. Cohen, S., Ward, L. M.: Sensation and Perception. HBJ, Orlando (1989)
5. DeLucia, A., Black, T. A.: Comprehensive Approach to Automatic Feature Generalization. Proc. 13th Int. Cartographic Conference (1987) 168-191
6. Goudie, A. (ed.): Encyclopedic Dictionary of Physical Geography. Blackwell, Oxford (1985)
7. ICA (International Cartographic Association): Multilingual Dictionary of Technical Terms in Cartography. Franz Steiner Verlag, Wiesbaden (1973)

8. Kato, Y., Yasuhura, M.: Recovery of Drawing Order from Single-Stroke Handwriting Images. IEEE Trans. Pattern Analysis and Machine Intelligence 22 (2000) 938 – 949
9. Lowe, D. G.: Perceptual Organisation and Visual Recognition. Kluwer, Boston (1985)
10. MacEachren, A. M.: How Maps Work. Guilford Press, New York (1995)
11. Palmer, S.: The Psychology of Perceptual Organisation: a Transformational Approach. In: Beck, J. (ed.): Human and Machine Vision. Academic Press, New York (1983) 269-339
12. Pfafstetter, O.: Classification of Hydrographic Basins: Coding Methodology. DNOS, Rio de Janeiro (1989)
13. Richardson, D. E.: Automatic Spatial and Thematic Generalization using a Context Transformation Model. PhD Thesis, Wageningen Agricultural University (1993)
14. Richardson, D. E., Thomson, R. C.: Integrating Thematic, Geometric and Topological Information in the Generalization of Road Networks. Cartographica 33 (1996) 75–83
15. Rusak Mazur, E., Castner, H. W.: Horton's Ordering Scheme and the Generalization of River Networks. The Cartographic Journal 27 (1990) 104-112
16. Thomson, R. C., Brooks, R.: Efficient Generalization and Abstraction of Network Data using Perceptual Grouping. Proc. 5th Int. Conf. on GeoComputation, Greenwich (2000)
17. Thomson, R. C., Claridge, E.: A Computer Vision'Approach to the Analysis of Crystal Profiles in Rock Sections. Proc 6th Scand. Conf. on Image Analysis (1989) 1208-1215
18. Thomson, R. C., Richardson, D. E.: The Good Continuation' Principle of Perceptual Organization Applied to the Generalization of Road Networks. Proc. 19th Int. Cartographic Conference (1999) 1215–1223
19. Thomson, R. C., Richardson, D. E.: A Graph Theory Approach to Road Network Generalization. Proc. 17th Int. Cartographic Conference (1995) 1871-1880
20. Weibel, R.: Generalization of Spatial Data: Principles and Selected Algorithms. In: van Kreveld, M., Nievergelt, J., Roos, T., Widmayer, P. (eds.): Algorithmic Foundations of Geographic Information Systems. Springer-Verlag, Berlin (1997) 99-152
21. Wertheimer, M.: Laws of Organization in Perceptual Forms. In Ellis, W. (ed.): A Source Book of Gestalt Psychology, Harcourt Brace, New York (1938) 71-88
22. Witkin, A. P., Tenenbaum, J. M.: On the Role of Structure in Vision. In: Beck, J., Hope, B., Rosenfeld, A. (eds.): Human and Machine Vision. Academic Press, New York (1983) 481-583.

Extraction of Contextual Information Existing among Component Elements of Origami Books

Takeyuki Suzuki, Jien Kato, and Toyohide Watanabe

Dept.of Information Engineering, Nagoya University
Furo-Cho, Chikusa-Ku, Nagoya 464-8603, Japan
{takeyu_s, jien,watanabe}@watanabe.nuie.nagoya-u.ac.jp

Abstract. This paper proposes a novel approach for extracting element clusters, associated with individual folding steps of origami, from a page-image of origami drill books. Unlike the documents investigated in other researches, origami books are usually not designed under a specific layout structure. Our approach is based on a Voronoi expression schema. Since the only clue to group related elements is spatial proximity relations among the elements, we utilize Voronoi regions, which are generated by using the centroids of main elements as characteristic points of Voronoi diagrams, to estimate the scopes of element clusters. This method provides necessary information to a recognition system that converts a folding process, described by a series of illustrations, instructive sentences and special symbols, into a 3D animation. The experimental results with good accuracy show the effectiveness of the proposed method.

Keywords: document analysis, origami, Voronoi diagram, drawing interpretation.

1 Introduction

Origami, the ancient Japanese art of paper folding, has attracted considerable attention as its merit in early childhood education is recognized by psychologists [1]. Every year a large number of new works of origami are presented. Among them, substantial works are collected in origami drill books and published to meet the needs of instruction and amusement. In recent years, origami works (origami for short) that employ high degree of technical skill trend toward increase. As a result, origami drill books (origami books for short) become harder to understand. The difficulty in understanding origami books leads to the growing interest in the approach of simulating paper folding by three-dimensional animation [2]. Animation techniques allow to see origami from different views, and moreover to return to any previous folding steps and repeat a certain folding operation at low speed. For instance, Miyazaki, *et al.* have developed an origami playing simulator in 3D virtual space [3]. That enables a user to pick edges or corners of a sheet of paper through a mouse, and perform folding operations in virtual space. On the other hand, Kato, *et al.* have proposed a recognition method that converts a sequence of origami illustrations printed in origami books

D. Blostein and Y.-B. Kwon (Eds.): GREC 2002, LNCS 2390, pp. 158–166, 2002.

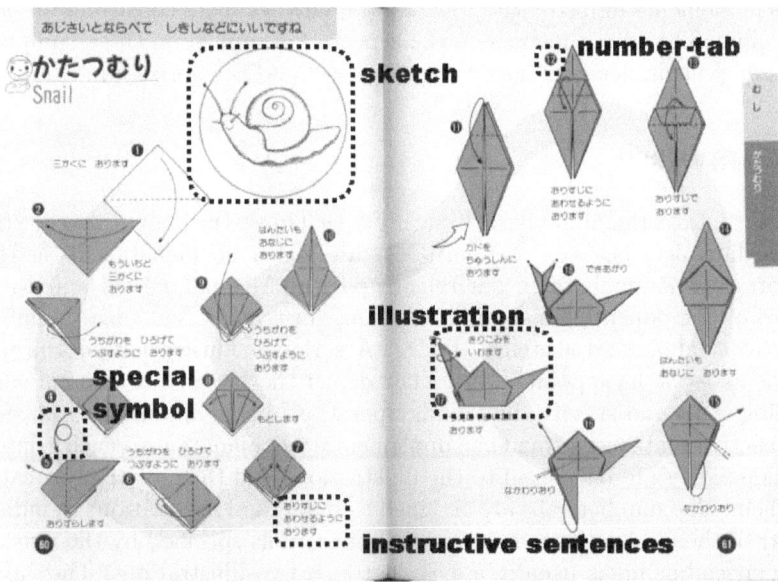

Fig. 1. An image of two facing pages taken from an origami book. Various component elements are included: illustrations, number-tabs, instructive sentences, special symbols and sketches

into 3D animation expression. With input of a sequence of the illustrations, the system interprets each illustration and generates appropriate folding operations that are consistent with succeeding illustrations step by step. A CG generator then simulates the resulting folding process corresponding to the input illustration sequence [4, 5].

As the extension of the researches addressed in [4, 5], this paper proposes a method for automatically extracting necessary input information for the recognition system from a page-image of origami books. In origami books, explanation for a single folding step mainly consists of one illustration, several instructive sentences and a number-tab (see Fig.1). The illustrations are the key elements. The number-tabs provide order information, while the instructive sentences give a supplementary explanation of folding operations respectively, for a related folding step. Detecting the three types of heterogeneous elements from a page-image and further grouping them into the units associated with individual folding steps are essential to automatically understanding origami books. We call the associations among the component elements and the correspondences between the element groups and the folding steps contextual information of origami books.

However, the component elements of a page-image in origami books are usually not designed under a specific layout structure. In order to extract the contextual information successfully, several procedures are needed: firstly to extract and categorize individual component elements, then to group the related component elements into stepwise units, and finally to find a consistent order among the units. This paper focuses on the aspect of grouping, on the assumption that the component elements have been extracted and categorized into three classes.

2   Approach

The layout of origami books is basically free. This is the main difference (also the main difficulty) between origami books and other documents such as technical reports and newspapers. In general, any page of origami books is a mixture of five types of component elements: illustrations, instructive sentences, number-tabs, special symbols and sketches (Fig.1). A series of illustrations is the principal elements as we have pointed out. They depict the states of origami at successive folding steps and specify the folding operations being applied at each step. The instructive sentences provide a supplementary explanation to each related illustration. They are incidental to the illustrations and thus relatively located near to them. The number-tabs are assigned to individual illustrations to indicate the order of them, also the order of folding operations specified by the illustrations. The special symbols usually appear between two illustrations. They alert to a view-change, namely, give a notice that origami will present in the following illustrations from a different angle. The sketches exhibit the real objects that origami attempts to express. With a view to understanding origami books and further generating corresponding animation, the former three elements, *illustrations*, *instructive sentences* and *number-tabs*, and the relationships among them are considered as the necessary input information for our recognition system.

Figure 2 shows our processing flow for extracting the contextual information existing among component elements. Except for preprocessing, it is composed of four phases: extracting and categorizing individual component elements from input page-images, grouping different types of the elements into stepwise units associated with individual folding steps and finding an optimal order among the element groups.

The extraction and categorization of component elements are essential since when different elements have been detected and classified, it at last becomes possible to investigate the relationships among them. Our extraction and classification procedures work effectively based on geometric and structural features of individual types of component elements. The illustrations are distinguished from sketches (an illustration is easily confused with a sketch) by use of distribution of line segments within a graph region, since illustrations include more line segments than sketches. The number-tabs are extracted by circle detection. The instructive sentences are extracted by using periodicity in vertical and horizontal directions appearing on images, because the font sizes included in a text region are basically same. The separate text strings which physically situate in neighboring positions with upper-lower relationships are then merged into one component element. Figures 3 (b) and (c) show extraction and classification results for a page-image given in Fig.3(a).

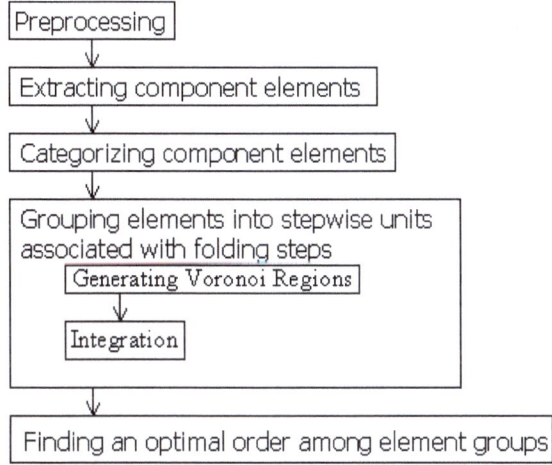

Fig. 2. Processing flow

We group the related component elements extracted and classified based on a Voronoi expression schema. The motive for utilizing a Voronoi diagram is because a page of origami books does not follow a specific layout. The only clue to group related elements into the units associated with individual folding steps is spatial proximity relations among the elements. Since illustrations are main elements, in both senses of significance and size, and moreover other two types of elements are incidental and thus relatively located near to the illustrations, the Voronoi regions generated according to the centroids of illustrations are able to roughly divide the component elements into appropriate groups. We go into details about grouping in the following section.

3 Grouping Component Elements

The grouping phase of individual component elements plays an important role in our approach. This phase, as you can see in Fig.2, is composed of 2 modules. We partition a page-image into several regions according to a Voronoi diagram in the first module. As to these elements that extend over two or more Voronoi regions, we decide to which group the elements likely belong in the second module.

3.1 Generating Voronoi Regions

Traditionally, a Voronoi diagram is used to partition an area into small regions based on some characteristic points. It draws the border lines of the regions so that the distances from the characteristic points of neighboring regions to the common border line are equal.

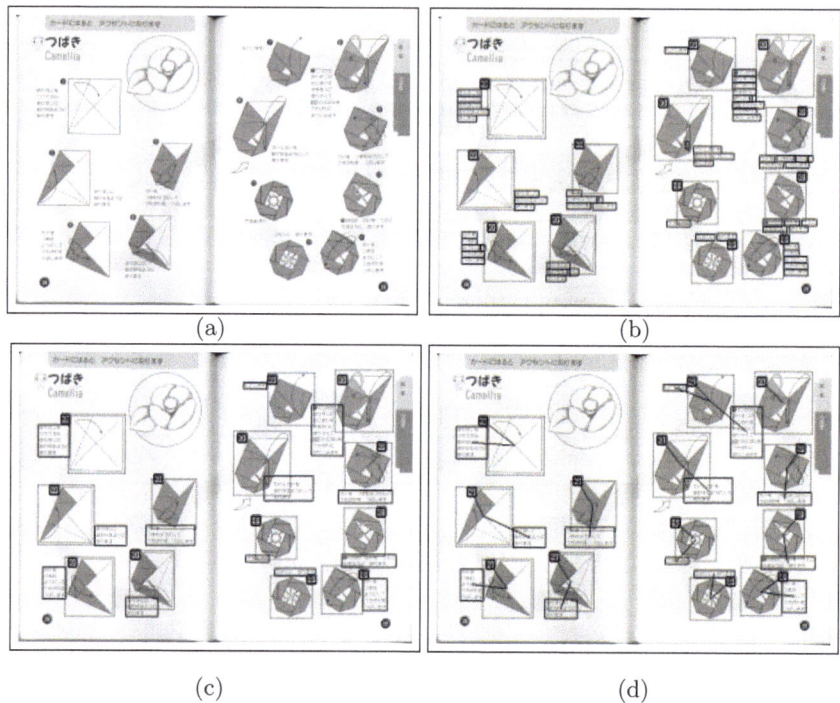

Fig. 3. Extraction of contextual information existing among component elements. (a) An original image, (b) extracting component elements, (c) categorizing component elements, and (d) grouping. In (b)-(d), illustration regions are surrounded with dot-lines, regions of instructive sentences are with thin solid-lines, and number-tab regions are with thick solid-lines. In (d), the elements grouped into one unit are connected by lines

Because an illustration in origami books always corresponds to one folding step, in our case we use the centers of individual illustrations as the characteristic points (see Fig.4). The generated Voronoi regions estimate the scope of element groups associated with individual folding steps. We call the scope obtained by such a Voronoi diagram *illustration scope*. By use of the illustration scopes, the relationships among component elements can be evaluated intuitively.

3.2   Obtaining Stepwise Units

From Fig.4 (b), we can observe that in many cases instructive sentences are located between two or more illustrations. In some cases, it is not very clear which illustration the text strings should be associated with. The illustration scopes are useful to make the subordination relationships between illustrations and instructive sentences, and also those between illustrations and number-tabs obvious. As to the elements extending over two or more illustration scopes, we

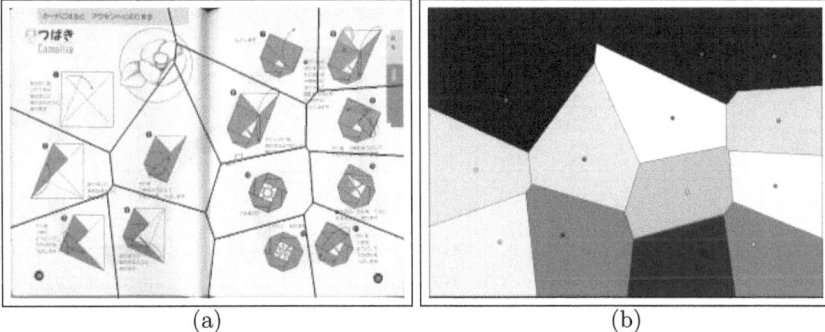

(a)   (b)

Fig. 4. Estimate the scopes of element groups using a Voronoi diagram. (a) Voronoi regions on a page-image. (b) Voronoi regions and the centers of illustrations

integrate them into a solitary group, corresponding to one stepwise unit. The integration is performed in two steps: the first is for number-tabs, and the second is for instructive sentences.

It is not difficult to integrate number-tabs into one stepwise unit because number-tabs always situate near the related illustrations explicitly. Actually, considerable number-tabs have been included in the scopes of related illustrations without integration. The way to know if one number-tab is included in an illustration scope is simply looking into whether the center of the number-tab comes in the illustration scope.

The judgment criterion for integrating the instructive sentences that extend over two or more scops is based on the spatial proximity relations among the elements, as same as the case of number-tabs. However, since two or more clusters of instructive sentences are possibly attached to one illustration, and moreover there are illustrations without any instructive sentence, the problem is much more difficult than the case of number tabs. In order to deal with this difficulty, we introduce some heuristic knowledge about spatial associations between instructive sentences and related illustrations. That is, instructive sentences should be located in the positions relative to a corresponding illustration according to the following priority: the lower, right, left and upper sides. With this heuristic knowledge, the relationships between instructive sentences and illustrations are estimated.

If instructive sentences relate to only one illustration, for example, the center point of the text strings is included in only one illustration scope, they are simply associated with this illustration. Otherwise, the equation below is used to calculate R_i, which measures how close text strings are to an illustration:

$$R_i = P_i + D_i. \tag{1}$$

In above equation, i is the identification number of illustration scope, P_i represents the percentage of the area, which is contained by the illustration scope i, of

Table 1. Results

| | Extraction of Component Elements | | | Categorization of Component Elements | | | Grouping |
	Illustrations	Instructive sentences	Number-tabs	Illustrations	Instructive sentences	Number-tabs	Group
Correct	478	6508	519	529	6524	526	368
Incorrect	90	——	168	55	——	34	24
Input	593	6868	593	593	6868	593	392
Accuracy (%)	69.99	94.76	68.20	81.64	94.99	95.12	93.88

instructive sentences, and D_i is a value that expresses the locational relationship between the instructive sentences and the illustration. D_i is heuristically specified. The right, lower-right, lower, lower-left, upper-right, upper, upper-left and left sides are respectively assigned as 0.45, 0.51, 0.60, 0.45, 0.30, 0.15, 0.21 and 0.30, based on analysis of a large number of examples in origami books. As the value of D_i becomes higher, the relationship between the instructive sentences and the illustration becomes closer. The instructive sentences are integrated into the illustration scope with the maximum value of R_i evaluated by Eq.1.

4   Experimental Results

Our method has been implemented and tested by 40 sample images of two facing pages, which are taken at random from an origami book [7]. The test images were acquired using a scanner with 256 gray-scale and 300 dpi resolution. The experimental results are summarized in Table 1 and an example can be found in Fig.3 (d).

In Table 1, "correct" and "incorrect" mean the numbers of correct and incorrect recognition results, respectively. For the extraction and categorization phases, by "incorrect" we especially mean misclassifying elements as incorrect categories. Accuracy in the extraction and categorization phases is calculated as

$$Ip = 100 \frac{In}{R + M}, \tag{2}$$

where In means the number of recognized elements, R and M stand for input number of elements and the number of misrecognized elements, respectively. For instructive sentences, misrecognized elements will rapidly increase in the particular case that there exit a lot of other text strings except instructive ones. Since the data in such case do not tell the truth, we did not include them into the recognition ratio and just set $M = 0$.

Accuracy in the grouping phase is calculated as

$$A = 100 \frac{C}{I}, \tag{3}$$

where C means the number of correctly extracted groups, I means the number of illustrations.

From Table 1 the improvement in accuracy by categorization phase can be observed. In the extraction phase, the component elements are roughly categorized by using local feature. On the other hand, in the categorization phase the

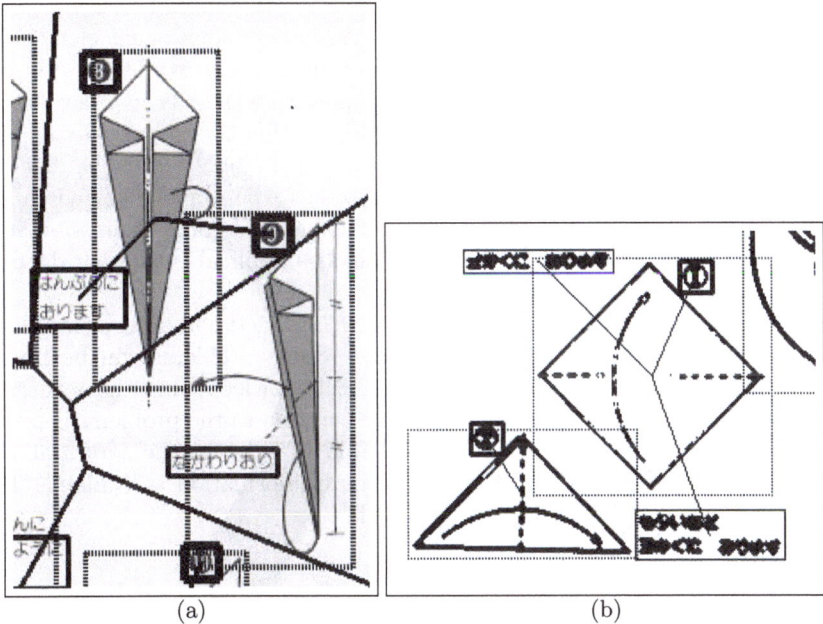

(a)   (b)

Fig. 5. Examples of grouping mistakes. (a) Associate a number-tab with a wrong group, and (b) associate instructive sentences with a wrong group

results are further refined by using the information obtained in the extraction phase, for example, the size and relative locations of component elements. Some mistakes yielded in the first phase can be modified in the second phase because more global features of elements are incorporated.

The grouping accuracy of 93.88% shows the effectiveness of the proposed method. However, some mistakes exists. Two typical examples are shown in Figure 5. One is a grouping mistake that occurs to integration between a number-tab and an illustration (Fig.5(a)): two number-tabs are associated with the same illustration scope. This mistake is because the distance between two illustrations is unusually small. The too close distance increases the proportion of the illustrations' areas that extend into neighboring Voronoi regions. The precision of scope estimation is thus reduced. Another mistake is caused by an exception of locational relationship between instructive sentences and an illustration (Fig.5(b)). The heuristic knowledge we introduced was not appropriate for this example. Recognition of the order of folding steps may provide useful information to solve the problems described above. We regard it as future work.

5   Conclusion

The ultimate objective of this research project is to develop an intellectual interface which converts a folding process, described by a series of illustrations, instructive sentences and special symbols, into a 3D animated expression. The extraction of contextual information existing among component elements, discussed in this paper, is no doubt one important step toward it.

It has to be pointed out that the proposed method is not developed for specific types of origami books. Because origami books are basically composed of common component elements and designed in similar styles recommended by Origami Society, the proposed method is able to be applied to general types of origami books. In addition, the idea of using a Voronoi expression schema to group meaningful component elements is also applicable to other documents without strict layout structures.

In the end of the previous section, we mentioned some problems that are considered as the causes for misextraction. Some problems can be improved somewhat by using more complicated heuristic knowledge, but the accuracy can not nevertheless be raised up to 100%. The solution to the problems is probably to build a flexible mechanism which is able to tolerate the errors at earlier processing steps and correct them when more information is available. This is the subject we will face in the further work.

References

[1] T.Okamura, "My Origami World," *http://www.oklab.com/origami/ori_wold.html*, 2000. 158

[2] T.Agui, T.Takeda and M.Nakajima, "An Origami Playing Folds by Computer," *Computer Vision, Graphics and Image Processing*, Vol.24, No.6, pp.244–258, 1983. 158

[3] S.Miyazaki, T.Yasuda, S.Yokoi and J.Toriwaki, "An Origami Playing Simulator in the Virtual Space," *J.Visualization and Computer Animation*, Vol.7, No.6, pp.25–42, 1996. 158

[4] J.Kato, T.Watanabe, T.Nakayama, T.Guo and H.Kato, "A Model-based Approach for Recognizing Folding Process of Origami," *Proc. of 14th Int'l. Conference on Pattern Recognition*, Brisbane, Australia, pp.1808–1811, 1998. 159

[5] J.Kato, T.Watanabe, H.Hase and T.Nakayama, "Understanding Illustrations of Origami Drill Books," *Trans.Information Processing Society of Japan*, Vol.41, No.6, pp.1857–1873, 2000. 159

[6] R.Kasturi and L.O'Gorman, "Techniques for Line Drawing Interpretation: An Overview," *Proc. of IAPR Workshop on Machine Vision Applications*, Tokyo, Japan, pp.151–160, 1990.

[7] M.Yamaguchi, "A Cyclopaedia for Paper Folding Art," *Seito Company*, Tokyo, P.324, 1990 (in Japanese). 164

Text/Graphics Separation in Maps

Ruini Cao and Chew Lim Tan

School of Computing, National University of Singapore
3 Science Drive 2, Singapore 117543
{caorn,tancl}@comp.nus.edu.sg

Abstract. The separation of overlapping text and graphics is a challenging problem in document image analysis. This paper proposes a specific method of detecting and extracting characters that are touching graphics. It is based on the observation that the constituent strokes of characters are usually short segments in comparison with those of graphics. It combines line continuation with the feature line width to decompose and reconstruct segments underlying the region of intersection. Experimental results showed that the proposed method improved the percentage of correctly detected text as well as the accuracy of character recognition significantly.

1 Introduction

The separation of overlapping text and graphics is a challenging problem in document image analysis [1][2]. This problem is found in many applications, including forms processing, maps interpretation and engineering drawings interpretation, where text and graphics are processed in fundamentally different ways.

Fletcher and Kasturi [3] developed an algorithm for text string separation from mixed text/graphics image. Taking account of curving labeled road names in the map, Tan and Ng [4] developed a system using the pyramid to extract text strings. Both methods, however, assume that the text does not touch or overlap with graphics.

In forms processing, the problem of touching text has been dealt with reasonably well [5][6][7][8]. The reason is that since forms contain only straight lines the problem is much simplified. While for engineer drawings [9][10][11], especially maps [12], the problem is much more complex since there are more types of data, various types of lines, possible curvature and even branching of the graphics.

The segmentation of connected text and graphics is in fact a chicken-and-egg problem. On the one hand, the purpose of segmentation is character recognition. On the other hand, correct segmentation may require the recognition of the characters. A recent paper [13] presented a method of cooperative processing of text and line-art in topographic maps and gave promising experimental results. Yet, the method assumed that all the text should be detected and grouped previously. The method as well as some other methods in the literature searched for graphics-connected characters in the neighboring areas of detected isolated characters. In this paper, we present an improved method of detecting, extracting

D. Blostein and Y.-B. Kwon (Eds.): GREC 2002, LNCS 2390, pp. 167–177, 2002.
© Springer-Verlag Berlin Heidelberg 2002

and grouping characters, including characters connected to graphics. The improvement is in that it can detect the strings whose constituent characters are all connected to graphics without the risk of introducing too many false alarms. The damages to the text image are also reduced by appropriate reconstruction of the strokes at the intersections.

2   The Data

The source maps are from a local Street Directory. The maps are printed in color. The focus here is on the black layer, which includes text, small icons, road lines, outlines of buildings and parks, and so on. Some icons in the maps are solid components, such as solid arrowheads, solid overhead pedestrian bridge. The road lines are either thick or thin lines, mostly curves. Some tracks and outlines of parks are dashed lines. Most outlines of buildings and parks are polygons.

The text strings in the maps are mainly road names, names of buildings and parks, and so on. The road names are labeled along the roads. That means the labels may be along curved lines. The touching of road names with road lines is somewhat similar to the case of strings touching the underlines or upper lines. On the other hand, the connection of the names and the outlines of the buildings is much more complicated. The reason is that the outlines of the buildings can be of any shapes.

3   The Proposed Method

3.1   Preprocessing and Initial Classification

The preprocessing includes three steps: sub-layer separation, solid graphical components removal, dashed lines removal. We extract the black layer in the following way: if the r, g, b values of one pixel are all less than a predefined threshold, then the pixel is considered black. The solid graphical components are removed by a morphological method [9]. Dashed lines are removed by detecting repeated similar bars as follows.

Dashed Lines Removal. Dashed lines in the maps mainly consist of short bars or dots. One dashed line either has only bars or dots, or has alternating bars and dots. First, we detect all the bars and dots and differentiate them from thin characters such as 'i', 'j' and 'l' that may be mistaken as dashed lines. This is done as follows. We perform total linear regression [15] on each connected component to obtain a straight line. A minimum band is set over and parallel to the straight line such that all points in the connected components are within the band. If the width of the band is smaller than a preset threshold, then the connected component is considered as a candidate dashed segment. Let the equation for the resultant line from total regression be:

$$x \sin \theta - y \cos \theta + \rho = 0 \tag{1}$$

Then, the width of the band is:

$$W_b = abs(\max_i(x_i \sin\theta - y_i \cos\theta + \rho) - \min_i(x_i \sin\theta - y_i \cos\theta + \rho)) \quad (2)$$

Fig. 1 shows some linear regression results. If we set the threshold to be 2.5, such thin characters as '1', 'i' and 'l' will not be detected as dashed segments due to the small hooks. Next, we group the candidate segments using the method described in sub-section 3.3. For each resultant group, we analyze the orientation distribution and the size distribution of the segments. If the orientation variance is small, and the size variance is small, or if the size variance is large, and the number of short segments is close to the number of long segments, and both are larger than 2, then the group is a dashed line. Fig. 2 gives an example of removing dashed lines.

After the preprocessing, the image mainly includes text, road lines, and outlines of buildings and parks. Now, we perform initial classification as described in [4]. In summary, connected components are generated first. Then, we use a size filter to roughly classify the components into text and graphics. Large components are identified as graphics and small components as text. Then, we get the minimum size S_{min} and the maximum size S_{max} of the text components to be used as refined size filters in the subsequent steps.

(a)   (b)

Fig. 1. Linear regression results. (a) samples of dashed segments. (left) slope − 52°, band width = 2.0; (right) slope = 87°, band width = 2.0; (b) samples of characters: (numeric character '1') slope = 97°, band width = 7.0; (lower case letter 'i') slope = 99°, band width = 3.0; (lower case letter 'l' slope = 95°, band width = 3.0

3.2   Detection and Reconstruction of Touching Characters

In the initial classification, text connected to graphics will be incorrectly classified as graphics since the text and the graphics form a whole connected component and this connected component is large. So, we need to further search the graphics image for these touching characters. It is observed that constituent strokes of characters are usually small in comparison with those of graphics.

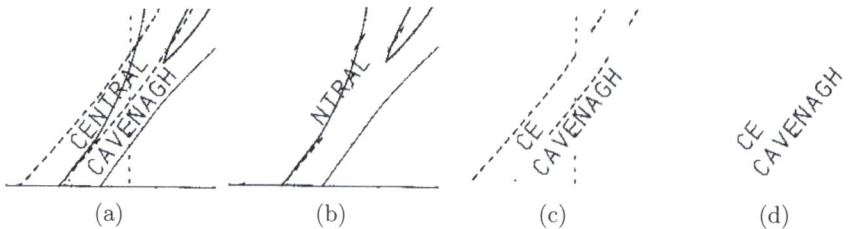

Fig. 2. Initial classification: (a) the original binary image; (b) the large components; (c) the small components; (d) the remaining text components after removal of dashed lines

Therefore, so long as we can decompose the graphics image with touching characters into appropriate segments in such a way that each segment is either a textual stroke or a graphical line, we can then use a size filter to separate them from each other.

We perform the decomposition process after thinning [16] since intersection points can be easily detected in the thinned graphics image. Intersection points are found at pixel locations where there are 3 or more neighbors. We first break the graphics image into small segments at the intersection points and then merge those segments that connected to the same junction by line continuation based on similar slopes of the adjoining segments within a thresholded difference. The slope of each segment is estimated by line fitting [17].

After decomposition, we use the size filters obtained in the initial classification to separate constituent strokes of text from graphical lines. We extract all the segments less than S_{max} and together with the intersection for which at least one segment connected to it has been restored. Then, we generate new connected components from the resultant image and remove those components smaller than S_{min} with some tolerance. Fig. 3 shows the separation results of touching characters and graphical lines of the image shown in Fig. 2(b). Fig. 4 gives another example.

Removal of large graphical segments may damage the touching characters in the overlapping area. The reason is that some parts of the removed segments possibly belong to both graphics and text. The damage may affect the subsequent recognition results severely. Fig. 5 gives such an example. We can see that the touching character 'o' is damaged as 'c' (Fig. 5c).

To achieve better recognition results, we should restore the damaged characters. It is observed that the line is usually thicker in the overlapping part. To reconstruct the damaged characters, we can restore the thicker parts of the removed segments. We choose the average width of the longest line segment in the image as the threshold. First, we pick the longest line segment in the thinned graphics image and represent it as a series of pixels: $P_1(x_1, y_1)$, $P_2(x_2, y_2)$, ..., $P_n(x_n, y_n)$, where P_{i-1}, P_{i+1} are the two neighbors of P_i, and P_1, P_n are endpoints. The line normal direction, which is perpendicular to the line slope di-

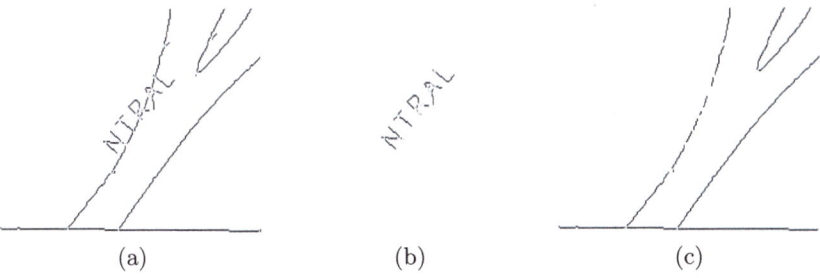

Fig. 3. Extraction of touching characters: (a) the thinned graphics image; (b) extracted touching characters; (c) graphical lines

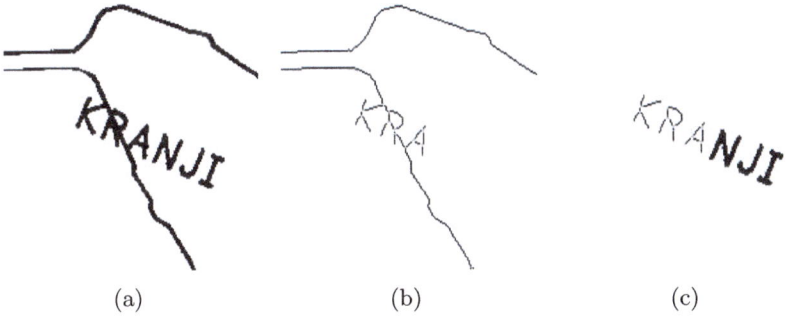

Fig. 4. Extraction of touching characters: (a) the original image; (b) the thinned graphics image; (c) extracted touching and isolated characters

rection, at pixel $P_i(x_i, y_i)$ can be estimated as the summation of the left 1-slope and the right 1-slope of this pixel:

$$\vec{N_i} = (\overrightarrow{p_{i-1}} - \vec{p_i}) + (\overrightarrow{p_{i+1}} - \vec{p_i}) \tag{3}$$

The line width W_i is roughly the distance between the two boundary points along the normal direction in the original image (Fig. 6). The average width of the segment is:

$$W = \sum_i W_i/L \tag{4}$$

where L is the length of the segment given by

$$L = \sum_i \sqrt{(x_{i+1} - x_i)^2 + (y_{i+1} - y_i)^2} \tag{5}$$

For each pixel at the removed segments, if the line width at this pixel is larger than the average line width W_i, then add it onto the text image. Fig. 5(d) shows the restoration results.

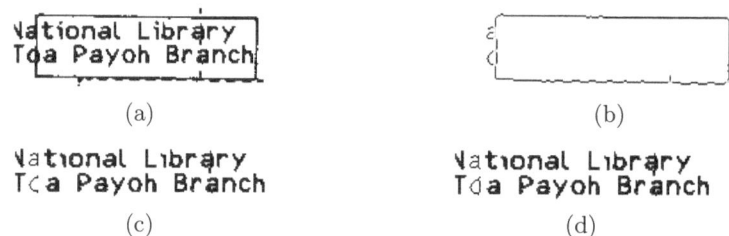

(a)   (b)

National Library
Tca Payoh Branch

National Library
Tda Payoh Branch

(c)   (d)

Fig. 5. Extraction of touching characters: (a) the original image; (b) the thinned graphics image; (c) extracted touching (without restoration) and isolated characters; (d) extracted touching (with restoration) and isolated characters

3.3   Characters Clustering

We assume that the gap between words (denoted as T_w) is larger than the gaps within words (denoted as T_c). We perform a dilation operation on the character images with a square structuring element larger than T_c and smaller than T_w. The characters of each word will form a connected component in this way. The coarse text-box can be drawn by finding the top, bottom, leftmost and rightmost side of each component in the resultant image.

This clustering method is very fast in comparison with the Hough Transformation based method in [3]. It also has the advantage of the method in [4], that is, it can group curving labeled road names. One disadvantage is that it may group two close words incorrectly if the assumption given at the beginning of this sub-section is violated.

4   Experimental Results and Discussions

We scanned 24 large maps at 300dpi and saved them in raw TIFF format. We cut out the neighboring area of overlapping text and obtained about 137 small

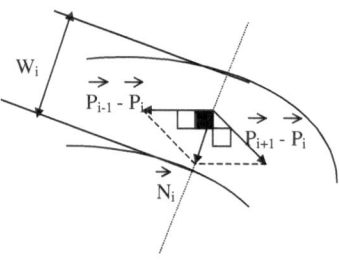

Fig. 6. Estimation of line normal direction and line width at each pixel, here P_{i-1}, P_i, and P_{i+1} are on the thinned line segments, W_i is calculated on the thick line images

Table 1. Statistic data of the test images

Total # of words	Touching words		Totally touching words	
464	184	39.7%	32	6.9%

Total # of characters	Touching characters	
2308	396	17.2%

images to do the experiments. Table 1 shows the statistic data of the testing images. We can see that the touching of words and graphics was serious (touching words: 39.7%; totally touching words: 6.9%). Here totally touching means all the characters within the word are connected to graphics.

The well-known IR standard measures, precision and recall, were used to measure the performance of the proposed method. Precision and recall are defined as follows:

Precision = # of correctly detected words / # of detected words,

Recall = # of correctly detected words / # of words in the image.

If all the constituent characters of a word are detected, then the word is considered as correctly detected. Table 2 shows the experimental results of the proposed method. The results from the initial classification are also shown to give some comparison. As we can see in Table 2, the proposed method detected almost all of the text in the test images. It improved the recall greatly (60.3% to 99.4%). At the same time, it acquired a comparable precision (87.5% to 87.1%). The increase of the false alarms was mainly due to small bars connected to graphics, some of which were dashed lines and some road line segments isolated by occlusion. Fig. 7 shows a larger segment of a map. Fig. 7(a) is the original image superimposed with the bounding boxes of detected text. Almost all the touching text has been detected successfully. Fig. 7(b) is the corresponding text image. Some small icons still remain in the text image since their characteristics are similar to text. Fig. 8 gives another example in which some occluded road line segments are wrongly detected as text.

The extracted words were also tested using an OCR system. The OCR system we used was the commercial system bundled with the HP scanner. The system can only recognize upright characters. Therefore, we rotated the characters into upright position before we conducted OCR experiments. Table 3 shows the testing results. As we can see, the recognition accuracy of both isolated characters and touching characters improved significantly after text extraction (71.7% to 94.3%, 30.2% to 64.8% respectively). Though our segmentation method only restores the touching characters and has no effect on the isolated characters, the recognition of the restored touching characters may have provided context information to the OCR system to improve the recognition of the nearby isolated characters.

There are several difficult cases for recognizing touching text. As we explained earlier, the proposed method merges line segments at the intersections by line continuation. However, the outlines of the buildings are usually polygon contours.

Table 2. Text detection evaluation

	detected words	false alarms	Precision	Recall
Before segmentation	280	40	87.5%	60.3%
After segmentation	461	68	87.1%	99.4%

Table 3. Text extraction evaluation by OCR

Accuracy	Isolated	Touching	Average
Before segmentation	71.7%	30.2%	54.7%
After segmentation	94.3%	64.8%	82.2%

Therefore, when text touches these outlines at the adjoining points, the line segments may not be merged correctly. As a result, there are still some short graphical segments remaining within the text boxes (Fig. 9). These segments usually cause confusion. Besides, some strokes of touching characters and lines coincide perfectly. The touching parts have the same line width as the free lines. In this case, the touching parts of the characters cannot be restored (Fig. 10). This also results in failure of recognition of the touching characters.

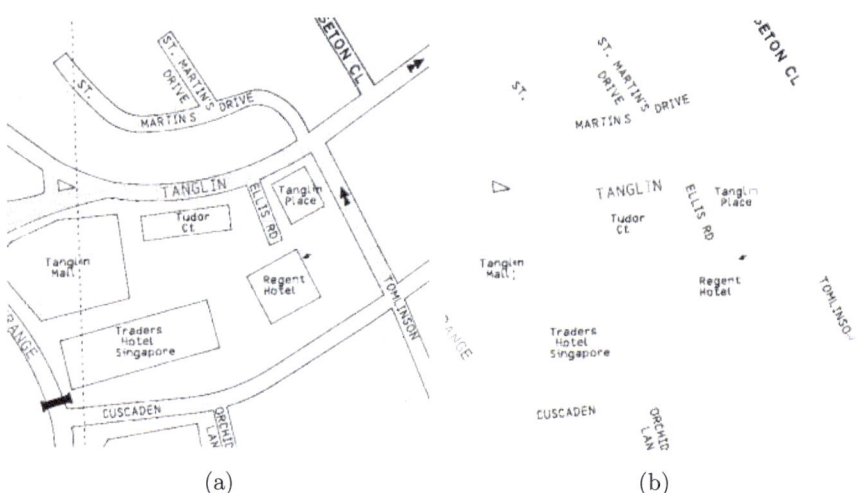

(a)   (b)

Fig. 7. Results of text detection (a) and extraction (b)

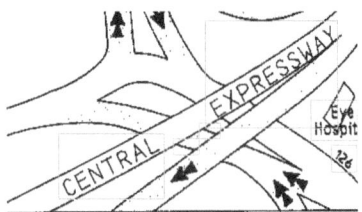

Fig. 8. Example of false alarms

(a)   (b)

Fig. 9. Example of text connected to polygon contours: (a) the original image superimposed with bounding box of detected text;(b) the extracted text image

5   Conclusion and Future Work

A method of detecting and extracting text connected to graphics has been proposed in this paper. The proposed method combines line continuation with line width to interpret connection of text and graphics. Experiments showed that the proposed method improved the percentage of correctly detected text as well as the accuracy of OCR significantly. However, the accuracy of recognizing touching text is still far too low to be usable. The integration of segmentation and recognition may be one promising approach to separate the touching characters completely from graphics. As the current precision and recall of text detection

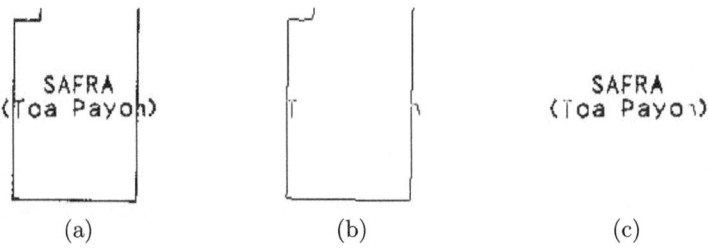

(a)   (b)   (c)

Fig. 10. Example of completely overlapping of text and graphical lines: (a) the original image; (b) thinned graphics image; (c) the extracted text

are high, the segmentation and recognition can be performed within the text boxes. This is more efficient than processing on the whole original images.

References

1. G. Nagy, Twenty years of document image analysis in PAMI, IEEE Transactions on Pattern Analysis and Machine Intelligence, Vol. 22, No. 1, pp. 38 - 62, January 2000   167
2. D. S. Doermann, An introduction to vectorization and segmentation, in Graphics Recognition: Algorithms and Systems, K. Tombre and A. K. Chhabra (eds.), Lecture Notes in Computer Science 1389, Springer, pp. 1 - 8, 1998   167
3. L. A. Fletcher and R. Kasturi, A robust algorithm for text string separation from mixed text/graphics images, IEEE Transactions on Pattern Analysis and Machine Intelligence, Vol. 10, No. 6, pp. 910 - 918, November 1988   167, 172
4. C. L. Tan and P. O. Ng, Text extraction using pyramid, Pattern Recognition, Vol. 31, No. 1, pp. 63 - 72, 1998   167, 169, 172
5. D. Wang and S. N. Srihari, Analysis of form images, in Document Image Analysis, H. Bunke, P. S. P. Wang, H. Baird (eds.), World Scientific, pp. 1031 - 1051, 1994   167
6. S. Naoi, Y. Hotta, M. Yabuki, and A. Asakawa, Global interpolation in the segmentation of handwritten characters overlapping a border, Proceeding of 1st IEEE International Conference on Image Processing, pp. 149 - 153, 1994   167
7. J. Yoo, M. Kim, S. Y. Han, and Y. Kwon, Line removal and restoration of handwritten characters on the form documents, Proceeding of 4th International Conference on Document Analysis and Recognition, pp. 128 - 131, 1997   167
8. K. Lee, H. Byun, and Y. Lee, Robust reconstruction of damaged character images on the form documents. In Graphics Recognition: Algorithms and Systems, K. Tombre and A. K. Chhabra (eds.), Lecture Notes in Computer Science 1389, Springer, pp. 149 - 162, 1998   167
9. R. Kasturi, S. T. Bow, W. El-Masri, J. Shah, J. R. Gattiker, and U. B. Mokate, A system for interpretation of line drawings, IEEE Transactions on Pattern Analysis and Machine Intelligence, Vol. 12, No. 10, pp. 978 - 992, October 1990   167, 168
10. D. Dori and Liu W., Vector-based segmentation of text connected to graphics in engineering drawings, in Advances in Structural and Syntactical Pattern Recognition, P. Perner, P. Wang, A. Rosenfeld (eds.), Springer, pp. 322 - 331, 1996   167
11. Z. Lu, Detection of text regions from digital engineering drawings, IEEE Transactions on Pattern Analysis and Machine Intelligence, Vol. 20, No. 4, pp. 431 - 439, April 1998   167
12. H. Luo, G. Agam, and I. Dinstein, Directional mathematical morphology approach for line thinning and extraction of character strings from maps and line drawings, Proceeding of 3rd International Conference on Document Analysis and Recognition, pp. 257 - 260, 1995   167
13. L. Li, G. Nagy, A. Samal, S. Seth, Y. Xu, Cooperative text and line-art extraction from a topographic map, Proceedings of 5th International Conference on Document Analysis and Recognition, pp. 467 - 470, 1999   167
14. D. Dori, Liu W. and M. Peleg, How to win a dashed line detection contest, in Graphics Recognition: methods and Applications, R. Kasturi and K. Tombre (eds.), Lecture Notes in Computer Science 1072, Springer, pp. 286 - 300, 1996

15. R. Jain, R. Kasturi, and B. G. Schunck, Machine Vision, MIT Press and McGraw-Hill, 1995   168

16. B. K. Jang and R. T. Chin, One-pass parallel thinning: analysis, properties, and quantitative evaluation, IEEE Transactions on Pattern Analysis and Machine Intelligence, Vol. 14, No. 11, pp. 1129 - 1140, November 1992   170

17. T. Pavlidis, Algorithms for graphics and image processing, Computer Science Press, 1982   170

Semantic Analysis and Recognition
of Raster-Scanned Color Cartographic Images

Serguei Levachkine[1], Aurelio Velázquez[1]
Victor Alexandrov[2], and Mikhail Kharinov[2]

[1] Centre for Computing Research-IPN
UPALM Zacatenco, Edificio-CIC, Cubo 2019 Ala Sur
07738 Mexico City, Mexico
{palych,avela}@cic.ipn.mx
[2] Institute for Informatics and Automation-The Russian Academy of Sciences
39, 14th line, 199178, St. Petersburg, Russia
{alexandr,khar}@mail.iias.spb.su

Abstract. Semantic analysis of cartographic images is interpreted as a separate representation of cartographic patterns (alphanumeric, punctual, linear, and area). We present an approach to map interpretation exploring the idea of synthesis of invariant graphic images at low level processing (vectorization and segmentation). This means that we ran "vectorization-recognition" and "segmentation-interpretation" systems simultaneously. Although these systems can generate some errors in interpretation, they are much more useful for the following understanding algorithms because its output is nearly recognized objects of interest.

1   Introduction

We have begun practical research on map recognition from global binarization followed by classical OCR-identification with artificial neural networks (ANNs), supervised clustering, knowledge-based recognition rules, and morphology-based vectorization. To overcome the problem of labor-intensive training, we designed simplified images. For this purpose, we used the linear combinations of color components or image representations (*false color technique*) and binary representation composing (*composite image technique*). These techniques are application-independent. However, in the frameworks of our approach, map recognition may be treated as a common (application-dependent) task [1]. We followed the concept that the important semantic information necessary to interpret an image is not represented in single pixels but in meaningful image segments and their mutual relations [15].

In the present work, a modified version of [9], both tasks, i.e., vectorization (Sections 3, 4) and segmentation (Sections 2, 3) are used for automatic interpretation of raster-scanned color cartographic images. An approach to separation of text from graphics, which involves detailed analysis of strokes and character components, is

D. Blostein and Y.-B. Kwon (Eds.): GREC 2002, LNCS 2390, pp. 178-189, 2002.
© Springer-Verlag Berlin Heidelberg 2002

described in Section 2. This approach can also be used at a low level: segmenting large connected components into constituent lines and arcs. In Section 3 we built up composite image representations based on object-fitting compact hierarchical segmentation [6,7]. In both techniques, artificial representations of source image form a "book" in which the objects of interest can be found and recognized on the appropriate page(s). Computer-based systems of alphanumeric, punctual, and linear cartographic pattern recognition (*A2R2V* system) are overviewed in Section 4. Finally, Section 5 contains our conclusions and summing-up of the paper.

2   False Color Technique

In this section, we present a new approach to color image segmentation [8], in which we aim to separate alphanumeric characters (for subsequent recognition) into raster-scanned color cartographic maps using model of color *RGB*. Note that this is a very difficult problem because there is either text embedded in graphic components, or text touching graphics (see *Remark 2*). The processed images only maintain pixels that truly belong to objects we wish to segment, eliminating all pixels not of interest, that are produced by noise or obtained due to erroneous selection of scan parameters (Figure 1).

Image thresholding can be seen as a segmentation step [2]. When applying thresholding to an image, we seek to isolate objects from background. In the case of a contrasted image (an image with objects contrasting with respect to background), the problem is easily solved using nearly any thresholding method; Otsu's approach is a good example [3]. In general, however, this is not the case because objects normally share the same gray-levels. Any simple thresholding method will not adequately isolate the desired objects. Intuitively, attempting to do the same with color images will be even more complicated.

We introduce an approach for thresholding color images as follows:

Preliminaries. Most existing color-based segmentation (thresholding) techniques use one of the *R, G,* and *B* components. Our approach utilizes this information plus the average of the combination of two of them, the average of all three, and transformation of the color image into an intensity image, transforming it into a *YIQ* image by selecting only the *Y* component. The eight images considered by our segmentation algorithm are the following: 1) M1 = "R"; 2) M2 = "G"; 3) M3 = "B"; 4) M4 = int(("R"+"G")/2); 5) M5 = int(("R"+"B")/2); 6) M6 = int(("G"+"B")/2); 7) M7 = int(("R"+"G"+"B")/3); 8) M8 = color_to_gray("RGB"). We call these images the *M-images* of a standard color image.

Methodology. The proposed methodology incorporates three main stages, which are:

Image pre-processing. During this stage, an image is first decomposed into its 8 M-images. Each M-image is subsequently processed to emphasize pixels of interest. In particular, we are interested in alphanumeric characters. This is carried out, on the one hand, by adjusting the contrast of the image to fit the range of 0 to 255 gray-levels. On the other hand, the same original 8 M-images are histogram-equalized. This results in 24 sub-images (8 original M-images, 8 normalized sub-images, and 8 equalized sub-images).

Image processing. This stage is divided into two steps: image pre-thresholding and image post-thresholding as follows:

Image pre-thresholding. The 24 images (not including the original image) are subsequently thresholded to obtain their 24 binary versions. Next, we describe how to select the desired threshold.

It is well known that threshold selection is not only a critical but also a difficult problem in general. One very well known method for obtaining threshold is by using the histogram [2] (or homogram [17]) of the image. We used both at the beginning of our research with insufficient results. Thus, we decided to use the following procedure:

To treat each sub-image independently, we applied the following modified Prewit edge masks:

-1	0	1
-1	2	1
-1	0	1

-1	-1	-1
0	2	0
1	1	1

in both the x and the y directions. Next instead of computing the magnitude of the gradient as is customary, we added, pixel-by-pixel, the two resulting images to again obtain one image, e.g., g. We then added all gray-level values of this image to obtain only one number. We finally divided this number by the total size of the image (the number of pixels of the image) to obtain the desired threshold, e.g., u. If g has M rows and N columns, then

$$u = \frac{1}{MN} \sum_{i=1}^{M} \sum_{j=1}^{N} g(i,j).$$

This threshold is computed and next applied to each of the 24 sub-images to obtain their binary version.

Remark 1. In the case of histogram-equalized images, the applied threshold is the original threshold u divided by seven. If the original threshold were applied directly to an equalized image, a completely black image would result. This factor was found empirically.

We used an edge detector to obtain the desired threshold, because letters and numerals are associated at their borders with abrupt changes with respect to image background. The use of an edge detector would thus accentuate the letters and numerals at least at their borders. An application of this type of method would also emphasize the presence of rivers, country borders, etc. These, however, could be eliminated later, for example, if only separating the alphanumeric layer of the image is required.

Image post-thresholding and region labeling. The white pixels in each of the 24 binary sub-images are representative of strokes of a letter or a numeral, the border of a river, the limit between cities, etc. They appear, however, fragmented and unclustered to form connected regions of letters or numerals. For example, the letter "i" in the word "Xichu" in some images appears incomplete, while in others appears complete. To solve this problem, we used the following heuristic method: We first obtain one image from the 24 binary sub-images by simple pixel-by-pixel addition. The value of a pixel in this image oscillates between 0 and 24. To again obtain a binary image, we verify whether the sum at a given (x,y) position is greater than 21 (a manually selected threshold). If this is true, we put a 1 into the buffer image, and zero otherwise. We thus have another binary image with 1s in the most probable positions representing letters, numerals and other symbols, and 0s in background areas. Figure 1(b) shows the resultant binary sub-image, that we call the *P-image*.

We next applied a labeling algorithm to this image to obtain adjacent regions representing desired elements (letters, numerals, and other symbols).

Image post-processing. On observing Figure 1(b), the adjacent connected regions in this P-image obtained with the procedure previously described nonetheless appear fragmented. Gaps between isolated regions must be filled to obtain the complete desired regions. These gaps are filled by means of the following procedure:

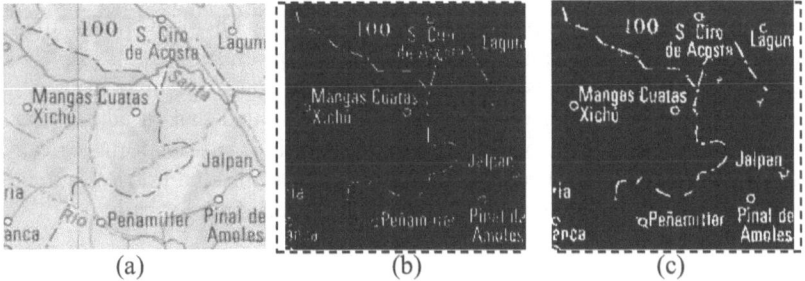

Fig.1. (a) – source image; (b) – *P-image*; (c) – thresholded image

Procedure FILLING GAPS

{The input is the *P-image*.

1. Select a connected region of pixels in the P-image. Dilate this region with a 3 x 3 structural element from 3 to 15 times, depending on the number of pixels ot the region. Each 10 pixels add one morphological-dilation to the process. This process results in square window the size of which is given by the number of dilations. We call this resulting square window the D-mask.
2. AND-Image-mask (see [4], p. 50, for details) the 8 M-images with the D-mask founded in step 1 to compute the average gray-level and the standard deviation of this region. Only pixel values under 1s in the D-mask are taken into account.
3. Turn off all pixels in the D-mask if corresponding gray-level value in any of the M-image is greater than the gray-level average plus the standard deviation value obtained in step 2. This allows, on the one hand, elimination of the undesired background pixels added to alphanumeric characters during dilation process. On other hand, this step permits aggregation of the missing pixels of the characters.

4. Apply steps 1-3 to all labeled regions inside the P-
   image.

The output is an image (*F-image*) with isolated characters
with gaps filled}.

Figure 1(c) shows the image resulting after gap-filling procedure previously described is applied to the image shown in Figure 1(b). Note the manner in which gaps
between isolated regions have disappeared, and letters and numerals now appear
complete.

From this image, we can now take each character with its position and file it for
further processing. Thus, the resulting images are now being analyzed by another
module to separate alphanumeric characters from other objects. The goal is to isolate
each alphanumeric character as much as possible and to determine its identity by
means of trained classifiers (see [14] and Section 4).

Comments. In this section a novel thresholding approach applicable for color images
has been proposed. This approach has been tested with many (more than 1,000)
raster-scanned color cartographic images, rendering promising results [8].

Our approach incorporates three main stages. The first stage involves generation of
artificial images with color components. We have found that in addition to natural
decompositions (e.g., *R, G,* or *B* – component-splitting), artificial images can also be
useful [8]. To use artificial representations also for semi-tone images or single-color
components, the following composite image technique is put forward:

3   Composite Image Technique

In this section, we set forth a novel conception of composite image representation
[7,9]. The main goal of image decomposition consists of the object linking by its associated names.

A new approach to image synthesis based on object-fitting compact hierarchical
segmentation [6,7] is proposed. Our approach provides composite representations (or
simply, *composites*) of the source image by means of a reduced number of color or
tone components and segments. In this manner, visually perceived objects are not
eliminated and the *image's semantics* is preserved. The *image's semantics* in this
context corresponds to the association of segment fields of different hierarchical levels being identified with identifying conceptions from the subject domain. For example, detection of a segment set identifying a coastline or highway becomes semantically meaningful. Further, this set of segments is renamed as "coastline", "highway",
etc.

Composite image representation is a sequence of binary images packed into different bit planes. These binary images are the result of two-valued classification of
source image by some feature (intensity, area, invariant moments, etc.). The composites detect local feature extrema, which are invariant with respect to hierarchical segmentation.

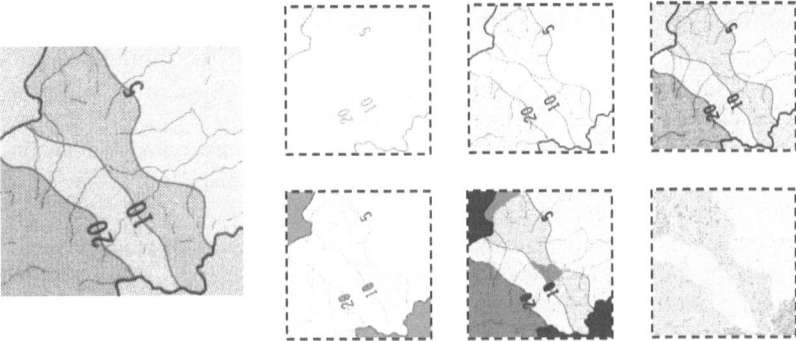

Fig.2. Source image (extreme left) and some "pages" of a "book" of composite images

Object-Fitting Compact Hierarchical Segmentation. Object-fitting compact hierarchical segmentation is a sequence of embedded partitions without repetition of composed segments in different partitions. A partition is obtained by iterative splitting or merging of segments [6,7]. In the merging mode, any segment in each iteration merges into an adjacent segment. Number 2^i bound total number of segments, where i is the number of iteration.

As stated previously, the composite images form a "book" in which the objects of interest can easily be found on appropriate page(s). Thus, a "page number" defines the method of thresholding and some tuning parameters (Figures 2 and 3).

Each composite image contains machine-treatable bit-planes for target object detection and purposeless bit-planes as well (Figure 3).

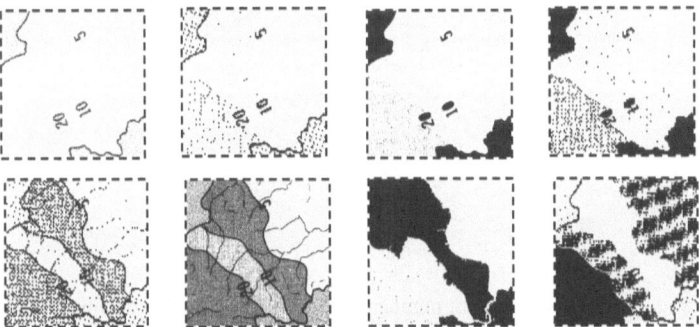

Fig.3. Bit-planes suitable for character recognition (above) and other purposes (below)

Composites. A bit component of composite image computes by means of thresholding of the current segmented image. Threshold is equal to the average all over image of intensity, geometric or otherwise, e.g., $\overline{I}$. To threshold the image, $\widetilde{I}$ is compared with its average over pixels of each segment, e.g., $\widetilde{I} : \widetilde{I} \geq \xi \overline{I}$ *or* $\widetilde{I} < \xi \overline{I}$, where ξ is the tuning parameter.

The bit components are packed in the resulting representation, where the extrema of intensity indicate pixels associated with unchanged binary feature.

Color Image Processing. Compact hierarchical segmentation of a color image is performed by each independent color component (R, G, and B) considering these as semi-tone images. In this way, coinciding intensities of the resulting R, G, and B composite images indicate segments of equal color with respect to using feature. This can be used for color invariant image description.

Segment number must be taken into account for automatic image analysis. When used for estimation of number of objects, this number characterizes the *image's semantics*. As a rule, compact hierarchical image segmentation implies that color segments are enlarged simultaneously. Their number decreases approximately as $(4 \div 5)^{-i}$, where i is the iteration number (Figure 4).

From our point of view, deviation from such exponential dependence leads to the image's semantics violation. Disclosed regularity can be useful for automatic analysis of color images. Due to the self-consistence of RGB-segmentation behavior, visual quality improvement in composite intensities becomes available (Figure 5a).

Applications of Composites. By applying composites, we are able to reveal cartographic data (Figure 5b). Note that false color technique and composite image technique do not contradict each another (compare Figure 5b with Figure 1).

Use of composite image representations improves existing computer systems for detection of polygons, contours, points, alphanumeric characters, etc. Based on object-fitting image segmentation, we proposed the method of parameterless image binarization [6,7]. Application of this method to composites also allows improvement of object detection.

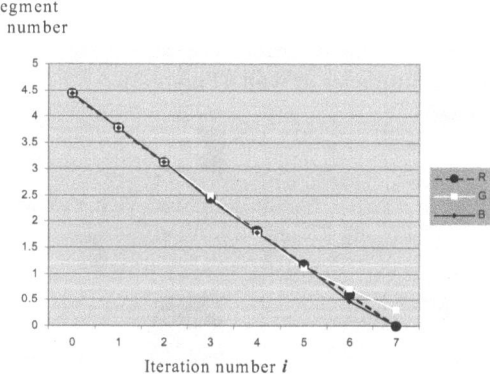

Fig.4. Linear dependence of segment number on iteration number i in decimal logarithm scale

Fig.5a. Left – map (1082 x 1406 pixels). Its visual improvement: right - by composite image technique; middle – by combining false color and composite image techniques

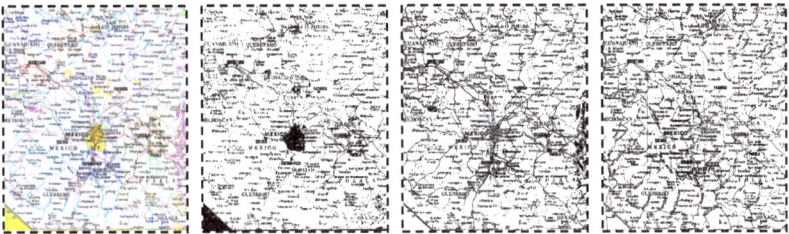

Fig.5b. Cartographic pattern page-out from a composite image (extreme left) and its *R, G,* and *B* components

Comments. The method requires significant operative memory space. To overcome this disadvantage, we used special data organization in the form of irregular dynamic trees, which provides optimal in memory space computing for successive scanning of image scales. Due to data organization, a practical use of our program package does not require further algorithmic development. The user needs only to make adequate choice to carry out task features (quantitative, qualitative or nominal) from the pre-scribed feature set. Specification of features associated with perspective representations is regarded as a type of *knowledge domain* [6,7].

4   Overview of Computer-Based Systems of Alphanumeric, Punctual and Linear Character Recognition

In this section, we overview single modules of our map interpretation system called *A2R2V* (from **A**nalogical-to-**R**aster-to-**V**ector) effected to date.

System of Alphanumeric Character Recognition (SACR). The following identification of segmented objects is particular previously prepared application. We first build the strings of characters and use a set of ANNs for identification of single letters and gazetteers (toponomy dictionaries) for word identification. Moreover, ANNs are previously trained on a set of artificially created characters of different fonts, sizes, inclinations, etc. (Table 1).

Table1. (a) - 16 selected fonts for testing previously trained ANNs; (b) – example of inclination of the characters Abadi MT Condensed Light font in the sizes of 13÷17 points

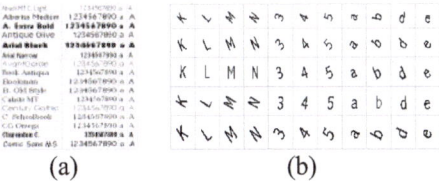

| (a) | (b) |

However, we apply these ANNs to identification of real characters into cartographic images using 1,025 alphanumeric characters (upper and low cases, different fonts, sizes and inclinations) extracted from real raster maps. 95.5% of which were successfully recognized by the system.

A gazetteer is used to verify either an identified string of characters forms meaningful word, or not [14].

System of Punctual Object Recognition (SPOR). This fully automatic system, which generates vector layer of punctual cartographic patterns (Figure 6), is described in details in [20].

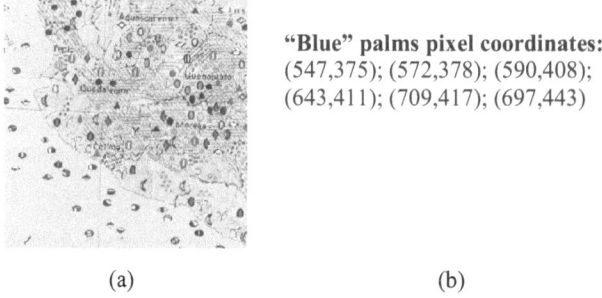

"Blue" palms pixel coordinates:
(547,375); (572,378); (590,408);
(643,411); (709,417); (697,443)

(a)   (b)

Fig.6. (a) – recognized "blue" palm restoration (in white) over source image; (b) – pixel coordinates of recognized palms

System of Linear Object Recognition (SLOR). This is a mainly semi-automatic system that provides full image treatment: pre-processing (image enhancement and segmentation), processing (a set of supervised identifiers and knowledge-based recognition rules), and post-processing (morphology-based vectorization). Thus, human-machine interaction is optimized [21].

A result of full processing of linear cartographic patterns is shown in Figure 7.

Remark 2. (a) We used OCR-identification with ANNs in SACR and SPOR systems only, training ANNs on sets of artificial characters, but applying them to real map characters; (b) We used perceptron multi layer-multi neuron feedforward to recognize alphanumeric and punctual characters. In the first segmentation, we selected and labeled the connected components to be analyzed by ANNs; (c) If characters touch graphics, we developed a method to separate and further recognize touching symbols [14]; (d) Size of training set is 95,000 samples; (e) ANNs have a special output to show shapes that do not appear to be characters of interest; (f) Linear character recognition does not use OCR; (g) Composites provide area character recognition because segmented area objects are labeled by the system in same, but are different for each type of objects, gray-level values.

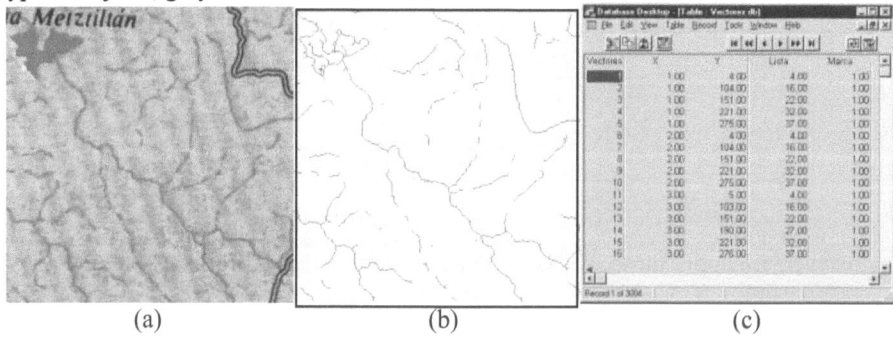

(a)   (b)   (c)

Fig.7. (a) – source image; (b) – vector image of rivers; (c) – related to (b) vector database

5   Conclusions

In this work, we presented a system for automatic interpretation of raster-scanned color cartographic maps. The highlight of our system is two color image segmentation techniques (Sections 2 and 3). They are followed by a set of identifiers (Section 4) that includes classical OCR-identification, supervised clustering, knowledge-based recognition rules, and morphology-based vectorization. The identifiers and vectorization worked well because they receive as input from the segmentation step nearly recognized objects of interest (alphanumerical, punctual, linear, and area).

The majority of color segmentation approaches is based on the monochrome segmentation approaches operating in different color spaces [11]. Gray-level segmentation methods can be directly applied to each component of a color space; thus, results can be combined in some way to obtain a final segmentation result. However, one problem is how to employ *color information as a whole for each pixel*. When color is projected onto three components, color information is so scattered that the color image becomes simply a multispectral image and color information that humans can perceive is lost. Another problem is how to choose *color representation for segmentation* [5]. Each color representation has its advantages and disadvantages. There is no single color representation that can surpass all others for segmenting all kinds of color images. In this work, we presented an approach that we call *the false color technique*. We aimed to separate different objects of interest into raster-scanned color cartographic images. We developed an alternative solution of the problems of defining image regions by means of additional quantitative color features (*false colors*). The novelty of the proposed approach is the use of linear and quasi-linear color spaces, such as *RGB* combination and *YIQ*, rather than nonlinear spaces, such as *HSI* and the normalized color space. Nonlinear spaces have essential, non-removable singularities and there are spurious modes in the distribution of values [13].

The problem of how and to what degree *semantic information* should be employed in image segmentation has led us to a novel conception of composite image representation for mutual object detection-recognition at low-level processing. We conjecture that modern segmentation systems must support mutual object detection-recognition-interpretation, starting at low-level processing, memorizing results at the intermediate level, and effectively communicating these results to the high level. The approach proceeding from this conjecture is called *composite image technique*. The idea is to prepare the source image as much as possible for subsequent high-level processing of image regions. In most existing color image segmentation approaches, definition of a region is based on similarity of color. This assumption often makes it difficult for any algorithms to separate objects with highlights, shadows, shadings, or texture, which cause inhomogeneity of colors of the object's surface. Using *HSI* can solve this problem to some extent, except that hue is unstable at low saturation. Some physics-based models have been proposed to solve this problem [11,12]. These models take into account color formation, but have too many restrictions which prevent them from being extensively applied. Fuzzy logic algorithms applied to the problem have also exhibited limitations [18]. We saw an alternative solution of the problem defining image regions by additional quantitative, qualitative, and nominal features (in addition to color feature), which on the whole render the *user's knowledge domain*. This is a kind of advanced simulation of human visual perception.

To our knowledge, this is one of the first attempts to design a segmentation-recognition computer system for color image treatment [10,11].

In our experiments, we used complex raster-scanned color cartographic images, which, being intermediate between drawings and natural images provide a nearly ideal model for testing because they have characteristics of both [9].

At the same time, automatic interpretation of color cartographic images presents certain difficulties for state-of-the-art in image processing and also artificial intelligence. A set of vector maps that one can expect as an output of such interpretation is very useful for Geographical Information Systems, new map production, and old map actualization. However, it appears unrealistic to obtain a fully automatic computer-based interpretation system free of errors [9,16]. Additionally, please note that high efficiency of interpretation is required for vector map production and actualization first. It seems reasonable to obtain in both cases 90÷95% successfully recognized objects. This is to avoid excessive work on corrections of errors produced by the computer system that can sometimes be greater than manual raster-to-vector conversion [19].

We believe that only an integrated approach to the problem can be fruitful. In the context of the present work, this means first, decomposition of source image by multiple hierarchical components to achieve a stable, accurate representation in the presence of degraded images. Second is segmentation with mutual recognition of appropriate primitives. Finally, there is the development of a unified knowledge-based learning and self-learning system with optimal human-machine interaction for color cartographic image treatment [7]. Our future research will be concerned with this approach.

References

1. Doermann, D. S.: An Introduction to Vectorization and Segmentation. In: Tombre, K., Chhabra, A.K. (eds.): Graphics Recognition Algorithms and Systems. Lecture Notes in Computer Science, Vol. 1389. Springer-Verlag, Berlin Heidelberg New York (1998) 1-8
2. Gonzalez, R. C., Woods, R. E.: Digital Image Processing. 3rd edn. Prentice-Hall PTR, NJ USA (2002)
3. Otsu, N.: A Threshold Selection Method from Gray-level Histograms. IEEE Transactions on Systems: Man and Cybernetics. 9(1) (1979) 62-66
4. Umbaugh, S. E.: Computer Vision and Image Processing: A Practical Approach using CVIPtools, Prentice-Hall PTR, NJ USA (1998)
5. Alexandrov, V. V., Gorsky, N. D.: Image Representation and Processing: A Recursive Approach. Mathematics and Its Applications, Vol. 261. Kluwer Academic Publishers, Dordrecht Boston London (1993)
6. Kharinov, M., Nesterov, M.: Intelligent Program for Automatic Image Recognition based on Compact Object-fitting Hierarchical Image Representation in terms of Dynamic Irregular Ramified Trees. In: Barulin, V.N. (ed.): Reports of International Academy for Informatics, Communications and Management, Special Issue 12-C. St. Petersburg, Russia (1997) 1-35 (Library of Congress Number: 98646239)
7. Alexandrov, V., Kharinov, M., Levachkine, S.: Conception of Hierarchical Dynamic Structure in Application to Audio and Video Data Recognition. In: Hamza, M.N. (ed.): Proc. IASTED International Conference on Artificial Intelligence and Soft Computing, ASC 2001, 21-24 May, Cancun, Mexico (2001) 348-353 (ISBN 0-88986-283-4; ISSN 1482-7913)

8. Levachkine, S., Velázquez, A., Alexandrov, V.: Color Image Segmentation using False Colors and its Applications to Geo-images Treatment: Alphanumeric Character Recognition. Proc. IEEE International Geosciences and Remote Sensing Symposium, IGARSS 2001, 9-13 July, Sydney, Australia (2001) (IEEE Catalog Number (CD-ROM): 01CH37217C; Library of Congress Number: 01-087978; ISBN CD-ROM: 0-7803-7033-3)
9. Levachkine, S., Velázquez, A., Alexandrov, V., Kharinov, M.: Semantic Analysis and Recognition of Raster-scanned Color Cartographic Images. In: Blostein, D., Young-Bin Kwon (eds.): Proc. 4th IAPR Int. Workshop on Graphics Recognition, GREC 2001, 7-8 September, Kingston, Ontario, Canada (2001) 255-266
10. Kumar, K. S., Desai, U. B.: Joint Segmentation and Image Interpretation. Pattern Recognition. 32 (4) (1999) 577-589
11. Cheng, H. D., Jiang, X. H., Sun, Y., Wang, J.: Color Image Segmentation: Advances and Prospects. Pattern Recognition. 34 (12) (2001) 2259-2281
12. Levachkine, S., Sossa, J.H.: Image Segmentation as an Optimization Problem. Computation and Systems. 3(4) (2000) 245-263 (ISSN 1405-5546)
13. Kender, J.: Saturation, Hue, and Normalized Color: Calculation, Digitization Effects, and Use. Computer Science Technical Report. Carnegie Mellon University (1976)
14. Velázquez, A.: Localización, Recuperación e Identificación de la Capa de Caracteres Contenida en los Planos Cartográficos. Ph.D. Thesis. Centre for Computing Research-IPN. Mexico City, Mexico (2002) (in Spanish)
15. Definiens Imaging GmbH e-Cognition: Object Oriented Image Analysis. http://www.definiens-imaging.com/ecognition/
16. Ogier, J.M., Adam, S., Vessaid, A., Bechar, H.: Automatic Topographic Map Analysis System. In: Blostein, D., Young-Bin Kwon (eds.): Proc. 4th IAPR Int. Workshop on Graphics Recognition, GREC 2001, 7-8 September, Kingston, Ontario, Canada (2001) 229-244
17. Cheng, H.D., Jiang, X.H., Wang, J.: Color Image Segmentation based on Homogram Thresholding and Region Merging. Pattern Recognition. 35 (2) (2002) 373-393
18. Chen, T. Q., Lu, Y.: Color Image Segmentation – an Innovative Approach. Pattern Recognition. 35 (2) (2002) 395-405
19. Wenyin, L., Dori, D.: Genericity in Graphics Recognition Algorithms. In: Tombre, K., Chhabra, A.K. (eds.): Graphics Recognition Algorithms and Systems. Lecture Notes in Computer Science, Vol. 1389. Springer-Verlag, Berlin Heidelberg New York (1998) 9-20
20. Levachkine, S., Polchkov, E.: Automated Map Raster Digitization by Cartographic Pattern Recognition. In: Muge, F., Ruíz Shulcloper, J. (eds.): Proc. 5th Iberoamerican Symposium on Pattern Recognition, SIARP 2000, 11-13 September, Lisbon, Portugal (2000) 81-96 (ISBN 972-97711-1-1)
21. Decelis-Burguete, J. O.: Digitalización automatizada de líneas en mapas ráster. M.S. Thesis, Centre for Computing Research-IPN, Mexico City, Mexico (2001) (in Spanish)

Structure Based Interpretation
of Unstructured Vector Maps

Manuel Weindorf

Institute of Photogrammetry and Remote Sensing
University of Karlsruhe, Germany

Abstract. This work presents an approach to map interpretation starting from unstructured vector data. For this task a map interpreter based
on PROLOG and grammatical object descriptions has been considered.
The major challenge in this approach is the definition of production rules
and grammars which are general enough to handle different data sets and
which are specific enough to discriminate the different object types.

1   Introduction and Motivation

During the last decades production as well as the utilisation of maps has developed from analogue to digital techniques. There is a large amount of digital
spatial data created by digitizing analog maps as well as resulting from direct
geodetic topographic measurments and image analysis. While the form of the
data itself has changed from analog to digital, the corresponding tools have been
improved in a way that they are able to model and represent spatial data. The
concept of geometry focused and layer oriented modelling has changed to object
oriented modelling techniques.

Today the interoperability of different geographic information systems (GIS)
is a major topic in research as well as in industry and the demand for well defined
data formats and data exchange procedures leads to some effort in standardization of data formats and services [7]. Most of the effort is taken in order to define
data formats used to exchange spatial information. Data is presently exchanged
mainly at the iconic level of representation (e.g. AutoCAD DXF, ESRI Shapefile,
Microstation DGN, etc.), whereas the exchange of semantics is rather rare. It is
stated in [4] that about 30% of the users work with graphics based (CAD) tools
for geoprocessing, and that about 40% of the data exchange is done via simple
graphics formats.

Map interpretation in this context is the transformation of raw map data
(raster or vector) from its physical to its logical representation, or in other words
to make the intrinsic geometical information explicitly available.

The approach presented in this work is based on the assumption that a human
map user is able to identify objects which are of interest to his application by
simply inspecting the map and using common sense. This work attempts to
emulate this behaviour with the purpose of data exchange by basically defining
the knowledge (the data model's requirements) on the target side of the data

D. Blostein and Y.-B. Kwon (Eds.): GREC 2002, LNCS 2390, pp. 190–199, 2002.

exchange chain. The basic idea relies on the assumption that, with methods of automatic map interpretation, only one definition of the target model could be sufficient to incorporate data from several sources. If one single description of the target data model is sufficient to select the interesting portions of data from a given dataset, then a corresponding system should be able to read data from many different sources. Furthermore, the target system should be able to decide on the content of a given dataset, if it is useful for a given application (content based retrieval systems).

The presented approach is based on a production system implemented in PROLOG. The models of the real world objects are formulated as grammatical descriptions and productions in the form of *'if condition then action'* rules.

The main challenge in this work is the definition of grammars and production rules, which are general enough to be applied to a large number of data sets. On the other hand, these rules have to be specific enough to discriminate the different object types in a reliable way.

Some work exists about automated map interpretation in the context of topographic maps at small scale [6, 8] as well as of cadastral maps at large scale [2, 1].

2 System Components and Processing

A system for map interpretation can be seen as a combination of several distinct components which have to cooperate with each other, aiming at a full description of a given map or scene.

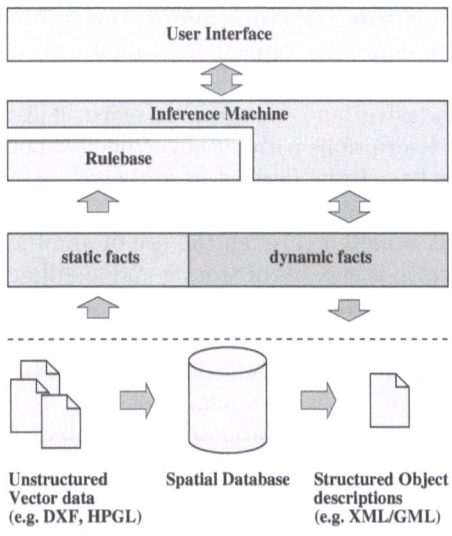

Fig. 1. System components

According to the system structure with several components and their interactions as displayed in fig. 1, the process is divided into several steps. First of all, the raw CAD data is transferred into the spatial database of the system. During this step all objects are reduced to simple lines and text elements. The reason for that is twofold: On the one hand, possible errors made while digitizing the data are eliminated. On the other hand, a single canonical data structure for all data sets can be assumed. After the import of the data, a feature neighbourhood graph concerning the lines is created within the database component. This graph structure (topological relations) together with the geometric primitives, is provided to the map interpreter as a set of geometrical and topological facts (PROLOG clauses). Based on this facts the map interpreter creates closed polygons which are stored in the spatial database where the corresponding spatial relations (containment, adjacency, etc.) are determined and provided to the map interpreter. Based on this structural knowledge about the data set, application specific object descriptions are inferred using rules in form of grammars and PROLOG predicates.

In the following sections these components are introduced and explained in more detail.

2.1   Spatial Database

The basic geometric primitives which may be used to define abstractions of real world objects are points, lines and polygons. These geometric primitives are used to express the structural information and lead to a pictoral description of the *real world objects*. Moreover, presentation attributes (e.g. line color, text fonts, line styles etc.) can be used to discriminate objects with different meanings or e.g. to visualize degrees of importance.

Within this work only the structural information (geometrical and topological properties) is used to determine objects from digital vector maps.

The basic data sets on which the presented processes operate are CAD drawings representing cadastral and topographic maps. This kind of data consists only of geometrical descriptions without any explicit topological relations. These topological relations have to be derived from the geometrical structure by algorithms of computational geometry. Therefore a fast and efficient spatial access is demanded. This is achieved through the use of an object relational database (PostgreSQL [5]), which is capable of storing and handling geometric primitives. This means the database supports geometric types like *point*, *line*, *polygon* and defines corresponding operators and functions to create, compare, store and retrieve these geometric objects. In addition, a spatial indexing method, which is able to index on two dimensions simultaneously, is necessary to gain efficient access to spatial data. In the proposed approach an indexing scheme called R-Tree [3], based on the bounding rectangles of the geometric primitives, is supported which leads to the ability to prove spatial predicates like *contains*, *intersects*, *meets*, etc. in an efficient way, even for a large set of primitives.

Additionally the functionality of the database is extended by rules to keep the data consistent. This means that the line segments, as they are stored in

Fig. 2. Example of a basic data set: Digital cadastral map-data based on CAD, just consisting of lines and text-elements

the system, are spread across two tables, one containing the line geometry and the other storing the start and end points as nodes of a graph. Furthermore, the database provides system-wide unique identifiers and ensures that only one node exists at a specific geometric location, which leads to the problem of clustering nodes within a small bufferzone, defined in the form of a threshold.

While adding new lines, the database rule system keeps track of the node table, rejecting new nodes which are up to be stored at the same geometrical location. In the same way, lines already existing in the database or already stored in inverse direction are rejected. The following SQL code fragments show how the rules, responsible for keeping track of double nodes and double lines, are defined.

```
CREATE FUNCTION point_is_already_there(point)
  RETURNS int4 AS
  'select count(*) from nodes where distance( $1, geometry ) < 0.01'
  LANGUAGE 'sql';

CREATE FUNCTION line_is_already_there(lseg)
  RETURNS int4 AS
  'select count(*) from lines where geometry_same( $1, geometry )'
  LANGUAGE 'sql';

CREATE RULE on_insert_node AS
  ON INSERT TO nodes
  WHERE point_already_there( NEW.geometry ) > 0
  DO INSTEAD NOTHING;

CREATE RULE on_insert_line AS
```

```
ON INSERT TO lines
WHERE line_already_there( NEW.geometry ) > 0
DO INSTEAD NOTHING;
```

After inserting all line segments into the database, the next step is the creation of a feature neighbourhood graph in respect to the lines. An additional set of edges is then created. Each edge corresponds to a specific line segment and is defined by an unique identifier and two nodes, corresponding to a start- and an end-node.

The spatial database is used to create an explicit topological description based on the geometric information of the raw graphics data. After this step it provides the topological and geometric facts which are demanded by the inference engine in the following interpretation process. The inference engine can not handle geometric operations efficiently. Therefore, it is convenient to use explicit topology to reduce the complexity of proving spatial predicates like adjacency and containment to simple table lookups instead of complex geometrical calculations.

2.2   Map Interpreter

The second component of the system is the map interpreter, which is responsible for inferring new information from the facts given by the database component. The map interpreters inference machine is based on knowledge formulated as rules and facts. The latter are subdivided in two categories, static facts and dynamic facts. The first kind is generated via database queries, describing the basic geometric primitives at the lowest level, namely line segments $line(x_1, y_1, x_2, y_2)$ and text elements $text(x_0, y_0, string, orientation)$ as well as *nodes* and *edges* of the feature neighbourhood graph. The dynamic facts are inferred during the process of map analysis and describe geometry of higher levels like arcs, closed polygons or application specific object descriptions like buildings, parcels and streets, etc. The map interpreter is organized in a hierarchical manner to construct new geometric features (e.g. polygons) from basic lines. Further application specific production rules are used to create semantic objects like buildings or parcels with their specific properties, for instance a buildings or a parcels number. A string parser is used to classify the text elements based on their syntactical structure.

The minimum cycles in the line graph are determined, which leads to a set of closed polygons. After adding these polygons to the spatial database, the database is used to create explicit topology for the polygonal features and to extend the base of facts of the map interpreter (facts concerning containment, connectivity or overlapping of polygons are added to the knowledgebase). The text elements are classified based on a grammar specifically defined for map texts. The map interpreter classifies polygons and assembles application specific objects (buildings, parcels, streets) based on production rules of a form like the one proposed in [9]. A production consists of a *condition part* and an *action part*. The condition part is a conjugation of predicates which relies on a configuration of geometrical and topological relations between the geometrical primitives (*input configuration*). If the features of the input configuration fulfill all the predicates of the condition part, the production rule creates a corresponding output configuration (in principal a new instance of a complex object). In fact, this corresponds to a grouping of geometric primitives forming a new object (eventually with additional properties like area, perimeter, orientation etc.).

Polygon Creation: Many objects of interest within a digital map are represented in polygonal form. Often these polygons are not represented explicitly in the data. In fact they are just a combination of distinct lines in the sense of being perceptable as polygons (visually). To deal with this problem an algorithm for the creation of polygonal features from a line graph was developed. This algorithm is formulated specifically to be implemented in a declarative way inside a system of production rules. The basic idea is that a closed polygon can be constructed from simple corner structures (called arcs) which themselves are defined by two adjacent edges. In fig. 3 and fig. 4 this idea is shown graphically.

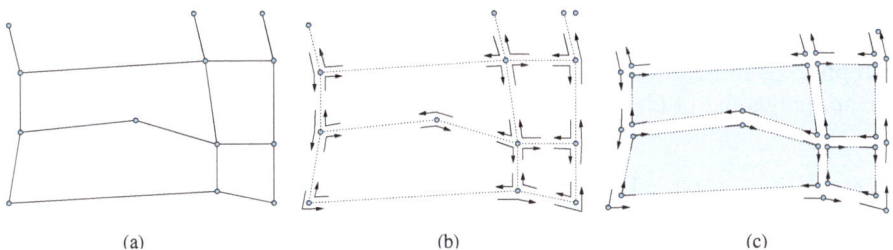

(a)   (b)   (c)

Fig. 3. Polygon creation, starting from a line-graph: a) Line graph $G_L(V, E)$, consisting of nodes and edges between this nodes (corresponding to points and lines in the map) b) Arcs (corner structures) associated to the nodes c) Graph $G_M(A, E')$ representing the closed polygons

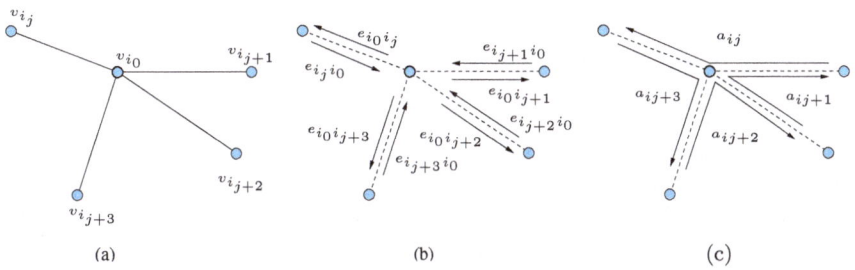

(a)   (b)   (c)

Fig. 4. Determination of the set of arcs related to a specific node v_{i_0}: a) Node v_{i_0} of graph $G_M(A, E')$. b) Directed edges E_i incident to v_{i_0}. c) Arcs A_i composed of directed edges

Geographical objects or phenomena are usually modelled within a digital vector map using the concept of boundary representations. Therefore these maps may be seen as the graphical representation of a planar cyclic graph $G_L(V, E)$ whose set of edges E is represented by the lines and whose set of nodes V is represented by the intersections of that lines. Each node v_{i_0} of degree d has a set of incident edges $E_i = \{e_{i_01}, e_{i_02}, \ldots, e_{i_0d}\}$ with a cyclic order, therefore for each node v_{i_0} a set of arcs $A_i = \{a_{ij}\}$ can be constructed (cf. fig. 4):

$$A_i = \{a_{ij}\}; \quad a_{ij} = \begin{cases} (e_{i_j i_0}, e_{i_0 i_{j+1}}) & \text{for } 1 \le j < d \\ (e_{i_j i_0}, e_{i_0 i_1}) & \text{for } j = d \end{cases} \tag{1}$$

$$e_{i_j i_k} = (v_{i_j}, v_{i_k}) \Rightarrow (e_{i_j i_0}, e_{i_0 i_k}) \Leftrightarrow (v_{i_j}, v_{i_0}, v_{i_k})$$

$$\text{with } k = \begin{cases} j - 1 & \text{for } 1 < j \le d \\ d & \text{for } j = 1 \end{cases} \tag{2}$$

Here the original feature neighbourhood graph $G_L(V, E)$ can be mapped into a second graph $G_M(A, E')$ where the set of nodes A is formed by the arcs, while the set of edges E' is constructed from two adjacent (subsequent) arcs. This graph G_M consists only of unconnected components representing the meshes (resp. polygons) of G_L (see fig. 3).

$$G_L(V, E) \longmapsto G_M(A, E') \text{ with } E' \subseteq E \tag{3}$$

$$A = \bigcup_{i=1}^{|V|} A_i \tag{4}$$

$$E' = \{e'_m = (a_{m_1}, a_{m_2}) = ((v_{m_11}, v_{m_12}, v_{m_13}), (v_{m_21}, v_{m_22}, v_{m_23}))\} \tag{5}$$
$$\text{with } v_{m_12} = v_{m_21} \wedge v_{m_12} = v_{m_23}$$

Creation of Application Specific Objects: Based on the closed polygons assembled in the step described in the previous section, the application specific objects are modelled. An example of an application specific object model is depicted in fig. 5 where, for example, a building is modelled as a composition of several building parts identified by a number.

To translate this application model of a building into a geometric model, several transformations have to be considered. The semantic relation *is_a* (which is used to model an inheritance of properties in terms of object oriented modelling) has to be translated into an spatial relation of the geometric model. Likewise the *has_a* relation of the semantic model needs to be translated into a spatial relation like *near_to*, *contains* due to cartographic common sense knowledge. For instance, the object type *building* is composed of a geometric boundary, represented by a closed polygon, and a label (inside the bounding polygon) representing the number of the building. The production rule for buildings may be formulated in the following way:

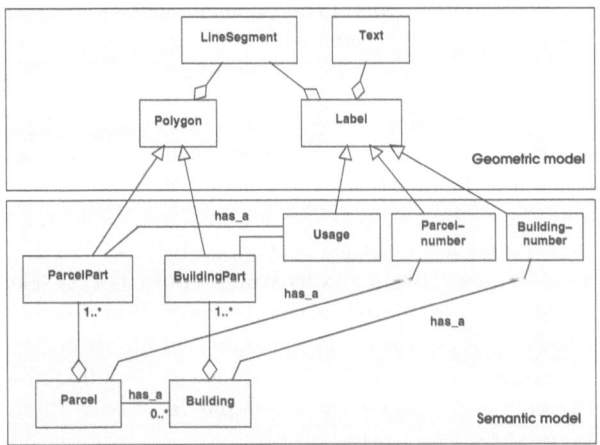

Fig. 5. Application specific object model, subdivided in a geometric and a semantic model

```
building_number( T, Number ) :-
  text( T, Number ),
  parse_text( building_number_type, Number).

building( Boundary, Number ) :-
  polygon( P, Boundary ),
  building_number( T, Number ),
  contains( P, T ).
```

The *building_number(T,Number)* predicate tests a given text string *Number*, identified by a system-wide unique identifier *T*, if it fulfills the grammatical requirements of a building number. The result represents a new instance of a *building_number* object which can be used as a basic element for further analysis or as a component for the construction of object descriptions. The grammatical requirements of a building's number are defined in the following way:

```
<digit>    := 1|2|3|4|5|6|7|8|9
<zero>     := 0
<number> := <digit>|
              <zero>|
              <digit>,<number>
<s_letter> := a|b|c...x|y|z
<c_letter> := A|B|C...X|Y|Z
<text_string> := <s_letter>,<s_letter>|
                  <s_letter>,<text_string>
<building_number_num> := <digit>|
                         <digit>,<digit>|
                         <digit>,<digit>,<digit>
<building_number_type> := <building_number_num>|
```

```
                              <building_number_num>,<s_letter>
<parcel_number>          := <number>|
                              <number>,/,<number>
```

A result of the process described above is displayed in fig. 6, where the interpreted building objects are visualized as shaded polygons. Not all of the buildings could be detected (cf. upper left part of fig. 2 and 6), because of geometrical defects in the basic dataset. These defects are for instance geometrically overlapping polygons (which should not exist as building-polygons). Furthermore the white parts of fig. 6 are caused by polygons which could not be closed completely. Therfore they were not considered in the interpretation process.

The interpretation is performed in a sequence of interpretation steps. Objects which can be discriminated with high confidence are handled first. In subsequent steps, when the knowledge about the scene has increased, these already known objects may be used to detect other objects (e.g. adjoining buildings rely on the existence of main buildings).

2.3 User Interface

The user interface is implemented as a separate component in the same way as the two components described above. It consists of a graphical map viewer, which is able to display the raw data stored in the spatial database as well as the interpretation results generated by the map interpreter. This component is essential to verify and validate the results inferred by the map interpreter but technical details are beyond the scope of this paper and are thus not addressed in particular.

Fig. 6. Building objects inferred from the data set displayed in fig. 2. Main buildings are displayed in dark grey, adjoining buildings are displayed in medium grey whereas light grey indicates unclassified buildings and parcel objects

3 Conclusions

The presented work has illustrated a system for automatic map interpretation. The components of the system have been introduced and explained while some implementation details have been demonstrated. Up to now the system is in an experimental state, but the approach has proven to be promising. Future work will be necessary in the field of quality assessment and automatic resolving of interpretation conflicts. Especially the determination of the rules mentioned in section 2.2 has to be supported in an automated way (e.g. via techniques of machine-learning). Although already tested on different data sets of comparable quality to the one displayed in fig. 2, the system has to be tested on different data sets of different scales and from several sources.

Other topics include the use of the system for interpreting GIS data in the sense of data mining, map update and map generalization tasks or as a general query tool for GIS analysis.

References

[1] Marlies de Gunst, Peter van Oosterom, and Berry van Osch. Modeling, computing and classifying topographic area features based on non-structured line input data. In Klaas Jan Beek and Martien Molenaar, editors, *Geoinformation For All, XIXth Congress of the ISPRS*, volume XXXIII, Sup. B4 of *International Archives of Photogrammetry and Remote Sensing*, pages 43–50, Groningen, The Netherlands, July 2000. ISPRS, Gopher Publishers. 191

[2] Jurgen den Hartog, Bernardus T. Holtrop, and Marlies de Gunst. Interpretation of geographic vector-data in practice. In Atul K. Chhabra and Dov Dori, editors, *Graphics Recognition - Recent Advances*, volume 1941 of *Lecture Notes in Computer Science*, pages 50–57. Springer Verlag, Berlin, 2000. 191

[3] Antonin Guttmann. R-trees: A dynamic index structure for spatial searching. In *ACM SIGMOD International Conference on Management on Data*, pages 47–57, Boston, June 1984. 192

[4] Jochen Kubinick, Bettina Barth, Gero Weber, Andreas Müller, and Alain Lefefre. Spatial Data Clearinghouse Saar Lor Lux (CLEAR). *Geoinformationssysteme*, 13(1):28–32, February 2000. 190

[5] Bruce Momjian. *PostgreSQL: Introduction and Concepts*. Addison-Wesley, 2001. 192

[6] Jan A. Mulder, Alan K. Mackworth, and William S. Havens. Knowledge structuring and constraint satisfaction: The mapsee approach. *IEEE Transactions on Pattern Analysis and Machine Intelligence*, 10(6):866–879, November 1988. 191

[7] Inc. Open GIS Consortium. Opengis simple features specification, 1999. http://www.opengis.org/techno/specs.htm. 190

[8] Hanan Samet and Aya Soffer. MARCO: MAp Retrieval by COntent. *IEEE Transactions on Pattern Analysis and Machine Intelligence*, 18(8):783–798, August 1996. 191

[9] U. Stilla and E. Michaelsen. Semantic modelling of man-made objects by production nets. In Armin Gruen, Emmanuel P. Baltsavias, and Olof Henricsson, editors, *Automatic extraction of man-made objects from aerial and space images*, pages 43–52, Basel, 1997. Birkhäuser Verlag. 194

Generating Logic Descriptions for the Automated Interpretation of Topographic Maps

Antonietta Lanza, Donato Malerba, Francesca A. Lisi,
Annalisa Appice, and Michelangelo Ceci

Dipartimento di Informatica, Università degli Studi di Bari
via Orabona 4, 70125 Bari, Italy
{lanza,malerba,lisi,appice,ceci}@di.uniba.it

Abstract. Automating the interpretation of a map in order to locate some geographical objects and their relations is a challenging task, which goes beyond the transformation of map images into a vectorized representation and the recognition of symbols. In this work, we present an approach to the automated interpretation of vectorized topographic maps. It is based on the generation of logic descriptions of maps and the application of symbolic Machine Learning tools to these descriptions. This paper focuses on the definition of computational methods for the generation of logic descriptions of map cells and briefly describes the use of these logic descriptions in an inductive learning task.

1 Introduction

Automating the interpretation of a map in order to locate some geographical objects and their relations [7] is a challenging task, which goes beyond the transformation of map images into a vectorized representation [3] and the extraction of single elements (symbol recognition), such as buildings [8] and roads [1]. In fact, though geographical information systems (GIS) store vectorized maps, information given as the basis of GIS models is often insufficient to recognize *geographical objects* relevant for a certain application. This deficiency is even more evident for *patterns* of geographical objects that are of interest to geographers, geologists and town planners. Map interpretation tasks, such as the detection of morphologies characterizing the landscape, the selection of both natural and artificial environmental elements, and the recognition of territorial organization forms require abstraction processes and deep domain knowledge that only human experts have.

In this work, we present an approach to the automated interpretation of vectorized topographic maps that is based on the application of machine learning tools to logic descriptions of maps. These descriptions are conjunctions of both attributive and relational features, which are automatically generated from vectorized maps. Two of the main problems in automating the map description process are the choice of an appropriate set of features and the definition of computational methods for their extraction. As to the former issue, the contribution of experts interested in automating the identification of some morphological elements in topographic maps is crucial. In a

D. Blostein and Y.-B. Kwon (Eds.): GREC 2002, LNCS 2390, pp. 200–210, 2002.

previous work [4], the collaboration of experts in geomorphology and territory planning allowed us to define a sufficiently general set of relevant features, which are *used by humans* in their map interpretation process.

In this work, we investigate the second issue, that is, the definition of computational methods for the generation of the chosen set of features. The next section is devoted to the presentation of some feature extraction algorithms, while Section 3 illustrates the use of the automatically generated logic descriptions in an inductive learning task. Conclusions are drawn and ideas for future work are presented in Section 4.

2   Generating Logic Descriptions of Map Cells

To simplify the localization process we follow the usual topographic practice of superimposing a regular grid on a map. This corresponds to a virtual segmentation of a map into square *cells*, whose size depends on the kind of geographical object or pattern we want to recognize in a cell.

The content of a map cell is described by means of a set of *features*. Here the term feature is intended as a characteristic (property or relationship) of a geographical entity. This meaning is similar to that commonly used in Pattern Recognition and differs from that attributed by people working in the field of GIS, where the term denotes the unit of data which represents a geographical entity in computer systems, according to the Open GIS Consortium terminology [16, 13].

Several spatial features can be extracted from vectorized maps. They can be distinguished on the basis of their *arity*, that is the number of arguments they can take. An *attribute* is a feature that predicates the property of one spatial object, and can be represented by a unary function or predicate. A *relation* is a feature that holds between two or more objects and can be represented by an n-ary ($n>1$) function or predicate. Spatial relations are actually conditions imposed on object locations.

According to their *nature*, it is possible to classify features as follows:

- *Locational* features, if they concern the location of objects. Locations are represented by numeric values that express co-ordinates.
- *Geometrical* features, if they depend on the computation of some metric/distance. Examples are area, perimeter, and length. Their domain is typically numeric.
- *Topological* features, if they are relations that are invariant under the topological transformations (translation, rotation, and scaling). Topological features are generally represented by nominal values.
- *Directional* features, if they concern orientation. Generally, a directional feature is represented by means of nominal values.

Geo-referenced objects also have aspatial features, such as name, layer, and temperature. Many other features can be extracted from maps, some of which are hybrid, in the sense that they merge properties of two or more feature categories. For instance, features that express parallelism or orthogonality between lines are both topological and geometrical. They are topological since they are invariant with respect to translation, rotation and scaling, while they are geometrical, since their definition is based on their angle of incidence. The relation of "faraway-west", whose definition mixes both directional and geometrical concepts, is another example of a hybrid spatial feature. Finally, some features might mix spatial relations with aspatial properties, such as the feature that describes coplanar roads by combining the condition of parallelism with the type of spatial objects (road).

Vectorized representations of topographic maps have been mainly used for rendering purposes in the field of GIS. Few works on feature extraction from vectorized maps are reported in the literature [2] and they refer only to cadastral maps. The first application of feature extraction algorithms to vectorized topographic maps can be found in the work by Esposito *et al.* [4]. For environmental planning tasks, fifteen features were specified with the help of domain experts (see Table 1). Since they are quite general, they can also be used to describe maps on different scales.

When features are explicitly modeled in a GIS, as is the case of the (sub-)type, color, and altitude of a geographical object, they can be easily computed. On the contrary, other features have to be extracted from vectorized maps. Actually, computational methods for feature extraction from vectorized maps are far from being a simple "adaptation" of existing graphics recognition algorithms. In fact, the different data representation (raster vs. vector) makes the available algorithms totally unsuitable for vectorized maps, as is the case of all filters based on mathematical morphology [17]. Each feature to be extracted needs the development of a specific procedure that relies on the geometrical, topological and topographical principles, which are involved in the definition of that feature. The problem is similar to that faced by some geographers who tried to capture the semantics associated with natural-language spatial relations through formal concepts of geometry, such as the 9-intersection model [15]. However, this model is appropriate for expressing only topological relations between linear and areal objects and cannot be used to define the semantics of some features, such as the relation of "parallelism" between contour slopes, which is deemed important for the recognition of some morphologies in the territory. In order to express the semantics of some features used in geometrical, topological and topographical reasoning, we propose the adaptation of 2D mathematical methods used for raster map processing [14].

For instance, the attribute *line_shape(O)* indicates the shape and the trend of the object O. It is a geometric attribute and has a nominal domain with values: straight, curvilinear, and cuspidal. It is extracted only for linear objects according to the following procedure. Let O be represented by n coordinate pairs (x_i, y_i); the angles of incidence w_i are:

$$w_i = \text{arctg} \frac{x_{i+1} - x_i}{y_{i+1} - y_i}, \text{ where } i = 1, 2, \ldots, n-1 .$$

Then, the differences dw_i are calculated as follows:

$$dw_i = w_{i+1} - w_i, \text{ where } i = 1, 2, \ldots, n-1 .$$

The *cuspidal* value is associated to *line_shape*, if the greatest difference between dw_i's exceeds a given threshold τ_{cuspidal}. If the cuspidality condition does not hold, then a check on a straight trend is performed. The *straight* value is generated if all differences dw_i are smaller than a threshold τ_{straight}, which depends on the examined territory. Otherwise, the *curvilinear* value is generated for the object O. The case of the cuspidal line is illustrated in Figure 1, where dw_3 is the greatest difference between dw_i and is greater than the given threshold.

A further example is the computation of the *distance* relation between two "parallel" lines. Let O and O' be two geographical linear objects represented by n and m coordinate pairs, respectively. Without loss of generality, let us assume that $n \leq m$. The algorithm first computes $dmin_h$ as the minimum distance between the h-th point of O and any point of O' (see Figure 2). Then, the distance between O and O' is computed as follows:

Table 1. Features extracted for the generation of map descriptions

Feature	Meaning	Type	Domain	
			Type	**Values**
contain(X,Y)	Cell X contains object Y	Topological relation	boolean	{true, false}
type_of(Y)	Type of Y	Aspatial attribute	nominal	33 nominal values
subtype_of(Y)	Specialization of the type of Y	Aspatial attribute	nominal	101 nominal values that are specializations of the type_of domain
color(Y)	Color of Y	Aspatial attribute	nominal	{blue, brown, black}
area(Y)	Area of Y	Geometrical attribute	linear	[0..MAX_AREA]
density(Y)	Density of Y	Geometrical attribute	ordinal	Symbolic names chosen by expert user
extension(Y)	Extension of Y	Geometrical attribute	linear	[0..MAX_EXT]
geo_direction(Y)	Geographic direction of Y	Directional attribute	nominal	{north, east, north_west, north_east}
line_shape(Y)	Shape of the linear object Y	Geometrical attribute	nominal	{straight, curvilinear, cuspidal}
altitude(Y)	Altitude of Y	Geometrical attribute	linear	[0.. MAX_ALT]
line_to_line(Y,Z)	Spatial relation between two lines Y and Z	Hybrid relation	nominal	{almost parallel, almost perpendicular}
distance(Y,Z)	Distance between two lines Y and Z	Geometrical relation	linear	[0..MAX_DIST]
region_to_region(Y,Z)	Spatial relation between two regions Y and Z	Topological relation	nominal	{disjoint, meet, overlap, covers, covered_by, contains, equal, inside}
line_to_region(Y,Z)	Spatial relation between a line Y and a region Z	Hybrid relation	nominal	{along_edge, intersect}
point_to_region(Y,Z)	Spatial relation between a point Y and a region Z	Topological relation	nominal	{inside, outside, on_boundary, on_vertex}

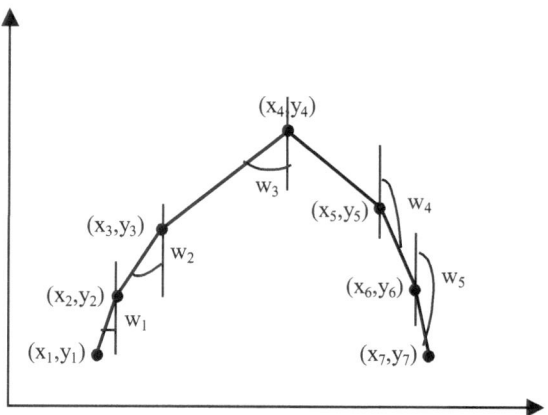

Fig.1. Graphic representation of a cuspidal line

$$\text{distance} = \frac{\sum_{h=1}^{n} d\min_h}{n} \ .$$

The complexity of this simple feature extraction algorithm is O(mn), though less computationally expensive solutions can be found by applying multidimensional access methods [6].

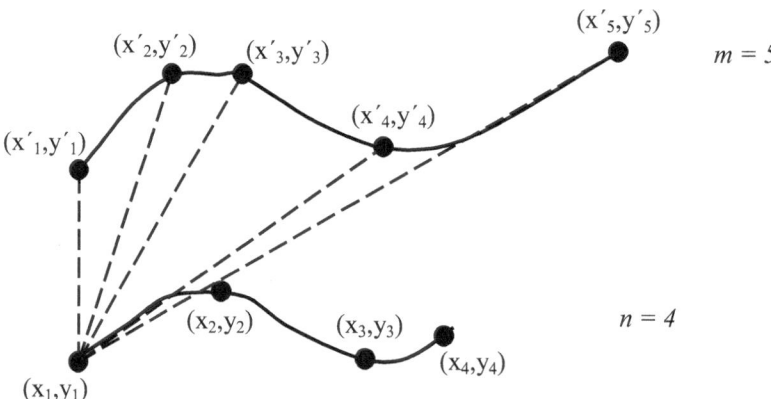

Fig.2. Computation of the average distance between two almost parallel lines

3   Learning from Logical Descriptions of Map Cells

Computational methods for the generation of logical descriptions of map cells have been implemented in the Map Descriptor module of INGENS, a prototypical GIS with inductive learning capabilities, that are used to discover geographic knowledge of interest to town planners [10]. The system manages vectorized topographic maps of the Apulia region of Italy and supports users in their map interpretation tasks. In this Section, we show an application of automated map interpretation to illustrate how logical descriptions of map cells are used. The problem is locating a "system of farms" in the large territory of his/her interest. This geographical object is not present in the GIS model, thus only the specification of its operational definition will allow the GIS to find cells containing a system of farms in a vectorized map. Who can provide such a definition? Asking the user to do that is not a feasible solution for a number of reasons.

Firstly, providing the system with operational definitions of some environmental concepts is not a trivial task. For example, the general description of a road given by an expert and reported in [11] is "a consolidated way, in the first place used for motor vehicle traffic, including over- and underpasses. Also dividing strips and roadsides … belong to roads." This declarative, abstract definition is difficult to compile into a query on a map repository.

Secondly, the operational definitions of some geographical objects are strongly dependent on the data model that is adopted by the GIS. For instance, finding relationships between the density of vegetation and the climate is easier with a *raster data model*, while determining the main orientation of some morphological elements is simpler in a *topological data model* [5].

Thirdly, different applications of a GIS will require the recognition of different geographical elements in a map. Providing the system in advance with all the knowledge required for its various application domains is simply illusory, especially in the case of wide-ranging projects such as those set up by governmental agencies.

A solution to these difficulties can be found in machine learning, a branch of artificial intelligence that investigates, among other things, how machines can be trained to recognize some concepts from a given set of examples [12]. The idea is that of extending a GIS with a training facility and a learning capability, so that each time a user wants to query the database about some geographical objects that are not explicitly modeled, he/she can prospectively train the system to recognize such objects. Training is based on a set of examples and counterexamples of geographical objects of interest to the user (e.g., road, ravine or steep slope).

Going back to our application, the user will train the system to recognize systems of farms. The only requirement for the GIS user is the ability to detect and mark some cells that are instances of his/her definition of a system of farms. Let us consider the cell shown in Figure 3, which corresponds to a square kilometer of the Apulian region reported on a map scale of 1:25,000. Suppose that the cell has been visually recognized as a system of farms by the GIS user, then it can be considered as an example of a "system of farms". Its logical description is automatically generated by the Map Descriptor (see Figure 3), which operates on the vectorized representation of the cell. The user obtains an operational definition of a "system of farms" by means of a machine learning tool embedded in INGENS, named ATRE [9]. Briefly, the learning problem solved by ATRE can be formulated as follows:

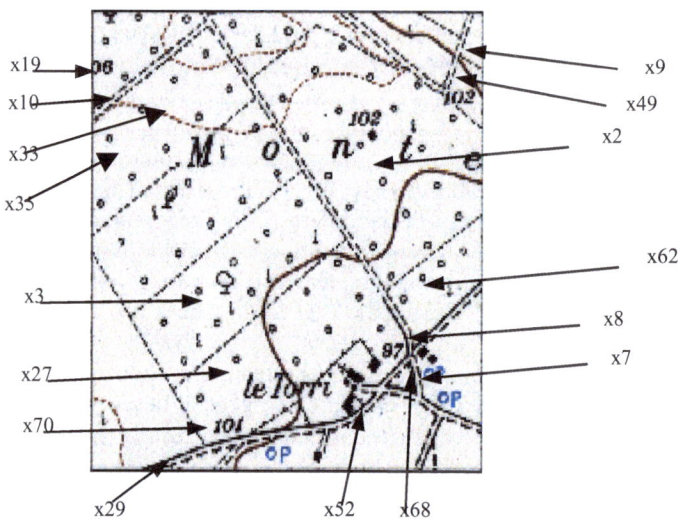

```
class(x1)=system_of_farms ←
  contain(x1,x2), contain(x1,x3), …,contain(x1,x70),
  type_of(x2)=parcel, type_of(x3)=parcel,…,type_of(x70)=quote,
  subtype_of(x2)=cultivation, subtype_of(x3)=cultivation, …,
  subtype_of(x68)=dry_wall, color(x2)=black,…,color(x70)=black,
  altitude(x19)=106.00, …, altitude(x70)=101.00,
  area(x2)=187525.00, area(x3)=99962.5, …, area(x62)=30250.00,
  density(x2)=high, density(x3)=medium, …, density(x62)=low,
  extension(x7)=111.018, extension(x8)=131.37, …,
  extension(x68)=101.119, line_shape(x7)=straight,
  line_shape(x8)=curvilinear, …, line_shape(x68)=straight,
  geo_direction(x7)=north, geo_direction(x10)= north_east, …,
  geo_direction(x68)=north, line_to_line(x7,x68)=almost_parallel, …,
  line_to_line(x52,x68)=almost_perpendicular,
  distance(x7,x68)=5.00,…,distance(x29,x52)=116.00,…,
  line_to_region(x8,x27)=adjacent, …,
  line_to_region(x33,x35)=intersect, region_to_region(x2,x3)=disjoint,
  region_to_region(x2,x9)=disjoint, …,
  region_to_region(x49,x62)=disjoint
```

Fig. 3. (*Above*) A cell containing a system of farms. (*Below*) The partial logical description of the cell. Constant *x1* represents the whole cell, while all other constants denote the sixty-nine enclosed objects. Linear values are expressed in (square) meters

Given
- a set of concepts C_1, C_2, …, C_r to be learned,
- a set of observations O,
- a background knowledge *BK*,
- a user's preference criterion *PC*,
  Find

a (possibly recursive) logical theory T for the concepts $C_1, C_2, ..., C_r$, such that T is complete and consistent with respect to O and satisfies the preference criterion PC.

The *completeness* property holds when the theory T explains all observations in O of the r concepts C_i, while the *consistency* property holds when the theory T explains no counter-example in O of any concept C_i. The satisfaction of these properties guarantees the correctness of the induced theory, with respect to the given set of observations, O.

In the context of map interpretation, each C_i is a geographical object not explicitly reported in the map legend, such as a "system of farms". Observations are logic descriptions of the cells, like that in Figure 3, while the background knowledge defines the relevant domain knowledge, such as the following rule for qualitative spatial reasoning:

$close\_to(X,Y)=true \leftarrow region\_to\_region(X,Y)=meet$
$close\_to(X,Y)=true \leftarrow close\_to(Y,X)=true$

which states that two adjacent zones are also close.

From a training set of twenty-nine observations, eight of which refer to the concept system of farms, ATRE induced the following two clauses:

$class(S1)=system\_of\_farms \leftarrow$
       $contain(S1,S2)=true, region\_to\_region(S2,S3)=meet,$
       $area(S2) \in [68437.5 .. 187525],$
       $region\_to\_region(S2,S4)=disjoint, region\_to\_region(S4,S3)=meet,$
       $type\_of(S1)=cell, type\_of(S2)=parcel, type\_of(S4)=parcel,$
       $type\_of(S3)=parcel$
$class(S1)=system\_of\_farms \leftarrow$
       $contain(S1,S2)=true, region\_to\_region(S2,S3)=disjoint,$
       $density(S3)=high, region\_to\_region(S2,S4)=meet,$
       $region\_to\_region(S4,S5)=meet, region\_to\_region(S2,S5)=meet,$
       $type\_of(S1)=cell, area(S2) \in [12381.2 .. 25981.2], type\_of(S2)=parcel$

The first clause states that "there are two pairs of adjacent parcels (S2, S3) and (S4, S3), one of which is relatively large (the area is between 68437.5 and 187525 m²)". This clause explains six training observations of a "system of farms". The second clause states that "there are three adjacent regions (S2, S4, S5), one of which is certainly a medium-sized parcel (the area is between 12381.2 and 25981.2 m²), and there is a fourth region (S3) with a high density (presumably vegetation), disjoint from the parcel S2". The second clause explains the remaining two training observations of a "system of farms".

This definition is complete and consistent, since it covers all observations classified as a system of farms and no other observation in the training set. In this example, ATRE is asked to generate more than 1200 consistent clauses (though not necessarily complete) before choosing the best. The preference criterion defined for the selection of the best consistent clause maximizes both the number of observations explained by the clause and the number of literals in the body of the clause.

INGENS can now recognize other examples of cells including a system of farms, such as that reported in Figure 4. This is done by matching the learned clauses against logical descriptions of other map cells. In the example shown in Figure 4, there is only one matching substitution that associates variables of the second clause with constants of the observation, namely S2 with x20, S3 with x57, S4 with x21 and S5 with x32. Generally speaking, INGENS recognizes complex geographical objects that have not been explicitly modeled by the computation of matching substitutions.

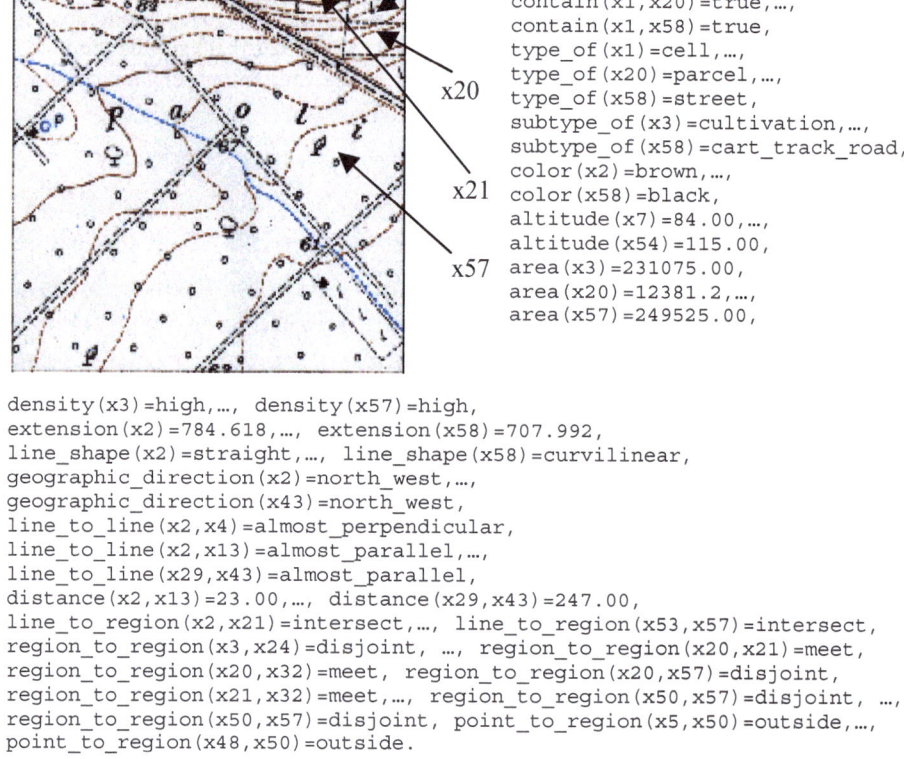

```
x32    contain(x1,x2)=true,…,
       contain(x1,x20)=true,…,
       contain(x1,x58)=true,
       type_of(x1)=cell,…,
       type_of(x20)=parcel,…,
x20    type_of(x58)=street,
       subtype_of(x3)=cultivation,…,
       subtype_of(x58)=cart_track_road,
       color(x2)=brown,…,
x21    color(x58)=black,
       altitude(x7)=84.00,…,
       altitude(x54)=115.00,
x57    area(x3)=231075.00,
       area(x20)=12381.2,…,
       area(x57)=249525.00,
```

```
density(x3)=high,…, density(x57)=high,
extension(x2)=784.618,…, extension(x58)=707.992,
line_shape(x2)=straight,…, line_shape(x58)=curvilinear,
geographic_direction(x2)=north_west,…,
geographic_direction(x43)=north_west,
line_to_line(x2,x4)=almost_perpendicular,
line_to_line(x2,x13)=almost_parallel,…,
line_to_line(x29,x43)=almost_parallel,
distance(x2,x13)=23.00,…, distance(x29,x43)=247.00,
line_to_region(x2,x21)=intersect,…, line_to_region(x53,x57)=intersect,
region_to_region(x3,x24)=disjoint, …, region_to_region(x20,x21)=meet,
region_to_region(x20,x32)=meet, region_to_region(x20,x57)=disjoint,
region_to_region(x21,x32)=meet,…, region_to_region(x50,x57)=disjoint, …,
region_to_region(x50,x57)=disjoint, point_to_region(x5,x50)=outside,…,
point_to_region(x48,x50)=outside.
```

Fig. 4. An example of cell classified as a "system of farms" by matching its logical description against the first learned rule

4   Conclusions and Future Work

Automated map interpretation is a challenging application domain for graphics recognition. Knowledge of the meaning of symbols reported in the map legends is not generally sufficient for recognizing interesting complex geographical objects or patterns on a map. Moreover, it is quite difficult to describe such patterns in a machine-readable format. That would be tantamount to providing GIS with an operational definition of abstract concepts that are often reported in texts and specialist handbooks. In order to enable the automation of map interpretation tasks in GIS, we propose a novel approach which is based on the generation of logic descriptions of maps and the application of machine learning tools to these descriptions. More specifically, every time a GIS user, such as a geographer, an urban planner, or a geologist, needs to "process" patterns of geographical objects of interest, he/she may generate the corresponding operational definitions by indicating the cell images which are instances of the patterns. This is done by invoking the computational methods for the generation of the logic descriptions of these cells, and then by applying an inductive learning algorithm to these descriptions. The induced rules act as operational definitions, and can be used to recognize new occurrences of the patterns in maps.

The proposed approach still presents some unresolved problems. In the data model for topographic maps, the segmentation of a map in a grid of suitably sized cells is a critical factor, since over-segmentation leads to a loss of recognition of global effects, while under-segmentation leads to excessively large cells with an unmanageable number of components. To cope with over-segmentation, it is necessary to consider the *context* of a cell, that is the neighboring cells, both in the training and the recognition phase. To solve problems caused by under-segmentation it is crucial to provide users with appropriate tools that hide irrelevant information in the cell description. Indeed, a set of generalization and abstraction operators will be implemented in INGENS to simplify the complex descriptions currently produced by the Map Descriptor module.

Acknowledgments

Thanks to Lynn Rudd for her help in reading the paper. This work is a partial fulfillment of the research objectives set by the IST project SPIN! (Spatial Mining for Data of Public Interest) funded by the European Union (http://www.ccg.leeds.ac.uk/spin/).

References

1. Barzohar, M., and Cooper, D.B. (1996). Automatic finding of main roads in aerial images by using geometric-stochastic models and estimation. *IEEE Transactions on Pattern Analysis and Machine Intelligence*, 18(7), 707-720
2. den Hartog, J., Holtrop, B. T., de Gunst, M. E., and Oosterbroek, E. P. (2000). Interpretation of Geographic Vector-Data in Practice. In A.K. Chhabra and D. Dori (*eds.*), *Graphics Recognition Recent Advances*, Lecture Notes in Computer Science 1941, Berlin: Springer, 50-57
3. Dupon, F., Deseilligny, M. P., and Gondran, M. (1998). Automatic Interpretation of Scanned Maps: Reconstruction of Contour Lines. In K. Tombre & A.K. Chhabra (Eds.), *Graphics Recognition: Algorithms and Systems*, Lecture Notes in Computer Science 1389, Berlin: Springer, 194-206

4. Esposito, F., Lanza, A., Malerba, D., and Semeraro, G. (1997). Machine learning for map interpretation: an intelligent tool for environmental planning. *Applied Artificial Intelligence*. 11(7-8), 673-695

5. Frank, A. U. (1992). Spatial concepts, geometric data models, and geometric data structures. *Computers & Geosciences*, 18(4), 409-417

6. Gaede V., and Günther O. (1998). Multidimensional Access Methods, *ACM Computing Surveys*, 30(2), 170-231

7. Keates, J. S. (1996). *Map understanding*. Edinburgh:Longman

8. Liow, Y.-T., and Pavlidis, T. (1990). Use of shadows for extracting buildings in aerial images. *Computer Vision, Graphics, and Image Processing*, 49, 242-277

9. Malerba, D., Esposito, F., and Lisi, F. A. (1998). Learning Recursive Theories with ATRE. In *Proceedings of the 13th European Conference on Artificial Intelligence*, ed. H. Prade, 435-439, Chichester (UK): John Wiley & Sons

10. Malerba, D., Esposito, F. , Lanza, A., Lisi, F. A., and Appice, A. (2002). Empowering a GIS with Inductive Learning Capabilities: The case of INGENS. *Journal of Computers, Environment and Urban Systems*, (to appear)

11. Mayer, H. (1994). Is the knowledge in map-legends and GIS-models suitable for image understanding? *International Archives of Photogrammetry and Remote Sensing* 30(4), 52-59

12. Mitchell, T. (1997). *Machine learning*. New York: McGraw-Hill

13. Open GIS Consortium (1996). The OpenGIS Abstract Specification. http://www.opengis.org/public/abstract.html

14. Pavlidis, T. (1982). Algorithms for graphics and image processing. Springer, Berlin

15. Rashid, A., Shariff, B. M., Egenhofer, M. J., and Mark, D. M. (1998). Natural language spatial relations between linear and aeral objects: the topology and metric of English Language terms. *International Journal of Geographical Information Science*, 12(3), 215-246

16. Sondheim, M., Gardels, K., and Buehler, K. (1999). GIS Interoperability. In P.A. Longley, M.F. Goodchild, D.J. Maguire, and D.W. Rhinds (eds.), *Geographical Information Systems, Principles and Technical Issues*, Volume 1, New York: John Wiley & Sons, 347-358

17. Yamada, H., Yamamoto, K., and Hosokawa, K. (1993). Directional Mathematical Morphology and Reformalized Hough Transformation for the Analysis of Topographic Maps, *IEEE Transactions on Pattern Analysis and Machine Intelligence*, 15(4), 380-387

Scan-to-XML: Using Software Component Algebra for Intelligent Document Generation

Bart Lamiroy[1] and Laurent Najman[2,3]

[1] LORIA – INPL / Projet QGAR
Campus scientifique – B.P. 239
54506 Vandœuvre-lès-Nancy cedex, France
[2] Océ Print Logic Technologies SA / CCR Department
1, rue Jean Lemoine – BP 113
94015 Créteil cedex, France
[3] Laboratoire A2SI, Groupe ESIEE
Cité Descartes, BP99, 93162 Noisy-le-Grand cedex, France

Abstract. The main objective of this paper is to experiment a new approach to develop a high level document analysis platform by composing existing components from a comprehensive library of state-of-the art algorithms. Starting from the observation that document analysis is conducted as a layered pipeline taking syntax as an input, and producing semantics as an output on each layer, we introduce the concept of a *Component Algebra* as an approach to integrate different existing document analysis algorithms in a coherent and self-containing manner. Based on XML for data representation and exchange on the one side, and on combined scripting and compiled libraries on the other side, our claim is that this approach can eventually lead to a universal representation for real world document analysis algorithms.

The test-case of this methodology consists in the realization of a fully automated method for generating a browsable, hyper-linked document from a simple scanned image. Our example is based on cutaway diagrams. Cutaway diagrams present the advantage of containing simple "browsing semantics", in the sense that they consist of a clearly identifiable legend containing index references, plus a drawing containing one or more occurrences of the same indices.

1 Introduction and Objectives

In this paper, we aim to validate a new approach to develop high level document analysis tools by composing existing components from a comprehensive library of state-of-the art algorithms, by developing a fully automated method for generating a browsable, hyper-linked document from a simple scanned image. The work presented here was conducted in collaboration with the QGAR research group of the LORIA laboratory, Océ Print Logic Technologies, and students from the École des Mines de Nancy [8].

The expressed need for composing existing algorithms naturally emerges from the observation that there exist a large number of papers describing ideas aimed

D. Blostein and Y.-B. Kwon (Eds.): GREC 2002, LNCS 2390, pp. 211–221, 2002.

at solving particular points of specific problems. Reusing this work for solving other applications seems to be quite a difficult task. In order to attain a functional level that can be used to tackle real-world problems, there is a pressing need to evaluate, experiment and combine existing approaches into an operational whole.

Moreover, the reuse issue is crucial, since the generic reasoning behind document analysis is to adopt a multi-level syntax+context=semantics paradigm. In other terms, most of the time, there is a loop where the semantics of a lower level algorithm, become the syntax of a higher level approach, and so on and so forth. The main problem is that the notion of the three terms: *syntax, context* and *semantics* tend to vary rapidly in function of a great number of parameters, one of the most important being the improvements of algorithms and advances in the state-of-the-art.

The syntax of a document tends to come from individual algorithms (vectorization, segmentation, skeletization, ...). These algorithms alone cannot decide on the *"truth"* (veracity, accuracy) of its output. It is only the whole treatment chain that can give information on that. Therefore, the question arises of how to use, to the best, imperfect results.

The context is given by the type of document and the problem that needs to be solved. It is generally expressed in terms of combination and conditions of syntactic expressions and expert knowledge. It is the correct formulation of this contextual information that gives the added value to a new approach that uses basic components to construct a higher level solution. Finding the correct contextual formulation is generally a process of trial and error, and requires a flexible framework for testing and reuse.

We insist on the fact that the paradigm we are describing is a general conception of how document analysis is conducted: a layered pipeline taking syntax as an input, and producing semantics as an output on each level. We are not referring to syntactic methods within the document analysis domain ([1,3,7,11] among others), but we are thinking of a more general framework.

Another important aspect that needs to be considered in the light of reuse and combination of existing components is the need to describe the results of one component in a way that can be easily translated (used) for further (more advanced) purposes. In other terms, *finding a flexible way to express the results.* There is some kind of intelligence already present in the way we express the result (there are two kinds of problems: the solved ones, and the ill-expressed ones). In today's world, the expression of choice is the XML format and its various derivatives (SVG, or in a way HTML). Most of the work in the XML context consists in using XSSL, XLT to transform the data, in a form easier to use for other purposes. We address this point in this paper.

As a summary we would like to emphasize the facts that

1. More than 20 years of advances in document analysis has given rise to an enormous pool of high performance algorithms. However, they tend to ad-

dress individual and specific problems, and (independently of any bench-marking or whatsoever) are more or less suited than competing approaches for particular tasks.

2. In order to solve higher level, real-world problems, there is a need for build-ing upon those existing components in an open manner that integrates the demand for flexibility and interchangeability.

   In that sense, we need:

   (a) a way to communicate results of one given treatment to another without being restricted to particular formats of their representation.

   (b) an experimentation and development framework that allows a flexible combination of different components, and facilitates their interchange. The principle objective is to form meta-components expressing an en-hanced context and opening the way to a straightforward assembly of intelligent components, and finally achieve a full treatment pipeline ar-chitecture.

The outline of the paper is as follows. First, we more thoroughly analyze the needs in terms of operational requirements for a platform that would suit the needs of flexible inter-component communication, and allow for efficient ex-perimental research. Secondly, we present a test-case for this environment, that consists in converting a cutaway diagram into a browsable document, and present the result of an automated treatment pipeline solving this problem.

2   The Model

In this section, we describe our approach to achieve the objectives expressed in section 1. At first, we refine the requirements induced by the introduction, then we propose our solution.

2.1   Requirements

The model for testing ideas in graphic analysis is the pipeline: an image is run through a series of operations, resulting in other images and other kinds of information extracted from those images and in various formats.

Our model should comply to the following requirements:

1. We want this pipeline model evolutionary, in the sense that it should possi-ble to be modified in a straightforward manner. This is necessary for several reasons. First, there is a need for flexible experimentation of new algorithms, combining existing components that are still in a non-finalized stage of de-velopment. Second, advances in the state-of-the-art may require fine-tuning of details or modifications and additions of new concepts.

2. We want to be able to add our home-made components as easily as possi-ble. If some (part) of those components play the same role, they need to be interchangeable for comparison purposes. Thus, our platform has to be modular.

3. Ideally, our platform should give us access to the greatest number of existing components, such components offering high-level services, like the ability to build a user-interface for demonstration purposes. Furthermore, the use of a console mode, i.e. a mode in which we can test step-by-step some idea, deciding only when seeing the result of the current step what will be the next step, is a must for debugging or fine-tuning of parameters.

2.2   Formal Analysis

If we formally represent the whole pipeline process as the application of a certain number of well chosen algorithms on a set of initial data, the usual way of seeing things is to represent algorithms as operators, and data as operands. There is, however, a more subtle way to represent the same situation, it has the particularity to be less stringent on the separation between what is algorithm and what is data.

Let us assume data is *truth*. Algorithms, therefore, must be seen as reasonings producing new truths from the initial data. In other terms, the initial data form *axioms*, subsequent data produced by the application of algorithms form corollaries or other theorems. In that sense, they are proofs. Re-expressing the previously cited requirements leads us to separate two classes of proofs: the *well established* ones, which are the experimented class of algorithms, found in the state of the art ; and the more experimental ones, built upon the previous ones by empirical trial and error until proven worthy enough to pass over to the first class.

This allows us to shed a new light on the syntax+context=semantics paradigm referred to in the introduction. The same equation can be rewritten into data + algorithm + meta-knowledge = data, where the meta-knowledge groups the following (non-exhaustive) list of items: choice of the algorithm (or combination of algorithms), parameterization of the algorithm, thresholding and selection of pertinent data, ... The individual algorithms are usually available, tend to come in multiple flavors and sorts, and are particularly specialized for a given context. The meta-knowledge itself is difficult to come by in an automated way[1], and is more generally the result of a fuzzy empirical process (which is a more snobbish way for saying "guesswork"). This meta-knowledge has the further tweak of being data-like without actually being part of the data, nor really part of the algorithm.

Solving complex problems thus consists of starting with the lowest possible syntax or semantics: a raw image, and progressively applying well chosen algorithms and adding context information (or meta-knowledge) as the process advances and intermediate data is produced, eventually converging to the final semantics.

This is not without raising a number of practical concerns:

---

[1] Which is rather obvious, once one thinks of it ... if it were, it would be part of the algorithm itself.

- Any intermediate algorithm solving a particular problem within the pipeline is probably the result of a choice between many existing alternatives. In the experimental context of finding the solution, it is absolutely necessary that this algorithm can be replaced by any other equivalent.
- Slightly changing the goal semantics should result in only a minimal change in the intermediate set of combined algorithms

2.3   A Component Algebra for Document Analysis

All the points discussed in the two previous sections naturally lead to what we'd like to call a Component Algebra:

- Valid data belongs to a well defined set.
- Components transform input data to output data belonging to the same, well defined set.
- Any intermediate result of a component is valid data.
- Data is self-explaining. Any component should take any data as input, and be able to determine if it is part of its application domain.
- Components are interchangeable.
- Context is easily represented and passed from one component to another as an attribute of the data itself.

In the following sections we shall develop the more practical choices that lead to a fully operational framework. XML is the most appropriate data representation format for the needs we have expressed. Determining the most efficient DTD to cover our ambitions is currently ongoing work. The glue for component construction and interaction is given by scripting. Combined with a compiled library it gives all latitude to realize our goal, as we are describing in the next section.

2.4   Implementation Proposal: Scripting Languages

It is now well known that object-oriented technology alone is not enough to ensure that the system we want to develop is flexible and adaptable. Scripting languages [9] go a step further than object oriented framework, as they integrate concepts for component-based application development. As it is stated in [9], *Scripting languages assume that there already exists a collection of useful components written in other languages. Scripting languages aren't intended for writing applications from scratch; they are intended primarily for plugging together components.* They thus tend to make the architecture of applications much more explicit than within an object-oriented framework [10].

- Scripting languages are extendible [2]: this means that new abstractions can easily be added to the language, encouraging the integration of legacy code.
- They are embeddable, which means that they offer a flexible way to adapt and extend applications. As it is stated in [10], *the composition of components using a script leads to a reusable component. Therefore, components and scripts can be considered as a kind of* component algebra.

- Scripting languages support a specific architectural style of composing components, namely the *pipe and filter* style supported by UNIX shells. As we have stated before, this is a good prototyping architecture, well adapted to image processing.

To obtain a fully operational experimentation platform, we build up a new component from our own graphic analysis library QGAR [4,5]. QGAR is a library written in C++, which implements basic methods for graphic recognition[2]: among the methods implemented, we found some low-level algorithms (binarization, segmentation, gradient methods, convolution masks, ...) and some vectorization methods. The use of SWIG [2] allows to automate the creation of the component: SWIG is a compiler that takes C/C++ declarations, turns them into the "glue" to access them from common scripting languages, including Python and Tcl.

3   An Example of an Application: A Browsable Cutaway

The most difficult part of succeeding this kind of generic architectures is the realization of a component algebra itself. The basic idea being the flexible combination of existing components to form new components, there is a crucial need of representing data for interchange between components. We have good conscience that the complete conception of such an algebra, and the representation of the data that is exchanged between components goes far beyond the scope of this paper. A thorough study of the formal requirements of such a construct will probably end at the fundamental difficulties of graphics analysis: what are the image semantics and how to represent them ? In our paper, we restrict ourselves to *a priori* known and very simple semantics. It is our ambition to continue this study in depth and beyond.

The application we have chosen for illustrating our ideas is to render a cutaway diagram browsable [8]. Cutaway diagrams present nicely identifiable image semantics, but extracting them requires the collaboration between a number of image treatments. They are composed of a graphical image, containing a number of *tags* (which we shall refer to as *indexes*) denoting zones of interest in their vicinity. A copy of all tags, together with an explaining text, is also present along the graphical image, and is called "the legend".

In what follows we take the following assumptions:

1. Tags are formed by alphanumeric characters we shall refer to as *strings*. There is no supplementary assumption concerning the tags, more particularly, we do not assume there are particular signs (such as surrounding circles, arrows, *etc. – cf.* Figure 2) that would allow easy detection of the tags.

---

[2] The complete source code can be downloaded or tested on-line at
http://www.qgar.org.

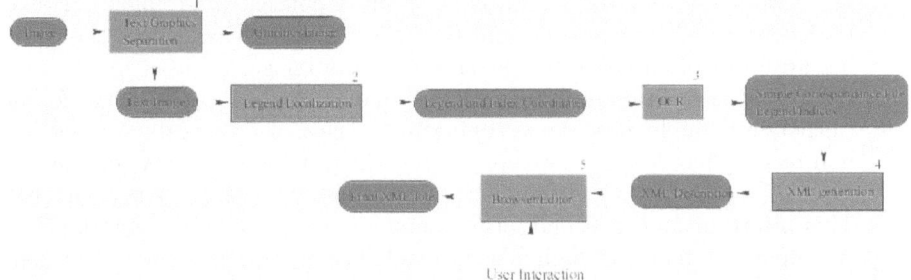

Fig. 1. General Software Architecture: A modular pipeline with easily inter-changeable components

2. The legend has an array structure, consisting of an undefined number of macro-columns, each of which contains two micro-columns: the leftmost containing the tag, the rightmost containing the explaining text.

Our goal is to detect the tags in the image, to detect the legend, and match the tags with the corresponding explaining text in the image by producing a mapping in the form of an XML document. A home-brew browser, will allow navigation through the document: clicking on a tag in the image or in the legend highlights the corresponding tag(s).

3.1   General Algorithm Outline

In order to achieve this goal, we use the pipeline architectured software model that takes a raw image as input, and sequentially executes the following treatment (Figure 1):

1. Separate text from graphics, and produce two new, raw images, one only containing text, the other only containing graphics.
   The text–graphics separation is an enhanced implementation of the FLETCHER–KASTURI algorithm [6]. It consists in an analysis of the connected components of the initial raw binary image. The algorithm is based on a statistical analysis of the distribution of the bounding boxes of all connected components. Given the fact that alphanumeric characters roughly have small squarish bounding boxes, a simple thresholding with respect to the bounding boxes' surface and ratio allows it to quite nicely separate text from graphics.
   It is noteworthy to mention that the quality of the Text/Graphic segmentation does not need to be absolutely perfect for our algorithm to work properly. As will be shown further on, we can cope with a certain level of badly classified glyphs.
2. Taking the text image as input, analyze it, and locate the position of the legend containing the needed references. Separate the text image into another two raw images: one containing the legend area, another containing the rest.

Since we know that the legend is generally a concentrated "blob" of texts (independent of its content), and that the referenced indexes in the drawing are most often sparsely scattered over the image, we us a simple algorithm that is general and robust enough to extract the location of the legend (although it might need some tuning in the case of awkwardly formed, *i.e.* non-rectangular, legends): we apply a Run-Length Smoothing Algorithm closure [12] on the image, with its parameter roughly the size of a text element. This will result in the legend form a homogeneous zone of filled black pixels. A simple search of very high density pixel clusters rapidly locates the legend with respect to the rest of the image or other textual elements.

3. Taking the legend area as input, analyze its structure and output a list of image zones, containing the references to be searched for in the rest of the image. These zones are fed to an OCR, and the final result of the treatment consists of a list of image zones and the corresponding references, as recognized by the OCR.

4. Taking the rest of the image as input, we also produce a list of the remaining text zones as well as the recognition results using the same OCR.

   This approach has the advantage to be rather robust. On the one hand, clutter in the Text layer will give at best a "not recognized" result from the OCR, or a random text label, at worst. The probability to encounter an identical text label (provided it contains a sufficient number of individual characters) is very low. Clutter therefore has very few chances to be matched with anything. A similar reasoning holds for the micro-analysis of the legend we mentioned before. Since the legend normally is composed of a tag, followed by a longer text string, there are few chances that the text string will be mismatched by a tag.

5. Both lists are analyzed in order to produce a coherent browsable structure, expressed in XML.

6. The XML file is then given as input for a customized browser-editor, written in Tcl/Tk, that allows navigation and editing of the final document.

3.2   XML for Data Exchange

In order to achieve our ambition of having complete interoperability, XML is missing link for collaborating modules. Indeed, in the context of a component algebra, any image treatment module can be either a final implementation, a component within a more complex treatment pipeline, or both. A sound, and universal knowledge representation scheme is absolutely necessary in this context, and XML [13] and its derivatives seem to be the perfect key to this framework. XML was designed to be an expression language in which it is possible to represent the underlying semantics of a document, and that that offers the possibility to easily extend and enrich its description.

XML Applied to Browsing Semantics Within the experimental context of this paper, we were principally concerned with expressing the content of a browsable document. We therefore decided on a very simple and straightforward DTD that would allow the interpretation of a cutaway tags, as detected by our algorithm. A browsable document (NAVIGABLE_IMAGE) contains three sub-parts:

- The path to a source image (IMG), mainly for displaying reasons.
- A list of rectangles (LEGEND), that contains a unique identifier as an attribute, and is supposed to be composed of two corners (UPPER_LEFT and BOTTOM_RIGHT) and an OCR recognition tag. This list represents the recognized items of the legend zone of the document.
- A similar list of rectangles (DRAWING, that also contains a unique identifier as an attribute, two corners (UPPER_LEFT and BOTTOM_RIGHT) and an OCR recognition tag. This lists enumerates the tags scattered over the drawing part of the document.

Our browser (*cf.* Figure 2) simply parses both lists and activates links between items of the DRAWING-list with those of the LEGEND-list, when they present the same OCR recognition tag. Unlinked zones can be highlighted as warnings for possible further fine-tuning and error correction methods. Moreover, the formal description of the document and the availability of a browser/editor, allows for a straightforward inclusion of human intervention: the browser/editor can allow for modification of the tag dimensions (merge/split/rezise) and tag labels (and indirectly the links) or even allow for creation and deletion of tags. Having a formal description of the document as well as a browser/editor nicely turns our experiment into a complete, exploitable platform for semi-automated hyper-linked document generation.

4   Conclusion and Perspectives

In this paper we have presented a first approach toward the creation of intelligent browsable documents, applied to cutaway diagrams. We identify the tags in the drawing and correlate them with the occurrences of the same tags in the legend adjoining the graphics.

Moreover, we identified the need for a real component algebra for document analysis applications and research and defined the major requirements for such an environment:

- flexibility and interchangeability, which we obtain through scripting basic image treatment components
- and interoperability, which we want to obtain by defining a sound XML-based description format for image and document analysis results.

The very first implementation we made of this component algebra, based on the Tcl/Tk scripting language and our own graphic analysis library QGAR, has demonstrated the power of such a design for rapidly obtaining a complete, exploitable platform for semi-automated hyper-linked document generation[3].

---

[3] All programming was done by 5 students, totaling roughly 300 man-hours, considering that they had no knowledge in image analysis, had to implement the pipeline, develop the XML browser/editor, compare different public domain OCR engines, *etc.*

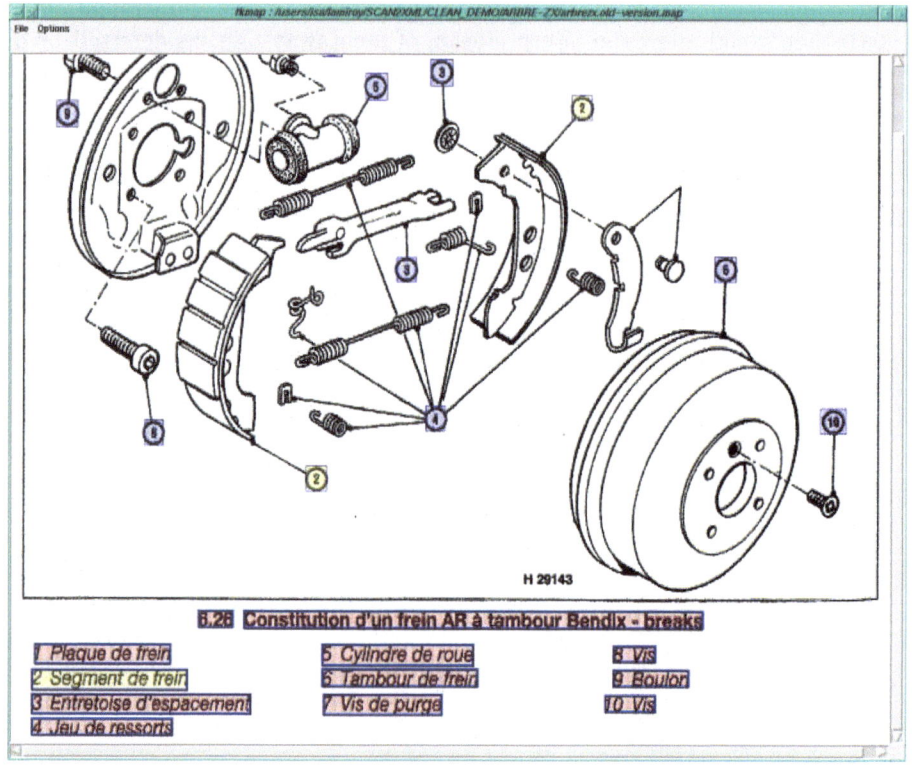

Fig. 2. Screenshot of Browser/Editor, showing a fully browsable cutaway diagram. Red and blue zones are "clickable". Highlighted zones are the the results of a selection

One of the major issues remains the data representation. In our test-case, we controlled the compiled software components, and their APIs were coherent with each other. Furthermore, we only used a limited part of our library, reducing the data types to be exchanged (finally, everything is reduced to their respective bounding boxes). Before testing the scalability of our approach (especially by integrating other sources for document analysis components), we need to formalize the data framework.

We are currently investigating the most appropriate way for expressing the data flow between different components. In our example, we developed an *ad hoc* DTD that is anything but generic and extendible. It is clear that, in order to achieve a realistic framework, the final representation should encompass most graphical entities and attributes. SVG is a very important initiative in that direction, and we are currently investigating its possibilities. However, at the printing of this document, it is still unclear whether this W3C recommendation remains suited for free, unlimited use, and other challengers, like MPEG7–VRML also need to be considered.

References

1. R. H. Anderson. Syntax directed recognition of hand-printed two-dimensional mathematics. In M. Klerer and J. Reinfelds, editors, *Interactive Systems for Experimental Applied Mathematics*. Academic Press, New York, 1968. 212

2. D. M. Beazley. SWIG and automated C/C++ scripting extensions. *Dr. Dobbs Journal*, (282):30–36, February 1998. 215, 216

3. D. Dori. A Syntactic/Geometric Approach to Recognition of Dimensions in Engineering Drawings. *Computer Vision, Graphics and Image Processing*, 47:271–291, 1989. 212

4. Ph. Dosch, C. Ah-Soon, G. Masini, G. Sánchez, and K. Tombre. Design of an Integrated Environment for the Automated Analysis of Architectural Drawings. In S.-W. Lee and Y. Nakano, editors, *Document Analysis Systems: Theory and Practice. Selected papers from Third IAPR Workshop, DAS'98, Nagano, Japan, November 4–6, 1998, in revised version*, Lecture Notes in Computer Science 1655, pages 295–309. Springer-Verlag, Berlin, 1999. 216

5. Ph. Dosch, K. Tombre, C. Ah-Soon, and G. Masini. A complete system for analysis of architectural drawings. *International Journal on Document Analysis and Recognition*, 3(2):102–116, December 2000. 216

6. L. A. Fletcher and R. Kasturi. A Robust Algorithm for Text String Separation from Mixed Text/Graphics Images. *IEEE Transactions on PAMI*, 10(6):910–918, 1988. 217

7. S. H. Joseph and T. P. Pridmore. Knowledge-Directed Interpretation of Mechanical Engineering Drawings. *IEEE Transactions on PAMI*, 14(9):928–940, September 1992. 212

8. B. Lamiroy, L. Najman, R. Ehrhard, C. Louis, F. Quélain, N. Rouyer, and N. Zeghache. Scan-to-XML for vector graphics: an experimental setup for intelligent browsable document generation. In *Proceedings of Fourth IAPR International Workshop on Graphics Recognition*, Kingston, Ontario, Canada, September 2001. 211, 216

9. John K. Ousterhout. Scripting: Higher-level programming for the 21st century. *Computer*, 31(3):23–30, March 1998. 215

10. J.-G. Schneider and O. Nierstrasz. Components, scripts and glue. In J. Hall L. Barroca and P. Hall, editors, *Software Architectures - Advances and Applications*, pages 13–25. Springer, 1999. 215

11. M. Viswanathan. Analysis of Scanned Documents — a Syntactic Approach. In H. S. Baird, H. Bunke, and K. Yamamoto, editors, *Structured Document Image Analysis*, pages 115–136. Springer-Verlag, Heidelberg, 1992. 212

12. K. Y. Wong, R. G. Casey, and F. M. Wahl. Document analysis system. *IBM J. Res. Develop.*, 26(2):647–656, 1982. 218

13. Extensible markup language (XML) 1.0 (second edition). Technical report, W3C, 2000. http://www.w3.org/TR/2000/REC-xml-20001006. 218

Interpreting Sloppy Stick Figures by Graph Rectification and Constraint-Based Matching

James V. Mahoney and Markus P. J. Fromherz

Xerox Palo Alto Research Center
3333 Coyote Hill Road, Palo Alto, CA 94304
{mahoney,fromherz}@parc.xerox.com

Abstract. Programs for understanding hand-drawn sketches and diagrams must interpret curvilinear configurations that are sloppily drawn and highly variable in form. We propose a two-stage subgraph matching framework for sketch recognition that can accommodate great variability in form and yet provide efficient matching and easy extensibility to new configurations. First, a rectification stage corrects the initial data graph for the common deviations of each kind of constituent local configuration from its ideal form. The model graph is then matched directly to the data by a constraint-based subgraph matching process, without the need for complex error-tolerance. We explore the approach in the domain of human stick figures in diverse poses.

1   Introduction

Programs for understanding hand-drawn sketches and diagrams must interpret common curvilinear configurations, such as arrows, geometric shapes, and conventional signs and symbols. These figures are often sloppily drawn and highly variable from one instance to the next. A recognition system should accommodate considerable sloppiness and variability in form, but without sacrificing matching efficiency or easy extensibility to new configurations. This paper proposes a framework for meeting these requirements, exploring it in the domain of human stick figures in diverse poses.

Figure 1 illustrates what we mean by "sloppiness". In the neat example, the lines meet precisely at junctions and corners. In the sloppy case, the lines often fail to be co-terminal, due to drawing overshoot or undershoot. Basic image analysis techniques alone (thinning, tracing, and detection of junctions and corners) would not provide an adequate basis for reliable recognition in this case.

Sloppiness is pervasive; people cannot draw a figure exactly the same way twice. In addition, there are several other pervasive sources of the variability with which matching must contend. Noise in sensing what has been drawn introduces gaps and extraneous marks. The class of figure to be recognized may be articulated and/or

D. Blostein and Y.-B. Kwon (Eds.): GREC 2002, LNCS 2390, pp. 222-235, 2002.
© Springer-Verlag Berlin Heidelberg 2002

flexible in shape (Fig. 3). Finally, there is interaction with context: the figure may overlap or intersect surrounding figures or markings (Fig. 4).

Because the configurations found in diagrams and sketches are often articulated or abstract, this recognition problem is well suited to a structural modeling approach. The configuration model and the input scene are represented as attributed graphs, with nodes representing figure parts (e.g., lines), and edges representing part relations (e.g., line connections). The attributes on nodes and edges represent geometric properties of the underlying lines and line relations. Recognition is cast as subgraph matching of the *model graph* to the *data graph*, wherein each line in the input is labeled with the name of the model part that it depicts. Matching subgraphs, rather than graphs, allows for background context in the input scene.

A stick figure configuration *model* is expressed in a simple syntax, illustrated below. Each `limb` statement defines a part of the figure. The `optional` modifier allows a part to be missing in the data. The `linked` statement asserts an end-to-end connection between two curvilinear parts. Two optional integer arguments allow the modeler to specify with which end of each part the link is associated. For example, the (default) values (2,1) indicate that the link goes from the second end of the first named part to the first end of the second, where "first" and "second" are assigned in an arbitrary but consistent manner. The syntax also allows constraints on part attributes to be specified. For example, the `minimize` statement in this example specifies optimal relative limb lengths.

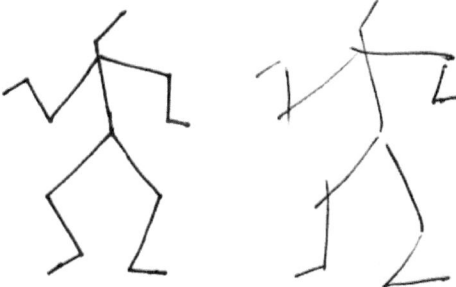

Fig.1. Neat and sloppy stick figures. In sketching, failures of co-termination are a frequent variation from the ideal form

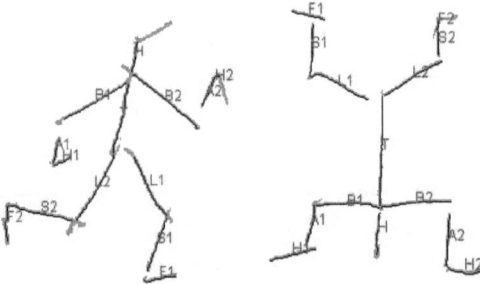

Fig.2. Example matching results. Labels denote model parts: Head, Torso, Biceps1, Arm1, etc

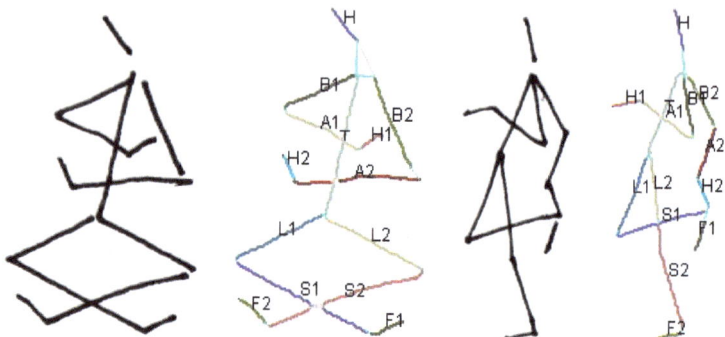

Fig.3. Matching must cope with variations, e.g., self-crossings, due to articulation

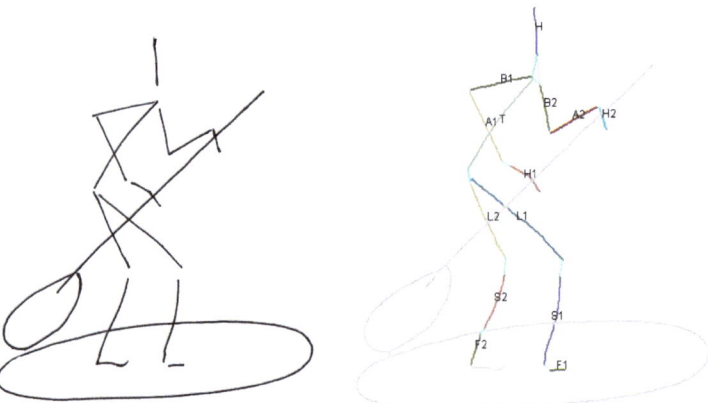

Fig.4. Matching must cope with variations, e.g., crossings, due to background context

```
model stick_figure {
  limb head, torso, biceps1, ...;
  optional limb hand1, hand2, ...;
  link(head, torso);
  link(torso, biceps1, 1, 1);
  ...
  minimize (torso.len-2*head.len)^2
   + (2*torso.len-3*biceps1.len)^2
   + ...;
  ...
} // end model stick_figure
```

Consider an initial data graph, G_D, created from an image of a line drawing like Fig. 1. Its nodes represent curve segments that result from applying standard operations of binarization, thinning, curve tracing, etc. Two nodes are linked if their curve segments terminate at the same junction or corner. Due to drawing variability and noise, the resulting graph would rarely contain a verbatim instance of the model as a subgraph.

The prevalent solution in the literature is to use error-tolerant subgraph matching to explicitly account for discrepancies in structure or attributes [14,11]. The matching process searches for a mapping that minimizes the so-called *edit distance* between the model and any data subgraph, which reflects predefined costs associated with particular discrepancies. However, this increases matching complexity; e.g., from $O(mn)$ to $O(mn^2)$ in the best case; from $O(m^n n^2)$ to $O(m^{n+1} n^2)$ in the worst case; m and n being the node counts of the data and model graphs respectively [11]. Typically, this added cost is incurred for every model matched.

As an alternative to error tolerant matching, we propose a two-stage approach to matching an ideal model graph to non-ideal data. The first stage, termed *graph rectification,* explicitly corrects data graph for each of the possible deviations of a constituent local relation from its ideal form, e.g., failures of co-termination (see Fig. 5). This process is an application of general *perceptual organization* principles, such as good continuation and proximity, to the specific goal of producing a data graph in which the model is much more likely to find a direct match.

We distinguish three classes of graph rectification operations: augmentation, reduction, and elaboration. *Augmentation* operations add new nodes or edges to G_D. E.g., in *proximity linking*, an edge is added where two free ends just failed to meet. *Reduction* operations remove elements from G_D. E.g., *spurious segment elimination* removes nodes representing short curve segments generated when other segments just fail to coincide at a common junction. *Elaboration* operations add subgraphs to G_D that constitute alternative descriptions of substructures already in G_D, to address locally ambiguous cases. E.g., if two segments satisfy a smooth continuation constraint, *continuity tracing* adds a new node to G_D that represents the alternative interpretation as a single curve.

The next section introduces methods for constructing and rectifying the data graph. Section 3 develops a constraint-based scheme for subgraph matching. Section 4 presents experimental results.

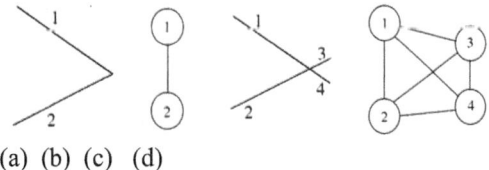

(a) (b) (c) (d)

Fig.5. Two lines just meeting at a corner (a) give data graph (b), but overshoot (c) results in graph (d). Graph rectification operations applied to (c, d) produce a graph identical to (b)

2   Data Graph Construction and Rectification

Graph Construction. The data graph, G_D, is constructed using standard image processing operations. The image is binarized and thinned; simple curve segments are extracted by tracing and then split into smooth segments at salient corner points.

(Corners are detected based on a technique in [12]; this involves a free parameter P_1.) For each of the resulting segments, its graph representation is added to G_D.

We now move to a more structured view of the graph representation of a segment. It consists of *two* nodes, one for each end point, connected by a type of edge we call a *bond*. For each pair of curves terminating at the same junction or corner, another type of edge, called a (*co-termination*) *link*, connecting their co-terminal end nodes, is added to the graph. The data graph then, consists of a single type of node, representing a curve segment end point, and two types of edges (links and bonds).

G_D is a natural description of a curvilinear input scene for matching purposes in that it mirrors the form of our configuration models; i.e., it makes explicit the end-to-end connectivity relations among line segments. Given noise, sloppy drawing, and background context, however, it may have three shortcomings. First, it may be *insufficiently connected* due to failures of one line to terminate precisely at an end or interior point of another line by a drawing undershoot, leaving a gap. Second, it may contain *spurious segments* arising from: (i) failure of one line to terminate precisely at the end of another line by a drawing overshoot, creating a spurious crossing or T-junction; (ii) failure of two lines to terminate precisely at the same point of a third line; and (iii) thinning artifacts. Third, it may be *over-segmented*: its segments may be in many-to-one rather than one-to-one correspondence with lines of the model, due to line intersections. Graph rectification operations are introduced below to address each of these problems. Free parameters of the method are enumerated as they arise.

Graph Augmentation. To deal with gaps, links are added to G_D based on proximity relations. One extreme possibility would be to simply make G_D totally connected, but this would make matching prohibitively costly. Ideally, the data graph would contain all links required by the model graph and as few extra ones as possible.

End-to-end proximity linking. A proximity link is an edge added to G_D based on proximity of two curve free ends. Links are added in increasing order of length until the graph becomes connected. As candidates for addition, we may use the set L of all n^2 possible links between n free ends. Various possible optimizations on this candidate set produce similar results. One is the set of edges of the Delaunay triangulation of the set of free ends, which can be computed in $O(n \log(n))$ time. Alternatively, we may use only those links in L between ends whose underlying image connected components are Voronoi neighbors. The results in this paper are based on this latter approach.

The process of adding links is a modification of Kruskal's algorithm for constructing the minimum spanning tree (MST) [4], resulting in a ʻclose-to-minimal spanning graph' of the base graph. (The MST itself would be insufficiently connected for matching: it would exclude some structurally important proximity relations, since it cannot contain a loop; see Fig. 6.). At each step a candidate link, l, is added either if (i) it connects two ends that are not yet connected (as in Kruskal's algorithm) or (ii) at least one of l's ends, e is, not associated with a previously added link, and the length of l is within a factor P_2 (a free parameter), of the length of the curve segment containing e. For n candidate links the complexity of this step is $O(n \log(n))$.

Fig.6. Minimum spanning tree (MST) and augmented MST of a configuration

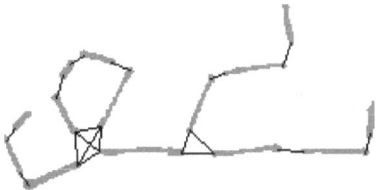

Fig.7. Link closure counteracts variability in relative link lengths

End-to-interior proximity linking (virtual junction detection). A virtual junction is a non-end point of a segment, *s*, that is sufficiently near a free end, *e*, and whose distance to *e* is a local minimum among points of *s*. The graph is modified to reflect a virtual junction *j* by splitting *s* into two segments *s1, s2*, and adding co-termination links relating *s1, s2* and the segment containing *e*. Virtual junctions may be detected using the Delaunay triangulation of the union of the free ends and evenly sampled non-end curve points, in a manner similar to the detection of proximity links.

Link closure. Even after proximity linking, G_D may still be insufficiently connected, due to differences in relative link lengths between the input instance and the model. E.g., the head and arms are each linked to the torso in the model, but in Fig. 6 (b) the head-torso link is missing. We address this problem by computing the transitive closure of links in G_D and then augmenting G_D with those links in the closure that it does not already contain (Fig. 7). (This operation was previously proposed in [13], in a somewhat different context.)

Graph Reduction. *Spurious segment elimination.* A spurious segment, *s*, results from two segments just failing to coincide at a common junction. This creates two nearby junctions in the skeleton, bridged by *s*. When *s* is removed from G_D, the junctions are merged into one, and appropriate new co-termination links are added. Deciding which segments are spurious is a classic scale selection or noise estimation problem that we have not addressed in a definitive way. Our current implementation is an *ad hoc local elimination* technique, in which a segment is removed if it is shorter than some factor (parameter P_4) of the length of the longest segment terminating at a common junction.

Graph Elaboration. Graph elaboration deals with local ambiguity in the input. We have implemented one elaboration process, *continuity tracing*, which addresses the ambiguity associated with smooth continuation of lines at crossings and T-junctions.

Continuity tracing. Suppose a local colinearity relation is established between co-terminal curve segments s_1, s_2 in G_D. (We do this by applying a threshold, free parameter P_5, to the angle between their co-terminal ends). A new segment, s_3, is formed by merging s_1 and s_2, and it is added to G_D along with links connecting s_3 to any other nodes to which s_1, s_2 are connected. Specifically, the graph representation of s_3 consists of copies of the two non-co-terminal nodes of s_1, s_2, connected by a new bond. The copy n_b of a node n_a is in turn assigned copies of all the links of n_a (but not of n_a's bond). Each link of n_a and its copy in n_b are related by mutual exclusion; i.e., a valid match of a model graph to G_D may not involve more than one of these edges. This fact is made explicit by assigning these links a common label: the unique identifier of the mutual exclusion set to which they both belong. This mutual exclusion relation applies transitively if n_b is itself copied later.

A step of *continuity tracing* involves applying the preceding operation to all pairs of collinear segments at every junction in the scene. By iterating this tracing step to convergence, all sequences of pair-wise collinear segments are added to G_D.

The idea of adding alternate subgraphs to the data graph, with a mutual exclusion relation holding among links that the alternates have to common nodes, is general. It may be applied to various other local relations where they are ambiguous, including co-termination at corners and colinearity at gaps.

3   Constraint-Based Subgraph Matching

Formulation as a CSP. Our subgraph matching problem may be formulated as a constraint satisfaction problem (CSP) as follows. Mirroring the data graph, the graph representation of a limb in the model is a pair of nodes, representing the limb's ends, connected by a bond. A link statement in the model specification gives rise to links between the specified nodes.

A CSP variable is defined for each node in the model graph. The initial domain of each variable is the set of data graph nodes plus the label *null* for a missing part, since our models may specify some parts as optional.

The primary constraint of the problem, termed *link support*, is that a link/bond, l, between two model nodes, m_1, m_2, requires the existence of a corresponding link/bond between the associated data nodes d_1, d_2. This requires some spelling out because of the possibility of missing parts. Specifically, if l is a bond, then the constraint is satisfied if (i) $m1$ and m_2 are both null; or (ii) d_1 and d_2 are connected by a bond. If l is link, then the constraint is satisfied if one of four conditions hold: either (i) m_1 and m_2 are both null; (ii) m_1 is null and d_2 has no links; (iii) m_2 is null and d_1 has no links; or (iv) d_1 and d_2 are connected by a link. (The reason for conditions (ii) and (iii) is that a model part should not be assigned null when there is support for it in the data.)

The *unique interpretation* constraint requires that each data node may be assigned to at most one model node. This is implemented as a global cardinality constraint.

These two constraints are sufficient to establish a purely topological match between the model and data graphs. This is often adequate for matching to an isolated instance, but spurious matches arise if the data contains noise, self-intersections, or background clutter. To find reliable matches in such cases, other criteria that rank topologically equivalent solutions, e.g., based on geometry, must also be applied. We have explored three such criteria.

The *minimal total link length* criterion prefers interpretations with smaller total length of links participating in the match. (A link participates in the match if it corresponds to a link in the model.) The *optimal part proportions* criterion prefers interpretations in which the length ratios of the segments closely approximate those specified in the model. The *maximal part count* criterion prefers matches that leave fewer model parts unassigned; it is needed because the constraints above allow optional model parts to be missing and make no distinction between solutions with differing numbers of missing optional parts. (The effect of progressively enabling these terms is illustrated in Fig. 12.)

To handle ambiguity, two global mutual exclusion constraints and an objective function term are added. The first constraint enforces mutual exclusion among links that are in the same mutual exclusion set. The second constraint enforces mutual exclusion among data subgraphs that have primitive segments in common; if a given segment is assigned to a model part, no other segment built from any of the same primitive segments may be assigned to any model part. The objective function term (whose weight is free parameter P_6) in effect preference ranks curve segment chains according to the number of segments they contain. This emulates the fact that long smooth curves are normally perceived as more salient than their constituent segments.

Implementation. One of our implementations for solving the above CSP employs a standard branch-and-bound state-space search framework. A state in the search space consists of an assignment to the set of CSP variables. (In the initial state, all variables are unassigned. In a goal state, all variables are assigned such that all constraints are satisfied.) The *successor function* of a search process is given a state taken from the search queue and returns a set of new states to be added to the queue. The successor function selects an unassigned CSP variable and creates a new state for each possible assignment of this variable, applying the specified constraints in the process to effect possible reductions in the set of new states generated.

In this design, the additional criteria may be applied by incorporating them into the objective function that is optimized by the search process. The three optimization criteria are combined into a single function as a weighted sum (the weights being parameters P_7, P_8, P_9). The link length and part proportion criteria have equal weights while the part count criterion has a much higher weight; i.e., the first two criteria only come into play among solutions with equal part counts.

This scheme will find only one instance of a given model in G_D. To find multiple instances, or to match multiple models, individual matches are done sequentially, removing the matched nodes and any associated links/bonds from G_D at each step.

We also implemented this basic matching scheme in a concurrent constraint programming formulation, expressed using the clp(FD) library and coroutining in SICStus Prolog [2]. A detailed description of this implementation is given in [7]. The two implementations give similar performance. The performance results given below are for the Prolog implementation running on a 600MHz Pentium III processor.

4   Experiments

We developed the matching scheme using five parallel sets of 24 example images each. Each set contains 20 images that contained an isolated stick figure in a distinct pose, as well as four images of a non-figure (see Figs. 8, 9). Non-figures are visually similar to figures with respect to local relations and number of lines. The example sets are parallel in the sense that the figure images are in one-to-one correspondence across them, with respect to posture, in the case of figures, or configuration, in the case of the non-figures. Three of the sets each emphasize a particular local relation: neat co-termination, line undershoot, and line overshoot, respectively. The other two sets combine all three relations at two degrees of sloppiness.

The free parameters of the matcher were tuned by hand to give roughly equal matching accuracy across all five example sets. Manually scored accuracy results, shown in Fig. 10, indicate that, with its parameters set favorably, the matching scheme can cope with highly variable configurations of a given model. Using this tuned performance as a baseline, our formal evaluations of the approach have focused so far on how runtime and accuracy scale with data graph complexity. (We plan to also compare matching accuracy on figures drawn by human subjects against these baseline results.)

Fig.8. The example set emphasizing line undershoot

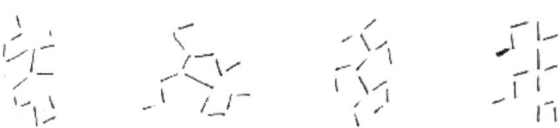

Fig.9. Non-figure examples from the example set of Fig. 8

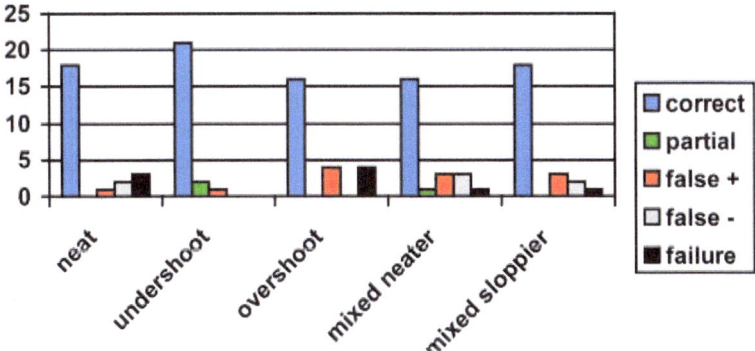

Fig.10. Tuned performance on development example sets. Vertical axis is number of examples with a given score. *Correct*: every line label correct; *partial*: at most 4 line labels incorrect; *false +*: matcher gave a solution for a non-figure; *false -*: matcher gave no solution for a figure; *failure*: more than 4 lines labels incorrect

Input complexity was varied in two ways: by adding distractor lines to images of isolated figures, and by composing multiple figure scenes from the figures in the example sets. In both cases, errors were counted automatically, in terms of the agreement between the altered and unaltered cases of the line labels assigned by the matcher. The figures used for these experiments were those of the undershoot example set, because its baseline accuracy results were nearly perfect.

Effect of Distractor Lines. Randomly generated distractor lines were added to the undershoot example images subject to two constraints: they were not allowed to cross the figure lines, since the figures were themselves disconnected; and their lengths were in a similar range as the figure lines. (Fig. 11 shows an example figure with distractor lines, its data graph, and the associated matching result.) A condition of the experiment was the number of distractors, which ranged from zero to 28. For each condition, the mean and standard deviation of runtime, in milliseconds, and match error, in number of mislabeled lines, were measured across all 24 example images at ten repetitions each (Fig. 13). A different set of distractors was generated for each image at each repetition. (The maximum of 28 distractors triples the number of nodes in the data graph, compared to an isolated stick figure.) For up to about ten neighboring distractors --a reasonable limit for realistic scenes -- the runtimes are clearly in a useful range. Overall growth in runtime is acceptable, considering the potentially exponential nature of this problem. The error rate appears to increase only linearly with the number of distractors.

Performance on Multiple Instances. Multi-figure scenes were made by selecting undershoot example images at random and composing them into a single new image subject to the constraint that figure lines should not cross. Fig. 15 shows an example composite image containing three stick figures, with the links and node identifiers of its data graph superimposed; Fig 16 shows the matching result. A condition of this experiment was the number of constituent figures, which ranged from one to five. The mean and standard deviation of runtime, in milliseconds, and match error, in number of mislabeled lines, were measured over twenty repetitions (Fig. 14). Runtimes are comparable to those in the previous experiment; these results suggest that performance should be in a useful range for realistic scenes containing a small number of target figures. For large data graphs, however, there is a trend toward rapid growth, and the error rate soon becomes unacceptably large. This indicates a clear need for additional computations to focus the matching process on appropriate subsets of a complex scene.

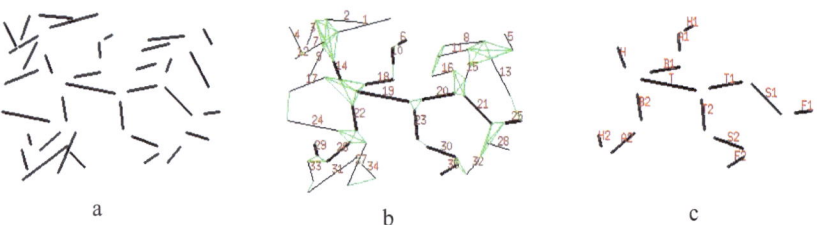

Fig.11. A stick figure with 20 distractor lines (a), the corresponding graph with labels and links produced by the image analysis stage (b), and the interpretation found by the matching process (c)

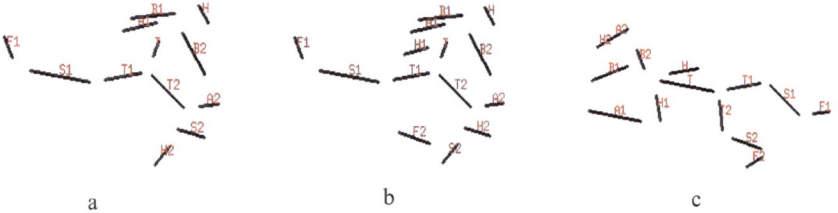

Fig.12. Three stick figure matches found in the data of Fig. 11 with progressively more objective terms: a) no optimization, b) maximal part count term only, c) maximal part count and optimal part proportion terms only

5   Discussion

Our research goal is a recognition system for simple, general curvilinear configurations of the kind routinely found in everyday graphic communications. Sloppiness and other types of variability characteristic of human drawing performance constitute a wide gulf between how people express themselves graphically and what current machine interpretation techniques can make sense of.

Where should the knowledge and computation needed to counter variability reside? The answer should not be the model, for that would hinder extending the system to new configurations. It should also not be the matcher, we feel, because that increases matching cost for a problem that is already dangerously complex computationally, and because it requires specialized matching machinery.

Instead, we located the knowledge about common local geometric variations and the processing to account for them in a dedicated prior process – graph rectification -- within a structural matching formulation. A constraint-based approach to matching suggests itself because it mates well with declarative modeling, and because constraint propagation and branch-and-bound search make effective use of the natural structure of this domain. But matching could be accomplished using various other technologies as well.

Graph matching is a well-established approach to recognition of structural models and there is previous work in which explicit correction of an initial graph has been employed prior to matching [8]. We are not aware of a previous application of this technique to the domain of hand-drawn sketches or diagrams. Messmer [11] gives an analysis of possible "distortions" in input drawings similar to our analysis of local variability, but applies it within the framework of error-tolerant subgraph matching. The work described in [6] also uses alternate subgraphs to represent alternative interpretations in a discrete relaxation framework.

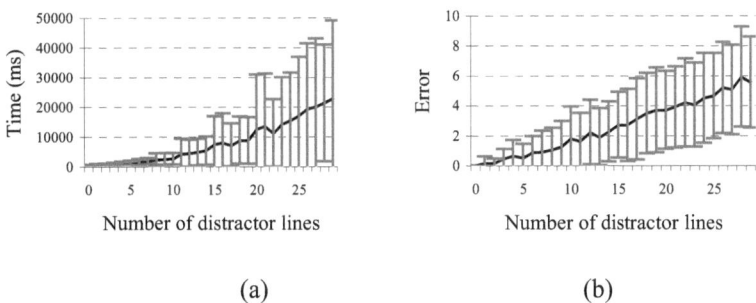

Fig.13. Mean match times and errors for 0 through 28 distractor lines. For each distractor-count, the standard deviation over 200 runs is shown as an error bar

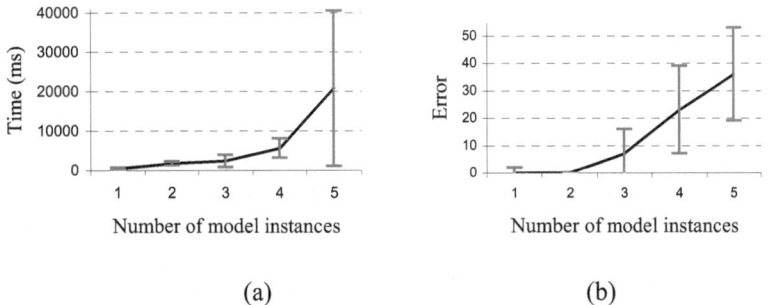

Fig.14. Mean match times and errors for 1 to 5 model instances. For each instance-count, the standard deviation over 10 runs is shown as an error bar

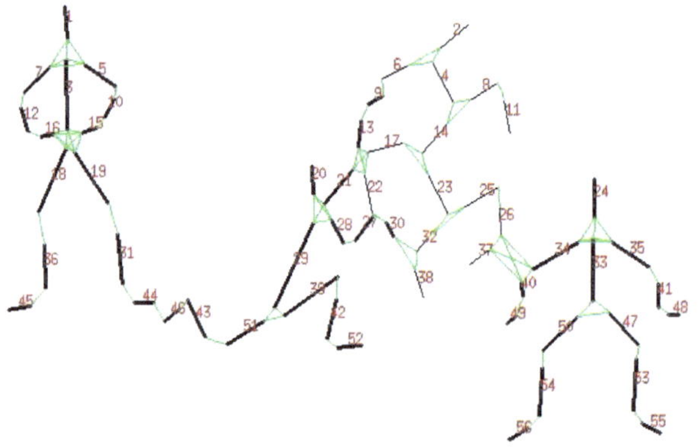

Fig.15. Sketch with three stick figures and a distractor figure with labels and links produced by the image analysis stage

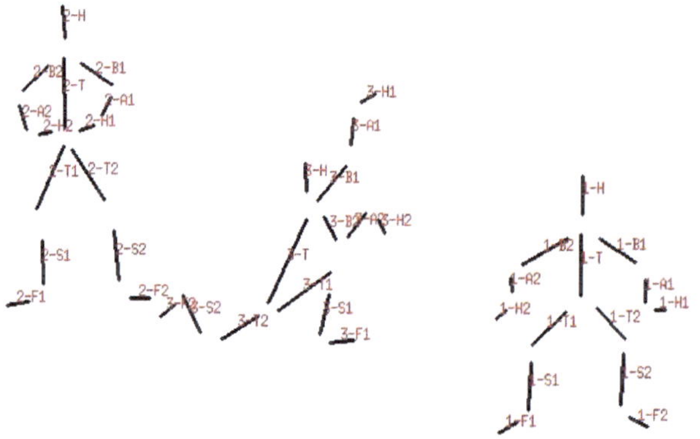

Fig.16. Interpretation of the data of Fig. 17, with part labels preceded by the index of the instance

Constraint-based pattern recognition approaches have previously been used mainly in domains with strong (visual) grammars, such as musical notation ([5],[1]) and state-transition diagrams [3]. The former two references extend Definite Clause Grammars to allow for the nonlinear composition of the graphical elements; the latter uses Constraint Multiset Grammars to similar effect. Other work has proposed dedicated forward checking and full lookahead search algorithms for subgraph matching [14][9]; these involved special-purpose algorithms that cannot be extended easily to user-defined constraints and objective functions.

There are three main avenues to be explored in our ongoing work. First, our assessment of the approach in this paper has been preliminary, limited to how it scales under varying scene complexity. Assessing the robustness of the scheme to variations in drawings collected from human subjects is a key next step. Second, the approach must be exercised on scenes containing instances of a variety of configuration models; the stick figure model was simply an entry point to this domain. Putting multiple models simultaneously in play raises issues of control and model indexing that we have not touched at all so far. Finally, the method involves a number of free parameters that would not in general be easy to tune by hand. It will be important to develop techniques for automatically setting them based on example data.

Acknowledgements

We are grateful to David Fleet, Eric Saund, and Dan Larner, of the Perceptual Document Analysis Area at Xrox PARC, for helpful discussions over the course of this research.

References

1. D. Bainbridge and T. Bell. An extensible optical music recognition system. In Proc. Nineteenth Australasian Computer Science Conf., 1996.
2. M. Carlsson, G. Ottosson, B. Carlson. An open-ended finite domain constraint solver. Proc. Programming Languages: Implementations, Logics, and Programs, 1997.
3. S. Chok, K. Marriot. Parsing visual languages. Proc. 18th Australasian Computer Science Conf., 27(1), 1995: 90--98.
4. T. Cormen, C. Leiserson, R. Rivest. Introduction to Algorithms. Cambridge, MA: M.I.T. Press, 1990.
5. B. Couasnon, P. Brisset, I. Stephan. Using logic programming languages for optical music recognition. Proc. 3rd Int. Conf. on Practical Application of Prolog, Paris, 1995.
6. H. Fahmy and D. Blostein. A graph-rewriting paradigm for discrete relaxation: application to sheet music recognition. Int. Journal of Pattern Recognition and Artificial Intelligence, Vol. 12, No. 6, Sept. 1998,pp. 763-799.
7. M. Fromherz and J. Mahoney. Interpreting sloppy stick figures with constraint-based subgraph matching. 7th Int. Conf. on Principles and Practice of Constraint Programming. Paphos, Cyprus: 2001.
8. D. Isenor and S. Zky. Fingerprint identification using graph matching. Pattern Recognition, vol. 19, no. 2, 1986, pp. 113-122.
9. J. Larrosa and G. Valiente, Constraint satisfaction algorithms for graph pattern matching. Under consideration for publication in J. Math. Struct. in Computer Science, 2001.
10. J. Mahoney and M. Fromherz. Interpreting sloppy stick figures by graph rectification and constraint-based matching. 4th IAPR Int. Workshop on Graphics Recognition: Kingston, Ontario: Sept., 2001
11. B. Messmer. Efficient graph matching algorithms for preprocessed model graphs. PhD thesis. Bern Univ., Switzerland, 1995.
12. E. Saund. Perceptual organization in an interactive sketch editor. 5th Int. Conf. on Computer Vision. Cambridge, MA: 1995: 597--604.
13. E. Saund. Perceptually closed paths in sketches and drawings. Submitted for publication.
14. L. G. Shapiro and R. M. Haralick. Structural descriptions and inexact matching. In IEEE PAMI, vol. 3, no. 5, Sept. 1981, pp. 504-519.

Using a Generic Document Recognition Method for Mathematical Formulae Recognition

Pascal Garcia and Bertrand Coüasnon

IRISA / INSA-Département Informatique
20, Avenue des buttes de Coësmes, CS 14315
F-35043 Rennes Cedex, France
couasnon@irisa.fr

Abstract. We present in this paper how to apply to mathematical formulae a generic recognition method already used for musical scores, table structure and old forms recognition. We propose to use this method to recognize the structure of formulae and also to recognize some symbols made of line segments. This offers two possibilities: improving the symbol recognition when there is a lot of symbols like in mathematics; and overcoming segmentation problems we usually find in old mathematical formulae.

1 Introduction

We presented in [3] DMOS (Description and MOdification of Segmentation) a generic recognition method for structured documents. This method is made of:

- the grammatical formalism EPF (Enhanced Position Formalism), which can be seen as a description language for structured documents. EPF makes possible at the same time a graphical, a syntactical or even a semantical description of a class of documents;
- the associated parser which is able to change the parsed structure during the parsing. This allows the system to try other segmentations with the help of context to improve recognition.

We have implemented this DMOS method to build an automatic generator of structured document recognition systems. Using this generator, adaptation to a new kind of document is then simply done by defining a description of the document with an EPF grammar. This grammar is then compiled to produce a new structured document recognition system. Each system can then use a classifier to recognize symbols which can be seen as characters.

By only changing the EPF grammar, and when needed by training a classifier, we produced automatically various recognition systems: one for musical scores [4] (figure 1a), one on recursive table structures whatever the number of rows or columns (figure 1c), and one on military forms of the 19th century (figure 1b). We have been able to successfully test this military forms recognition system on 5,000 pages of forms even if they were quite damaged [6].

D. Blostein and Y.-B. Kwon (Eds.): GREC 2002, LNCS 2390, pp. 236–244, 2002.

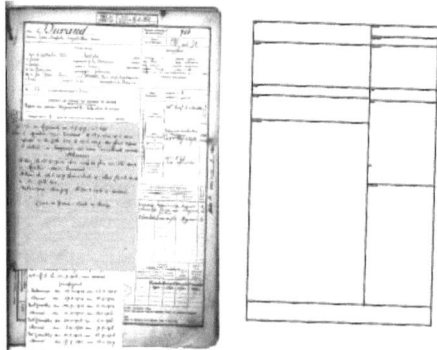

(a) On musical scores: the original image (up above) and the reconstructed score from the recognized structure (down below)

(b) On a military form of the 19th century: the original image and his well recognized structure in spite of the added sheets of paper

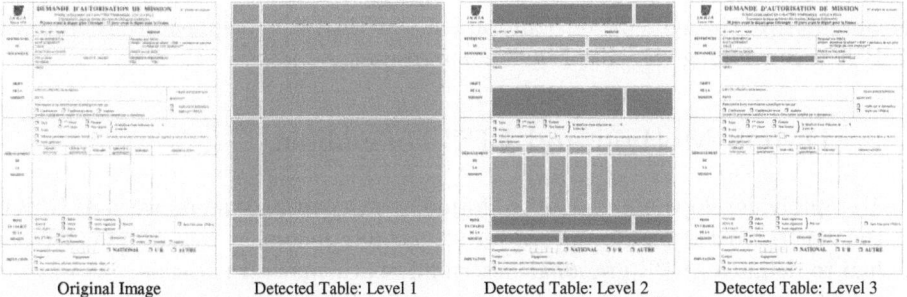

| Original Image | Detected Table: Level 1 | Detected Table: Level 2 | Detected Table: Level 3 |

(c) On table structure: recognition of the table hierarchy

Fig. 1. Applications of DMOS, a generic document recognition method

We present in this paper the way we used this generic method to automatically produce a mathematical formulae recognition system. Thanks to DMOS we have been able to strongly reduce the development time which is one of the main interests of using a generic method. Moreover, we could propose an answer to some unsolved problems of mathematical formulae recognition:

- by describing some symbols with EPF, we could limit the number of mathematical symbols recognized by classifiers. This is very important as mathematical notation can easily uses 250 symbols and sometimes a lot more (LaTeX documentation presents around 500 symbols). Besides, some symbols can have very different sizes in a same document. All this makes difficult the building of good classifiers for mathematical symbols. The grammatical description we propose on some symbols can appreciably decrease the number of symbols classes handled by classifiers and therefore increase their quality;

– the use of the DMOS method offers the possibility to deal with some of the over and under segmentation problems we can find in old mathematical books.

Even if we can find in the literature quite a lot of work done on the mathematical formulae recognition [8,2], only few of them tried to deal with an important number of symbol classes. On the formulae structure recognition, various grammatical methods have been proposed (for example [7,9]), but none of them are able to use the context to improve segmentation problems and therefore to improve the recognition quality.

The system we propose in this paper is limited to isolated and printed formulae with the following vocabulary: basic arithmetical expressions with subscript and superscript, trigonometric expressions, relational expressions, sums, products, square roots and integrals. The generated system produced by the EPF grammar compilation is able to produce the recognized formula in LaTeX.

We will start this paper by a fast presentation of the EPF formalism. Then we will see how we used it to define a description of mathematical formulae in a way to recognize their structure. Section 4 will show that this description can improve the symbols recognition and can deal with segmentation problems. Before concluding we will present some results.

2   DMOS Method and EPF Formalism

To rapidly develop a mathematical formulae recognition system, we used DMOS, the generic method we proposed for structured document recognition. We only had to define an EPF grammar describing mathematical notation.

The EPF formalism (morely presented in different papers [5,3]) can be seen as a grammatical language to describe a structured class of documents. From an EPF grammar we can automatically produce an adapted recognition system by using the automatic generator we developed.

EPF can be seen as an adding of several operators to mono-dimensional grammars like:

Position Operator (encapsulated by AT):

    A && AT(pos) && B

means A, and at the position pos in relation to A, we find B.
Where && is the concatenation in the grammar, A and B represent a terminal or a non-terminal. The writer of the grammar can define as much as necessary position operators as well as he can for non-terminals.

Factorization Operator ## (in association with the position operators):

    A && (AT(pos1) && B ##
          AT(pos2) && C)

means
(A && AT(pos1) && B) and (A && AT(pos2) && C)

Space Reduction Operator (`IN ... DO`):
EPF offers also an operator to optionally reduce the area where a rule should be applied:

`IN(aera_definition) DO(rule)`

This is very useful, for example, to make recursive descriptions. Thus we can describe what a square root notation is (where `::=` is the constructor of a grammar rule):

```
squareRoot ::=
    termSqrtSign &&
    AT(rightSqrt) && IN(areaUnderSqrt)
                    DO(expression).
```

The `termSqrtSign` describes the symbol of square root and the `IN DO` operator allows to limit the recursive description in the part of the image under the square root. The `expression` is positioned relatively to `termSqrtSign` by the position operator `rightSqrt`.

$$\int_0^x y^2 \, dy + (x+y)^2 \qquad\qquad \sum_{0 \le i \le n} i^2$$

$$1 + \left(\frac{1}{1-x^2}\right)^3 \qquad\qquad \prod_{p+q=n} x_p + y_q$$

(a) The beginning symbol of each formula is the leftmost

(b) The beginning symbol of each formula ($\sum$ and $\prod$) is NOT the leftmost

Fig. 2. The beginning symbol

3   Grammar of Mathematical Formulae

With the help of EPF we defined a grammatical description of mathematical formulae. This description follows the natural reading order of a formula. The grammar needs to be able to define three kinds of descriptions:

- the beginning of a formula;
- the alignment, subscript and superscript;
- the recursivity of mathematical formulae.

3.1 Description of the Beginning of a Formula

As we work on isolated formulae, we can describe the symbol starting the formula by the leftmost symbol in the image except when a $\sum$, a $\prod$ or a fraction line starts the formula (see figure 2). When a fraction line is the beginning of a formula, it is not necessary the leftmost symbol because of the scanning skew. We can then describe the beginning symbol of a formula by this EPF grammatical rule:

```
formBegin ::= sumBegin.
formBegin ::= prodBegin.
formBegin ::= fracLineBegin.
formBegin ::= leftMostSymbol.
```

This means that the beginning of a formula can be a $\sum$, a $\prod$ or a fraction line under certain conditions and if none of the conditions are found it is the leftmost symbol in the image. The description of the conditions for a sigma to be the beginning symbol is defined by:

```
sumBegin ::= termSigma && (
                AT(leftSide) && noSymbol ##
                AT(topBottom) && noFracRule).
```

Where `termSigma` is the non-terminal for the symbol $\sum$.The position operators `leftSide` and `topBottom` define areas - non specifically closed - like in figure 3, relatively to `termSigma` with the factorization notation. This grammar rule explains that a sigma is the beginning of a formula if :

- the symbol $\sum$ exists;
- at the left side there is no symbol;
- above and below there is no big fraction line.

The description of a formula is then defined using EPF relatively to this beginning symbol (`formBegin`).

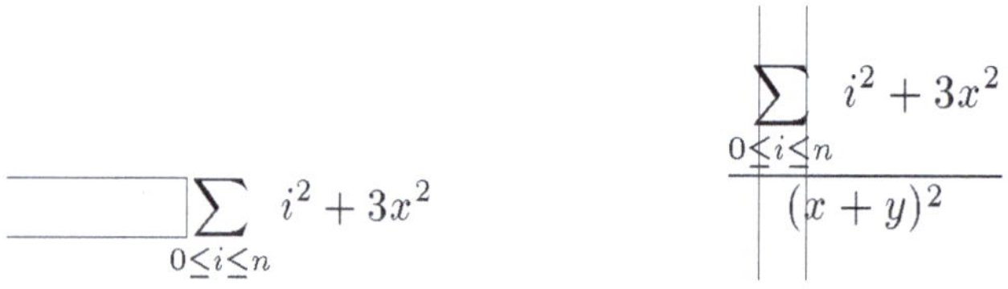

(a) Area defined by `leftSide`

(b) Area defined by `topBottom`

Fig. 3. Position operators for the description of $\sum$ as the beginning symbol

3.2   Description of Alignment, Subscript and Superscript

To describe the relative position (alignment, subscript or superscript) of symbols in a mathematical expression, we use position operators (defined by AT) to find the next symbol after the current one. Then we add a condition on the positions of base lines of these two following symbols.

3.3   Description of the Recursivity

Mathematical notation is very recursive. For example, we can find a mathematical formula in superscript, in a denominator or under a square root. Using the IN DO operator of EPF, we can easily describe this recursivity, like it is done in the squareRoot rule previously presented.

For example, a mathematical expression can contains a variable with subscript or superscript. It can then be described by:

```
expression ::= ... | atomExpr | ...

atomExpr ::=
   variable && (
      AT(subPos) && subExpressionOrNot ##
      AT(supPos) && supExpressionOrNot
      ).

supExpressionOrNot ::=
   IN(areaSup) DO(expression).
supExpressionOrNot.
```

The position operator supPos defines the area where a superscript can be found. supExpressionOrNot uses the IN DO operator to recursively call expression. As for the beginning of the formula, recursivity will look for the beginning of the formula inside the areaSup (see figure 4). The beginning symbol will be used to check conditions of being superscript.

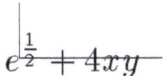

Fig. 4. Recursivity in superscript : area defined by areaSup

4   Mathematical Symbol Recognition

In this mathematical formulae recognition system we need to develop a mathematical symbol classifier.

To do so, we decided to use the classifier (able to deal with reject notions) we presented in [1], but for time reasons we have not been able to make the learning phase yet. However, we propose to use the EPF grammar to recognize symbols made of line segments.

4.1   Symbol Recognition Using Line Segments

Mathematical symbol recognition is quite difficult because of the large number of symbol classes (LaTeX uses around 500 symbols for mathematics). It is there important to reduce the number of mathematical symbol classes given to classifiers, in order to get better recognition rates. This is even more crucial as redundancy is quite weak in the mathematical notation [8].

We can notice that quite a lot of mathematical symbols are made of line segments. As with the DMOS method it is relatively easy to describe (and then recognize) graphical objects made of line segments, we propose to describe those symbols with EPF. For the notation we implemented we can recognize those symbols with EPF: $+$, $-$, $\times$, $/$, fraction line, square root symbol, $>$, $<$, $\geq$, $\leq$, $=$, $\neq$, $\sum$, $\prod$ and the absolute value. The classifier then will only have to recognize those symbols: 0 to 9, a to z, the dot, (,) and $\int$.

This reduces the number of classes recognized by the classifier from 54 to 40. This reduction can be even more important when we will increase the vocabulary (to add Greek letters...). For example there are 60 symbols in LaTeX (AMS) for negation of binary relations, and on those 60 symbols 44 can be described with EPF. Moreover those descriptions are size invariant.

4.2   Resolving Some Segmentation Problems

Using the DMOS method to recognize symbols made of line segments offers the possibility to deal with over and under segmentation problems which can be found mostly in old versions of mathematical formulae.

$$\gamma^{k}_{\nu,\,k+1} = -\sqrt{\frac{(\nu + k + 1)\,(k - \nu + 1)}{(2k + 1)\,(2k + 2)}}\,\beta^{k}_{k+1,\,k+1}, \qquad (27)$$

Fig. 5. Square root over segmented

Over Segmented Symbols: Figure 5 presents an example of an over segmented square root symbol found in a book of 1962. This is not a problem for the system we propose because a description of a graphical object with EPF (like for the square root symbol) produces a recognition system which is not linked to connexity.

$$P_n(x) = f(0) + \frac{f'(0)}{1!}x + \frac{f''(0)}{2!}x^2 + \frac{f'''(0)}{3!}x^3 = x + x^2 + \frac{1}{3}x^3$$

Fig. 6. Fraction lines touching symbols ("x" and ")")

Under Segmented Symbols: Figure 6 is from a book of 1970 and shows a fraction line touching some symbols. As the DMOS method can deal with these kinds of segmentation problems [5] by automatically changing the parsed structure during parsing, all the symbols described by EPF can touch other symbol - made of line segments or not - without any problem to be recognized.

5   Results

This description of mathematical notation done with EPF was compiled by the generator of structured document recognition systems. This automatically produced the recognition system for mathematical formulae.

Due to time constraints, we have not been able yet to integrate a classifier for non line segments symbols. Therefore we used a manual labelling for those symbols to validate the grammatical part. Thus we show that it is possible to use the generic method DMOS for mathematical formulae recognition. We present in figure 7 some examples of recognized formulae taken in more than 60 formulae (a LATEX representation of the formula is produced by the recognition system).

6   Conclusion

The grammatical and generic DMOS method have been used to automatically produce three recognition systems of document structure: one for musical scores,

$$(1) \quad \frac{4x^2 + 3x + 4}{5x^3 + 4x + 2} \qquad \sum_{0 \leq i \leq n} i = \frac{n \times (n+1)}{2} \quad (2)$$

$$(3) \quad |a + b| \leq |a| + |b| \qquad \cos^2(x) + \sin^2(x) = 1 \quad (4)$$

$$\prod_{i=0}^{n} \frac{2}{3} - \frac{\int_0^{\frac{x}{2}} y^2 + 3dy}{\frac{2}{3} + 56} - 1 + 2 \times \sum_{i=0}^{n} i^2 / \frac{1 + \frac{1+4}{3+x+yz}}{2xy + \frac{\frac{5}{2}+6}{4}} \times e^{\frac{\frac{1}{2}}{a+\frac{1}{3}}} \quad (5)$$

Fig. 7. Five examples of recognized mathematical formulae

one on recursive table structures and one on military forms of the 19th century. We have presented in this paper how to use the same DMOS method to automatically produce a mathematical formulae recognition system. Due to the genericity of the method we have been able to strongly reduce the development time. Moreover by using EPF (the grammatical formalism associated to DMOS) to recognize symbols made of line segments, we can limit the number of symbols recognized by a classifier. This grammatical description of symbols can also deal with size variation. Thanks to the DMOS method it is also possible to overcome some segmentation problems on symbols.

We still have to integrate a classifier for symbols without line segments and increase the number of classes to be able to validate the whole system on a real test set. We will also study the possibility to recognized non isolated formulae.

References

1. E. Anquetil, B. Coüasnon, and F. Dambreville. A symbol classifier able to reject wrong shapes for document recognition systems. In Atul K. Chhabra and Dov Dori, editors, *Graphics Recognition, Recent Advances*, volume 1941 of *Lecture Notes in Computer Science*, pages 209–218. Springer, 2000.   242

2. K. F. Chan and D. Y. Yeung. Mathematical expression recognition: a survey. *International Journal on Document Analysis and Recognition*, 3:3–15, 2000.   238

3. B. Coüasnon. Dmos: A generic document recognition method, application to an automatic generator of musical scores, mathematical formulae and table structures recognition systems. In *ICDAR, International Conference on Document Analysis and Recognition*, pages 215–220, Seattle, USA, September 2001.   236, 238

4. B. Coüasnon and J. Camillerapp. Using grammars to segment and recognize music scores. In L. Spitz and A. Dengel, editors, *Document Analysis Systems*. World Scientific, 1995.   236

5. B. Coüasnon and J. Camillerapp. A way to separate knowledge from program in structured document analysis: application to optical music recognition. In *ICDAR, International Conference on Document Analysis and Recognition*, volume 2, pages 1092–1097, Montréal, Canada, August 1995.   238, 243

6. B. Coüasnon and L. Pasquer. A real-world evaluation of a generic document recognition method applied to a military form of the 19th century. In *ICDAR, International Conference on Document Analysis and Recognition*, pages 779–783, Seattle, USA, September 2001.   236

7. A. Grbavec and D. Blostein. Mathematics recognition using graph rewriting. In *ICDAR, International Conference on Document Analysis and Recognition*, volume 1, pages 417–421, Montréal, Canada, August 1995.   238

8. D. Blostein A. Grbavec. Recognition of mathematical notation. *Handbook of character recognition and document image analysis, pp. 557-582*, 1997.   238, 242

9. S. Lavirotte L. Pottier. Mathematical formula recognition using graph grammar. *Electronic Imaging*, 1998.   238

Interpreting Line Drawing Images:
A Knowledge Level Perspective

Tony P. Pridmore, Ahmed Darwish and Dave Elliman

Image Processing & Interpretation Research Group
School of Computer Science and Information Technology
University of Nottingham, Nottingham, UK
{tpp,axd,dge}@cs.nott.ac.uk

Abstract. Image understanding systems rely heavily on a priori knowledge of their application domain, often exploiting techniques developed in the wider field of knowledge-based systems (KBSs). Despite attempts, typified by the KADS/CommonKADS projects, to develop structured knowledge engineering approaches to KBS development, those working in image understanding continue to employ unstructured 1st generation KBS methods. We analyse some existing image understanding systems, concerned with the interpretation of images of line drawings, from a knowledge engineering perspective. Attention focuses on the relationship between the structure of the systems considered and the KADS/CommonKADS models of expertise, sometimes called generic task models. Mappings are identified between each system and an appropriate task model, identifying common inference structures and use of knowledge. This is the first step in the acquisition of models of the expertise underpinning drawing interpretation. Such models would bring significant benefits to the design, maintenance and understanding of line drawing interpretation systems.

1   Introduction

Early knowledge-based, sometimes called expert, systems typically comprised large collections of simple, supposedly independent, rules. These were applied to input data by a monolithic inference engine. It was often argued that this architecture allowed expert systems to make flexible use of their knowledge, be developed incrementally and/or provide explanations of their reasoning. In practice the unstructured nature of first generation expert systems made them extremely difficult to develop, maintain and understand. In recent years significant attempts have been made to replace the ad hoc tools and techniques underlying classic expert systems with a more structured knowledge engineering approach. A number of knowledge engineering methodologies have appeared, though KADS/CommonKADS [1-4] is arguably the

D. Blostein and Y.-B. Kwon (Eds.): GREC 2002, LNCS 2390, pp. 245-255, 2002.
© Springer-Verlag Berlin Heidelberg 2002

best developed. Central assumptions of the model-based, knowledge-engineering approach are:

- the simple "rules + inference engine" model masks, rather than describes, the true structure of knowledge-based systems
- system and inference structures are naturally closely related to task type.

Key to these methodologies is the notion of the generic task model or model of expertise, which describes the knowledge, inference and control structures that may be used to perform instances of particular classes of task. An early step in the development of a system using KADS/CommonKADS is to identify which of a library of task models is most applicable to the problem at hand. Subsequent design decisions are then made with reference to this template, which may be modified if the task does not fit an existing model with sufficient accuracy. Access to a suitable model of expertise allows the system developer to be clear about the function of each of the available knowledge sources. System failures (and successes) can therefore be more easily attributed to specific system components. User interfaces are also easier to develop when the inference structure of the task is known and it is clear what knowledge the user is required to provide at each point in the system's operation.

Knowledge engineering has had a significant impact on KBS development and subsequent industrial uptake. While KADS/CommonKADS is very much a European methodology, similar issues have been extensively investigated in the US by, e.g. Chandresakaran [5]. Knowledge-based image understanding systems, however, continue to rely upon 1st generation techniques and design methodologies; unstructured rule bases remain the norm. Crevier and Lepage [6], for example, use the standard rules + inference engine architecture as the basis of their review.

In the longer term, the research reported here aims 1) to examine the possibility of applying knowledge engineering to the development of image understanding systems and 2) to produce line drawing interpretation systems based on sound knowledge engineering principles. We begin by analysing existing drawing understanding systems from a knowledge engineering perspective; identifying common inference structures and use of knowledge. In this we follow Clancey's [7] seminal work on medical diagnosis systems. Attention is focused throughout on the relationship between the structure of the systems considered and the existing models of expertise.

2   Models of Expertise: Classification vs. Configuration

Line drawing interpretation comprises two closely related tasks. A successful system must both determine the type of each entity present and recover a sufficiently accurate spatial description, ensuring that each reported entity is correctly located, both in absolute coordinates and in relation to neighbouring entities. These two issues of classification and spatial coherence are, for example, both explicitly addressed by current [8] performance evaluation schemes. Initial examination of the KADS/CommonKADS model library suggests that, given these goals, models describing classification and configuration tasks are the most relevant.
' Classification models are perhaps the most commonly used elements of the KADS/CommonKADS library, and have already been used to some extent in image analysis. Ton et al [9] explicitly based their Landsat image segmentation method on a

form of classification known as systematic refinement, though a recent review [10] of aerial image interpretation systems suggests that heuristic classification [3] is more commonly adopted, albeit implicitly, in that domain. Configuration models are intended to describe design tasks in which the goal is to assemble some artifact given a catalogue of available components and a design brief in the form of a set of requirements on and constraints between components. To apply this model to drawing interpretation would be to take an analysis by synthesis approach.

In the remainder of this section we introduce generic task models of classification and configuration. CommonKADS grew from a series of EU projects active throughout the 1980s and 90s. Over this period a number of notations and formalisms were developed and used to describe models of expertise. While the UML-based notation described by Schreiber et al [4] is the most recent, we adopt, for their simplicity and clarity, the conventions employed by Tansley and Hayball [3].

2.1 Heuristic Classification

A generic task model comprises an inference structure, task structure and strategies. Few KADS/CommonKADS task models, however, include detailed strategy layer; strategy knowledge, though high level, is often also strongly domain dependent. The inference structure is a diagram in which rectangles represent domain roles - the knowledge and data available to and manipulated by the system - and ovals denote inferences over those roles. Arrows highlight possible chains of inference. These diagrams are supplemented by descriptions of each domain role and inference type. In heuristic classification (Fig. 1), for example, the role observables contains "any observable phenomenon". Variables represent "the value placed on the observed data, from the system's perspective". Solution abstractions are "an abstract classification of a problem (or other concept)" and solutions are "a specific, identified solution".

Each inference type description contains a brief textual description of its inputs, outputs, knowledge used/required and possible implementation methods. During Abstract "observable data are abstracted into variables". This requires definitional knowledge and may be achieved via definitional abstraction, based on essential and necessary features of a concept, qualitative abstraction, a form of definition involving quantitative data, or generalization in a subtype hierarchy [3]. Match operates "heuristically by non-hierarchical association with a concept in another classification hierarchy". This may be achieved using heuristic rules. Finally, Specialise refines abstract solutions into more specific ones, again exploiting heuristic knowledge [3].

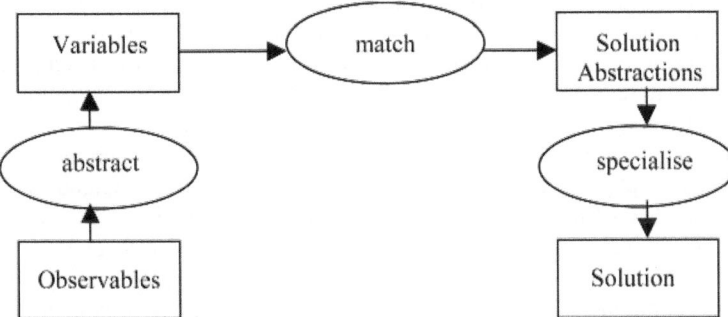

Fig. 1. The inference structure of Heuristic Classification [3]

The task structure component of a generic task model is a pseudo-code description of ways in which the inference structure might be traversed. The task structures for the two most common forms of heuristic classification are [3]:

```
/* Forward Reasoning */
Heuristic Classsification (+observations, - solutions) by
    Obtain data (for observables)
    Abstract (+ observables, -variables)
    Match (+variables – solution abstractions)
    Specialise(+ solution abstractions, - solutions)

/* Backward Reasoning */
Heuristic Classsification (+observations, - solutions) by
    Specialise(-solutions, - solution abstractions)
        Match (-solution abstractions, -variables)
            Abstract(-variables, +observables)
                Obtain data (for observables)
```

Heuristic classification is well suited to line drawing interpretation: it relies on hierarchical representations of both data and solutions and employs heuristic knowledge in the match inference. The model was, however, designed with symbolic, not iconic data in mind. If it is to be employed within line drawing interpretation some way must be found to incorporate the spatial dimension of the problem.

2.2 Incremental Configuration

Configuration is a simple form of design in which components must be selected from a fixed catalogue and placed into an empty or partially filled framework so as to satisfy a set of a priori requirements and constraints. In systems based on incremental configuration, initial requirements are broken down to form descriptions of the resources required by each design element and the constraints on how they can be arranged within the system. Element demands and constraints are then used to identify grouping contexts, usually functional areas or sub-systems of the design, which in turn are used to select a set of elements from the current design to which additional elements need to be connected. In a separate branch of the inference structure (Fig. 2), element demands and constraints are also used to identify components that would satisfy the requirements if included in the design. The required design elements are then compared to those identified as needing extension, any additional required design elements are identified and an Assemble inference employed to merge the new elements into the existing design.

As a model upon which to base drawing interpretation, incremental design is attractive. It neatly separates spatial issues (which entities to extend and in which grouping context) from the classification problem of deciding what type of design element (drawing entity) should be included. One would expect a drawing interpretation system based around this task model to similarly identify and separate these key components. The effective use of constraints is a key issue in scene formation; incremental design provides one template for their use. One criticism that might be made of the model, however, is that it treats design elements/drawing entities as flat sets of unrelated constructs. The hierarchical data and solutions often found in both classification tasks and drawing interpretation has no place in Fig. 2.

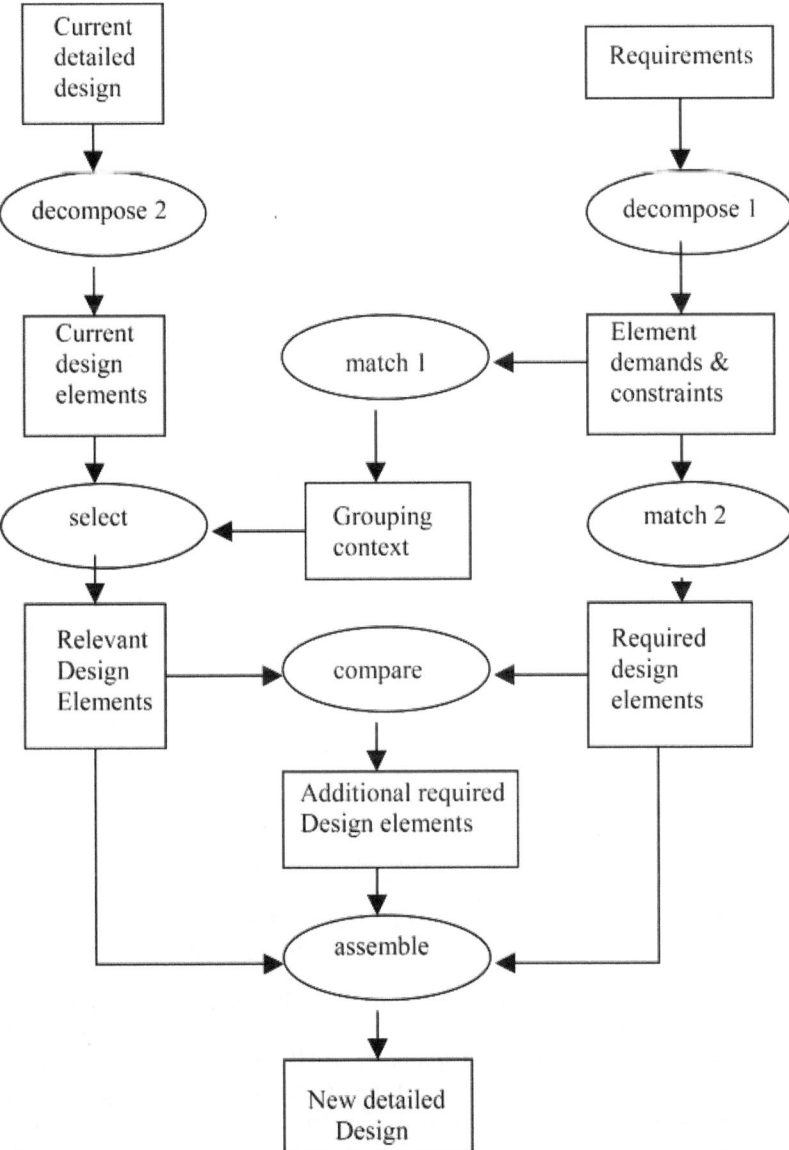

Fig.2. The inference structure of Incremental Configuration [3]

3   Mapping Rule-Based Systems to Expertise Models

One of the most direct attempts to exploit the rule-based expert system paradigm in drawing interpretation is the SAFE system of Goodson and Lewis [11]. SAFE aims to extract extended curvilinear features from images of sea charts. The system is semi-automatic, but relies heavily on a rule-based KBS, only querying the user when the KBS cannot provide a satisfactory response.

The initial image processing operations (Gaussian filtering, thresholding) may be thought of as a KADS/CommonKADS Transform inference. While Transform is defined by Tansley and Hayball as converting "one structure into another", it is commonly used to model general algorithmic operations. The subsequent extraction of line segments best corresponds to the Aggregate inference, which "takes an unstructured or partially structured collection of objects and results in a more or completely structured arrangement of the objects". In this context the objects are pixel locations and the structure imposed aggregates those objects into vector segments.

SAFE's KBS is a backward chaining rule-based system operating over a working set of symbolic statements about lines and their relationships. Each statement is assigned a certainty value that is updated, using a method similar to MYCIN, as interpretation progresses. As is often the case with 1st generation expert systems, the available written description of SAFE considers the statement and rule formats and certainty propagation method, but does not discuss the operation of the system at the knowledge level, i.e. in terms of inference and task structures, etc. The examples given strongly suggest, however, that the IKBS performs heuristic classification, matching the line segments provided by the Aggregate inference to a set of (probably implicitly represented) drawing entities. It is not clear whether any Specialisation occurs, though this seems likely.

Described at the knowledge level, SAFE's KBS performs a variant of heuristic classification in which the Abstract is replaced by an Aggregate and an additional Transform inference is included. The task structure, assuming a Specialise stage, is

> SAFE (+image, -features) by
> Transform (+image, -binary image)
> Aggregate(+binary image, -line segments)
> Specialise (+feature, -feature class)
> Match(+line segments, +feature class)

the result of the match being to test for the presence of a particular feature type.

Although less obviously influenced by classic expert systems such as MYCIN, MARIS [12] also takes a rule-based approach. MARIS extracts building lines, contour lines and lines representing railways, roads and water areas from images of maps. The input image is vectorised by a process which labels pixels with line widths, thins, creates and prunes a graph structure and approximates pixel strings with straight lines. Rules are then applied, in fixed order, to identify the constructs sought.

The inference structures of MARIS and SAFE are very similar. There is no initial Transform operation, in the MARIS software at least, as the scanner employed produces a binary image. The Transform is, however, implicit in the complete system. Examination of MARIS' rules shows clear examples of both heuristic Match and

Specialisation being performed. Buildings are identified by thresholding their internal areas, border lengths etc. Continuous lines are extracted and specialised to contours.

Both SAFE and MARIS identify drawing features by collecting together and performing tests on groups of line segments. It is therefore tempting to view the Match inference in both systems as also comprising some measure of Aggregation. The key operation of Match, however, is to make a decision; to label each vector as contributing to one of a predefined set of drawing constructs. This decision is made after consideration of groups of vectors, in the same way as a medical diagnosis system considers the existence and severity of a range of symptoms when deciding whether or not the patient is suffering from a virus. Any vector grouping that occurs during this stage is therefore contained within and secondary to the Match inference. It is often difficult to see the boundary between Aggregate and Match in rule-based systems like SAFE and MARIS, but in inference terms the separation is clear.

Recognition that these two systems are instantiations of the same generic task model allows them to be compared and contrasted, in terms of inference and task structure, in a systematic fashion. The methods used to achieve aggregation are different, for example. MARIS and SAFE are both semiautomatic. In SAFE, the user is asked to indicate line segments that should be included in a specified feature, effectively including the user only in the Aggregation needed to support Match. MARIS, however, allows the operator both to influence Aggregation, by directing line tracking algorithms to particular data items, and correct classification errors by moving lines between the various map layers. At the task level, MARIS employs the standard forward reasoning approach, while SAFE uses the structure described above. To construct reliable knowledge-level descriptions, however, requires detailed and accurate knowledge of the system(s) concerned; note that Clancey [7] had access to the source code of many of the systems he considered.

4   Comparing Structured Systems at the Knowledge Level

Some developers have sought to produce more structured knowledge-based drawing interpretation systems. Two drawing interpretation systems which comprise similar components communicating in similar ways are the mechanical drawing interpretation system of Joseph and Pridmore [13] and the utility map interpretation system of den Hartog, ten Kate and Gerbrands[15].

ANON [13] extracts 2D graphical elements from grey level images of piece-part drawings. At the heart of ANON is a hierarchically organised set of schema classes describing prototypical drawing constructs. Schema instances contain a geometrical description of the construct they represent, state variables noting the current condition of that representation and a number of C/C++ functions. The latter both manipulate schema descriptions; creating new instances, accessing and adding components, performing tests on and modifying state variables, etc. and interface to the system's low-level image analysis operations; all image analysis in ANON is performed under the control of some schema. ANON's control structure is based upon the cycle of perception model. The controlling schema both directs image analysis and interprets the result. A high-level control system is informed of this interpretation and responds by modifying the controlling schema. ANON's control system comprises rules written in an LR(1) grammar and applied by a parser generated using the Unix utility yacc.

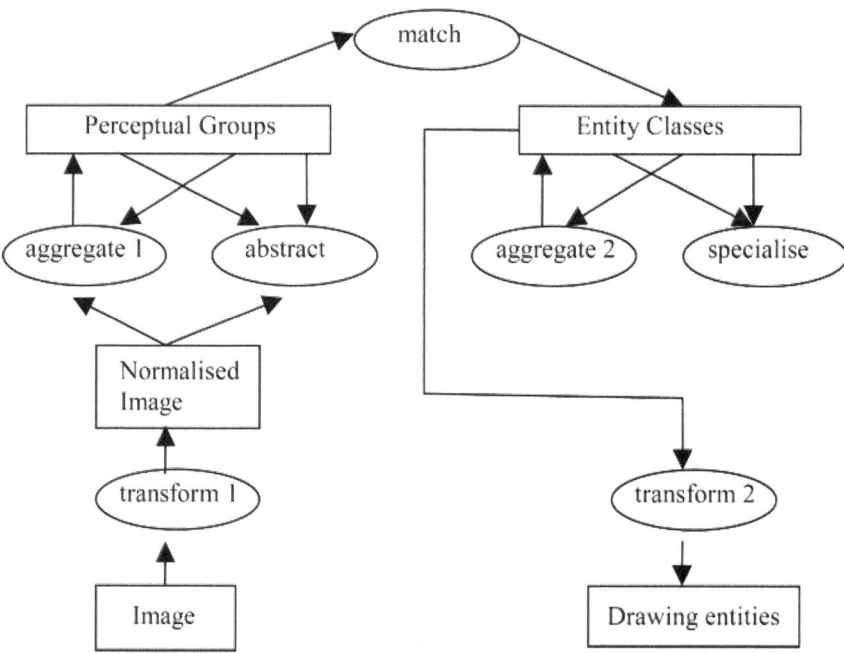

Fig.3. The inference structure of ANON

ANON's rules take two forms. The first describe relations required between constructs. Within this rule set two sub-types can be identified; both Aggregate data, (e.g. adding new segments to a broken line schema) but one also creates new, higher-level instances, bringing about an Abstract inference. Members of the second rule set perform tests on the current schema. These often Specialise the controlling schema but can also perform a Match. ANON often pursues lines of inquiry that do not lead to formation of acceptable drawing entities. A decision must be made as to whether a developing schema should be retained or discarded. This is achieved by Matching the current schema against the requirements for some drawing construct(s).

It is unfortunate to find that individual rules in ANON can perform more than one inference, but as the system was developed without any reference to knowledge engineering methodologies it is not surprising. The mapping between ANON's high level rules and KADS/CommonKADS inferences remains comparatively simple.

ANON performs a variant of heuristic classification (Fig. 3) in which Aggregation and Abstraction are iterated until a schema instance is created which can be Matched to a drawing entity class. That entity is then extended by further Aggregation and/or Specialised until no further operations are possible, at which point a specific entity description is produced. It will be noted that an initial Transform 1 inference appears, as in SAFE While ANON performs all image analysis (thresholding, etc) under schema control, the grey level image is first normalised and various local image statistics produced. The Transform 2 inference performs administrative operations on the final schema description, computing, for example, means and standard deviations of line and break lengths. ANON's task structure is difficult to specify in detail as the pattern of calls to the Aggregate 1/Abstract and Aggregate 2/Specialise pairings varies as interpretation progresses., the process is, however, essentially forward reasoning.

The MDUS system of Dori and Liu [14] adopts a similar structure. MDUS operates on an initially unstructured vector representation of the input drawing, grouping raw vectors into higher-level constructs. At the heart of the system is a hierarchy of graphic object classes, each containing specialised object recognition algorithms based upon the same generic procedure. In MDUS, identification of the "key component" of a drawing entity triggers the creation of a hypothesis, an instance of the appropriate graphic object. That object then seeks to verify itself through application of an object-specific "extend" method that searches for further object components. When no further extension is possible the object is tested and, if credible, added to the database. Objects that fail the test are deleted. Objects are recognised class by class. As the various recognition methods are "relatively independent" of each other the order adopted is not very critical [14] though it is accepted that determining the best order in which to search for objects is a difficult problem. MDUS avoids this by allowing searches for one object class to be dynamically triggered during the search for another. A search for arrowheads might, for example, be triggered while seeking leader lines.

MDUS implements a variant of heuristic classification similar to ANON's, though it is less easy to isolate the match inference as this appears to be embedded within the class-specific code. The initial Transform (which might be better described as Aggregation) creates a vector map. A hypothesis (Perceptual Group) is then generated and extended (Aggregated). At some point top-down searches for other entities begin to appear as the class is specialised.

The map interpretation system of den Hartog et al [15] also appears very similar to ANON. It employs an object-oriented, semantic net representation of drawing entities and their interrelationships and is based around a three-component, cyclic control structure. A top-down search generation process requests a search for a specific entity in a specific area of the image, much as ANON's current schema performs a search determined by its class and state variables. In den Hartog's system the result is passed to an inconsistency detection module. In ANON, inconsistency is detected occurs by the yacc-generated parser. Both systems use the classified object returned by the search to determine the next search action, repeating the process until acceptable representations are produced or no more possibilities remain to be explored.

Despite these surface similarities the inference structures of the two systems are very different. Den Hartog et al's system maps neatly, not to heuristic classification, but to a form of incremental configuration (Fig.4). Note first that the two Decompose inferences are not necessary, as all the representations are already available in component form. The semantic net provides the "element demands and constraints" found in incremental configuration and is used to determine what should be searched for (required design elements) in what part of the drawing (grouping context). The semantic net also determines, via the grouping context, which parts of the developing interpretation need to be considered during the inconsistency detection (Assemble) phase. Only den Hartog et al's object recognition step does not fit naturally into incremental configuration, though incremental design does include the Compare inference to check that the design elements chosen are sufficient. One could consider "current design elements" to include low level geometric primitives and the Compare inference to represent object recognition. Instead we prefer to explicitly Compare that the required entities with geometrical primitives Aggregated from the input image.

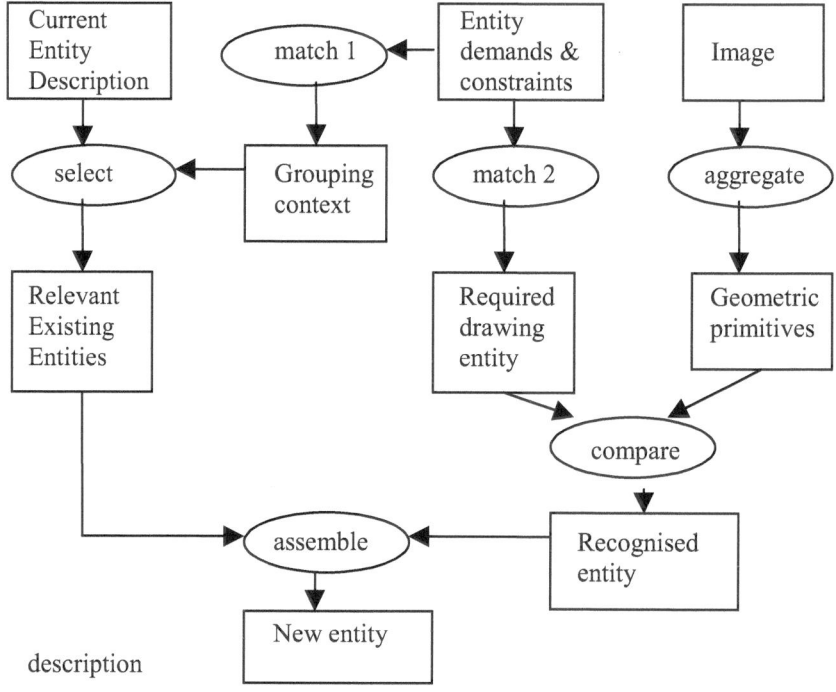

Fig. 4. The inference structure of den Hartog's map interpretation system

5   Conclusion

Reviews of various aspects of image interpretation are common in the literature and, though they often provide useful guides to those new to the field, it can be argued that most do not advance the state of their particular art. We would like to stress that the present paper is less a review than an analysis of knowledge-based line drawing understanding systems. Although much has been made of the importance of knowledge in drawing understanding, we believe that to date the field has lacked the tools required to best determine, make use of and even understand the role of the available knowledge. While almost certainly not perfect, we further believe that the tools and techniques developed under the banner of Knowledge Engineering have significant potential in this area. The analyses presented here have shown how generic task models can be used to describe, compare and distinguish between knowledge-based image interpretation systems. They have also generated first versions of models that might in time be used to inform the design of new systems. This work is ongoing; a more comprehensive knowledge-level analysis of line drawing interpretation systems is in preparation. The use of generic task models to drive the design of new systems will be the subject of future reports. In the meantime we would urge colleagues working in knowledge-based drawing understanding systems to give some consideration to the generic task models, and particularly inference structures, underlying their systems.

References

1. B. Wielinga, A. Th. Schreiber and J. A. Breuker, "KADS: A Modeling Approach to Knowledge Engineering," Knowledge Acquisition, Vol. 5, pp. 5-53, 1992.
2. G. Schreiber, B. Wielinga and J. Breuker, (ed.), KADS: A Principled Approach to Knowledge-Based System Development. London: Academic Press Ltd., 1993.
3. D. S. W. Tansley and C. C. Hayball, Knowledge-Based Systems Analysis & Design: A KADS Developer's Handbook. Prentice Hall International (UK) Ltd., 1993.
4. G. Schreiber, et. al., Knowledge Engineering and Management: The CommonKADS Methodology. Cambridge, Mass.: MIT Press, 1999.
5. Chandresakaran, B., Generic Tasks in Knowledge-based Reasoning: High Level Building Blocks for Expert System Design, IEEE Expert, pp 23-30, Fall 1986.
6. D. Crevier, and R. Lepage, "Knowledge-Based Image Understanding Systems: A Survey," Computer Vision & Image Understanding, Vol. 67, no. 2, pp. 161-185, 1997.
7. W.J. Clancey, Heuristic Classification, Artificial Intelligence, Vol. 27, pp 215-251, 1985.
8. I. Phillips and A.K. Chabbra, Empirical Performance Evaluation of Graphics Recognition Systems, IEEE Transactions PAMI, Vol. 21, No. 9, pp 849-870, 1999.
9. J. Ton et. al.: "Knowledge-Based Segmentation of Landsat Images", IEEE Transactions on Geoscience & Remote Sensing, Vol. 29, No. 2, 1991.
10. A. Darwish, T.P. Pridmore and D. Elliman, Interpreting Aerial Images: A Knowledge-Level Perspective, Proc. ES-2001, pp 169-182, Cambridge, UK, 2001.
11. K. J. Goodson and P. H. Lewis, A Knowledge-based Line Recognition System, Pattern Recognition Letters, Vol 11, pp 295-304, 1990.
12. S. Suzuki and T. Yamada. Maris: map recognition input system. Pattern Recognition Vol. 23, pp 919-933, 1990.
13. S. H. Joseph, S. H., and T. P. Pridmore, Knowledge-directed interpretation of mechanical engineering drawings, IEEE Trans. Pattern Analysis and Machine Intelligence, Vol. 14, No. 9, pp. 928-940, 1992.
14. D.Dori and W. Liu, Automated CAD Conversion with the Machine Drawing Understanding System, IEEE Trans. Systems, Man and Cybernetics, Vol. 29, No. 4, pp 411-416, 1999.
15. J. E. den Hartog, T. K. ten Kate and J. J. Gerbrands, Knowledge-based Interpretation of Utility Maps, CVIU, Vol. 63, No. 1, pp105-117, 1996.

Smoothing and Compression of Lines Obtained by Raster-to-Vector Conversion

Eugene Bodansky, Alexander Gribov, and Morakot Pilouk

Environmental Systems Research Institute, Inc – ESRI
380 New York St., Redlands, CA 92373-8100 USA
(ebodansky,agribov,mpilouk}@esri.com

Abstract. This paper presents analyses of different methods of post-processing lines that have resulted from the raster-to-vector conversion of black and white line drawing. Special attention was paid to the borders of connected components of maps. These methods are implemented with compression and smoothing algorithms. Smoothing algorithms can enhance accuracy, so using both smoothing and compression algorithms in succession gives a more accurate result than using only a compression algorithm. The paper also shows that a map in vector format may require more memory than a map in raster format. The Appendix contains a detailed description of the new smoothing method (continuous local weighted averaging) suggested by the authors.

1   Introduction

The result of raster-to-vector conversion of a black and white line drawing can be represented as borders of connected components and as centerlines. Some definitions for the term centerline can be found in [1]. The simplest vectorization consists of building piecewise linear curves or polylines that are borders of connected components. It has a unique solution, which can be obtained without any control parameters.

There are two standard methods of building borders of connected components. The first method [1] results in borders as orthogonal polylines, lines that consist only of horizontal and vertical segments. The second method [2] results in a set of digital lines that consist of border pixels. In this case, borders in the form of polylines can be obtained with special approximation algorithms.

Post-processing borders should increase accuracy, smooth and generalize contours, defragment straight lines, arcs, and circles, and compress data. Very often, it is impossible to separate these tasks. For example, increasing accuracy while suppressing a high frequency noise gives the effect of data compression. The same effect can be achieved by defragmentation or generalization, and data compression can be used for defragmentation and generalization.

D. Blostein and Y.-B. Kwon (Eds.): GREC 2002, LNCS 2390, pp. 256-265, 2002.

It is necessary to explain what we mean by increasing accuracy: Figure 1 shows a piece of a contour map that was converted into a raster format with a resolution of 300 dpi. Any differences between contours on a source map and corresponding contours of the raster image of this map will be called errors of digital description: errors that appear because of scanning, binarization, and discretization. For error correction, we can use some *a priori* information about depicted objects.

It is known that isolines (in our case contours) cannot intersect. Hence any intersections on Figure 1 are errors. These are most likely caused by scanning and binarization. Only an operator can correct these errors. To ascertain this, it is enough to see Figure 2a. Should it be two contours (Figure 2b) or only one (Figure 2c)?

Fig. 1. A piece of a contour map scanned with a resolution 300 dpi

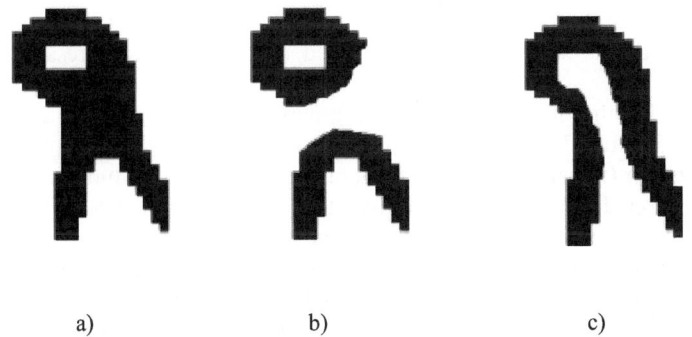

a)   b)   c)

Fig. 2. A fragment of Fig.1. a) An error; b) Two contours; c) One contour

Another type of error can be corrected automatically. It is known that contours are smooth. The main cause of failure to carry out this condition is an error in discretization. This error can be corrected with a smoothing algorithm after choosing a value of a control parameter. In this paper we will analyze only the second type of error.

The importance of the pre-processing goals listed above change as time goes by. Not long ago it was very important to compress data. We can read in the literature that this was one of the main goals of raster-to-vector conversion [3, chapter 8], but now the situation has changed. Progress in computer technology has resulted in huge increases of available computer memory and processing speed. This is why data compression is not as critical as it used to be. Progress in computer technology has also led to another very important result: we can now use very sophisticated algorithms for smoothing and generalization that were considered unrealizable not long ago.

2   Post-processing Algorithms

All existing algorithms that can be used for post-processing can be divided into three groups:

a) Deletion of some specific features of curves (for example, arms and narrow peninsulas from a coast line).
b) Deletion of some vertices of polylines.
c) Approximation of lines with some mathematical functions.

For a rigorous solution of generalization tasks, the best algorithms are those of the first group. Because these algorithms are essentially recognizing algorithms, they are very complex. In [4], the development of such algorithms is characterized as a "fantastically complex problem." Moreover, they have a very high computational complexity. The generalization of different objects may require different algorithms, as it is unlikely that the algorithm used for cost line generalization can be used to generalize roads, contours or building silhouettes. There is some progress in developing such algorithms, but at the beginning of the sixties, the difficulties of development and implementing such algorithms seemed so insuperable that the algorithms of the second group were used not only for compression but also for generalization.

Obviously, algorithms of the second group have to be used for data compression and straight-line defragmentation. But until recently, they were also used for generalization (in cartography) and smoothing. Many algorithms of this group [4-7] were developed. These algorithms result in a new polyline with vertices (critical points) that are a subset of vertices of the source polyline. Critical points are the points that determine the shape of the processed polyline. The distances of other vertices (noncritical points) from the resulting polyline have to be less than some threshold h, chosen by an operator. The algorithms of this group differ from each other in inessentials.

The author of [8] wrote that the Douglas-Peucker method is "mathematically superior" and the author of [9] suggests that "the generalizations produced by the Douglas algorithm were overwhelmingly – by 86 percent of all sample subjects – deemed the best perceptual representations of the original lines." But it is necessary to note that all these algorithms, including the Douglas-Peucker algorithm, have a common shortcoming: errors of vertices of resulting polylines are equal to errors of critical points of the source polylines, and therefore they cannot be reduced by decreasing the control parameter h. Because of this, resulting polylines are uneven, orientation of segments of resulting polylines depends on errors of the critical points' coordinates, and these algorithms cannot be used to filter random noise.

The lower the resolution of the image, the more noticeable the errors are. These shortcomings cause undesirable consequences when these algorithms are used for map generalization, defragmentation of straight lines, or noise suppression. Figure 3 shows a connected component and Figure 4 shows the borders of this connected component after processing with the Douglas-Peucker algorithm, with values of the threshold h = 1 and h = 0.75.

Fig. 3. Raster connected component

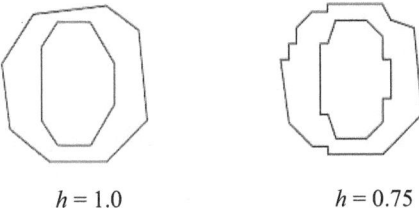

$h = 1.0$ $h = 0.75$

Fig. 4. Border of the connected component after compression with different values of a threshold ($h)$

We have already said that contours have to be smooth. The source image (Figure 1) contains a large amount of binarization noise. Comparing borders before and after compression, it is easy to conclude that it is easier to smooth source borders manually than to smooth compressed borders manually. The same conclusion may be drawn for straight lines. To define the correct direction of straight segments after compression is more difficult than to define direction of these segments before compression. So we can conclude that compression algorithms at least sometimes result in irreversible loss of information about a noise.

3   Smoothing and Accuracy

The most appropriate algorithms for suppressing a discretization noise (at least for contours) are those of the third group, which use mathematical functions to approximate the original polylines. They include algorithms that define straight lines and polynomials with the least square method and algorithms of local smoothing, which are especially important for cartography. An approximation can be done for a whole polyline (global approximation), or sequentially, for small pieces of polyline (the local approximation). In [4], several such methods are described and their shortcomings are listed:

a) The resulting curves are smoother than real objects.
b) These methods look for an average behavior of polylines and consequently corrupt the critical points that define a polyline shape.
c) Processed polylines contain more points than source polylines.

We have objections to all these conclusions:

a) Methods of this group use a control parameter, which defines a power of smoothing. An operator can choose a value for this parameter that gives the required power of smoothing. Figure 5 shows polylines obtained with the same algorithm of smoothing but with different values of the smoothing parameter (d = 3 and d = 5). Developed by A.Gribov, this algorithm is described in the Appendix.
b) There are different methods for looking for critical points [9], based on evaluation and analysis of a curvature and the first derivation of a curvature. If critical points are found, it is possible to split a curve at critical points before smoothing. Each of the new curves can be smoothed with fixed end points.
c) Let us try to understand why this algorithm can produce too many points. These algorithms can calculate smoothed curve points with any given step. Usually this step is constant, i.e. distances between any two consecutive points are almost equal. The appropriate number of points required for a good approximation of the curve with a polyline depends on the nature of the processing curve and is defined by the operator. Extra points can appear if the chosen step was too small or if curves contain long segments with a very small curvature. It is possible to delete extra points by increasing a step or by using compression algorithms. We explain in the next section why using smoothing and compression algorithms consecutively gives a better result than simply using the same compression algorithm.

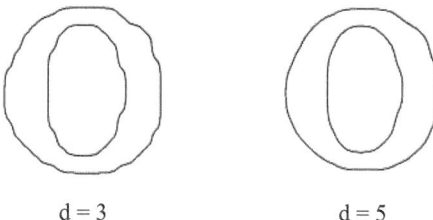

d = 3   d = 5

Fig. 5. Border of the connected component after smoothing with different values of a smoothing parameter (d)

4   Using Smoothing and Compression Consecutively

Consider one essential difference between the algorithms of the second (compression) and the third (approximation) groups. For the second group, we have already noticed that all vertices of the processed polyline are a subset of the vertices of the source polyline. That means that along with the vertices we inherit their errors so that these algorithms are unsuitable for noise suppression. Methods of the third group produce polylines with vertices that do not coincide with vertices of the source polyline; they can, therefore, be used for accuracy enhancement.

Assume that a smoothing algorithm produced points located on the original smooth curve without any errors. Let us plot a polyline using these points as vertices. If the vertices were calculated with a small step, the polyline would be a good approximation of the smooth curve. If we process this polyline with a compression algorithm, the vertices of the compressed polyline would be free of error. The maximum distance between the source and resulting polylines would be less than the threshold h, i.e. we could build the approximation with any given accuracy. Of course, smoothing algorithms produce the result with some error. But if the control parameter of smoothing is selected correctly, errors of the smoothed polylines will be much less than errors of the original polyline. Processing this smoothed polyline using a compression algorithm with small enough threshold h will not increase the error enough to matter.

Figure 6 shows the border of the connected component after smoothing with the algorithm described in the Appendix and with the parameter of smoothing $d = 5$, followed by processing with the Douglas-Peucker algorithm with the threshold h equal to 0.25 and 0.5, respectively. Comparing the polylines shown in Figures 6 and 4 visually, it is easy to arrive at the conclusion that accuracy increases if you use smoothing before compression. The polylines ($h = 0.5$) shown on Figure 6 (smoothing with compression) have only 5% more vertices than the polylines ($h = 0.75$) shown on Figure 4 (only compression).

d = 5, h = 0.25   d = 5, h = 0.50

Fig. 6. Border of the connected component after smoothing with the same value of a smoothing parameter (d) and compression with different thresholds (h)

5   Different Formats and Requirement for Memory

Fairly often, you can read that vector representation of maps, engineering drawings, and other line drawings is much more economical than raster representation, and that this is one of the main incentives for raster-to-vector conversion [3]. Maybe this is true for drawings consisting mainly of straight lines and circles, such as engineering drawings and electrical schematics, but it is not true for maps and other documents consisting mainly of irregular curved lines. To show this, we evaluated the memory required for storing the piece of a contour map shown in Figure 1 in both raster and vector formats.

The raster image of this fragment consists of 293 rows and 879 columns of 257547 pixels. Because raster-to-vector conversion systems usually process black and white images, only 1 bit is required for each pixel. So for storing the image in uncompressed raster format, approximately 32 Kbytes will be required. If the image will be converted into the run-length encoding (RLE) format, the required memory will decrease to 24 Kbytes.

The exact borders of all connected components shown on Figure 1 include 14169 points. To store information about each point will require 16 bytes (2 coordinates in the double format). So for the exact borders in the vector format, more than 220 Kbytes will be required, or almost 7 times more than for the raster format. A vector database saves not only coordinates but also other information about the geometry and topology of image features: area of polygons, length of perimeters of the polygons, number of holes, and so on. After smoothing (for the smoothing algorithm, see the Appendix) with the power of smoothing $d = 5$ and $d = 3$, the number of points increased to 26921 and 48174, respectively.

If, after smoothing with $d = 5$, the resulting image is processed with the Douglas-Peucker algorithm with parameter $h = 0.25$, then the number of points will be reduced to 5872, but all the same a vector representation of the map fragment will require much more memory (about 92 Kbytes) than the raster representation.

If the image consists of linear elements (as Figure 1), it can be represented in a vector format as a set of centerlines with their width. That will require almost two times less memory in comparison with borders. If the float or short format is used instead of the double format, the requested memory will be even less, but it will still be comparable with the memory required for the raster image.

The analyzed example allows one to conclude that memory compression cannot be the main reason for raster-to-vector conversion of line drawings (at least for maps). Obviously this transformation is performed for other, more important goals. That is why the importance of this problem was not reduced by the enhancement of computers and the tremendous increase in available computer memory.

6   Conclusion

The raw results of raster-to-vector conversion of linear drawings often require post-processing, usually by generalization, smoothing, and compression. Smoothing can be used to increase accuracy by depressing discretization noise. Increasing accuracy is especially important for processing images with a low resolution. Smoothing can be implemented with the algorithm of local averaging. The continuous local weighted averaging algorithm used in this paper is given in the Appendix. This algorithm has some advantages; particularly that it does not require preliminary densification and that it attains a worst-case running time of $O(n)$, where n is a number of vertices of a source polyline.

Compression algorithms do not increase the accuracy of the vectorization, but they can be used after smoothing for reducing the required memory with little diminution of accuracy.

A vector format of map representation may require more memory than a raster one, but raster-to-vector conversion is performed not for saving memory but for other more important goals. That is why the significance of vectorization was not reduced by the enhancement of computers and the tremendous increase of available computer memory.

References

1. Bodansky, E., Pilouk, M.: Using Local Deviations of Vectorization to Enhance the Performance of Raster-to-Vector Conversion Systems. IJDAR, Vol. 3, #2 (2000) 67-72.
2. Quek, F.: An algorithm for the rapid computation of boundaries of run-length encoded regions. Pattern Recognition, 33 (2000) 1637-1649.
3. Ablameyko, S., Pridmore, T.: Machine Interpretation of Line Drawing Images. (Technical Drawings, Maps, and Diagrams.) Springer (2000).
4. Douglas, D., Peucker, Th.: Algorithm for the reduction of the number of points required to represent a digitized line or its caricature. The Canadian Cartographer, Vol.10, #2 (1973) 112-122.
5. Ramer, U.: Extraction of Line Structures from Photographs of Curved Objects. Computer Graphics and Image Processing. Vol.4 (1975) 81 – 103.
6. Sklansky, J., Gonzalez, V.: Fast Polygonal Approximation of Digitized Curves. Pattern Recognition, Vol.12 (1980) 327-331.
7. Curozumi, Y., Davis, W.: Polygonal approximation by minimax method." Computer Graphics and Image Processing, Vol.19 (1982) 248-264.
8. R.B.McMaster Automated line generalization. Cartographica, Vol. 24, #2, 1987, pp. 74-111.
9. E.R.White. "Assessment of line-generalization algorithms using characteristic points." American Cartographer, Vol. 12, #1, 1985, pp. 17-27.
10. H.Asada, M.Brady. "The curvature primal sketch." IEEE Transactions on Pattern Analysis and Machine Intelligence, Vol.8, #1, 1986, pp. 2-14.

Appendix: Smoothing with Continuous Local Weighted Averaging

We analyze curves that are approximated with piecewise linear polylines. The smoothing algorithm is based on an approximation of a source curve with a mathematical function. A local averaging algorithm calculates points of a smoothed curve, taking into account only a few neighboring points of the source curve. In [10], coordinates of points of the smoothed polyline are calculated as the convolution of the coordinates of points of the source polyline and the Gaussian.

Our smoothing method differs from them by the convolution kernel. Instead of the Gaussian we used function $e^{\frac{|t-\tau|}{d}}$, $d > 0$, where (τ) is the length of a path along source polyline. It gives us the possibility to build the algorithm that attains a worst-case running time of $O(n)$, where n is a number of points of the smoothed polyline. Besides, a local approximation is performed with the second order polynomial

$$u_\tau(t) = \alpha_0^u(\tau) + \alpha_1^u(\tau) \cdot (t-\tau) + \alpha_2^u(\tau) \cdot (t-\tau)^2$$

This equation represents two equations, which can be obtained by substitution x or y for u.

Values of coordinates $x(\tau)$ and $y(\tau)$ are calculated as

$$x(\tau) = \alpha_0^x(\tau),$$
$$y(\tau) = \alpha_0^y(\tau),$$

where $\alpha_0^u(\tau)$ is the solution of the next equation

$$\sum_{i=0}^{2} c_{i+j}(\tau)\alpha_i^u(\tau) = b_j^u(\tau), j = \overline{0, N}, \text{ where } c_k(\tau) = \int_0^L e^{-\frac{|t-\tau|}{d}}(t-\tau)^k \, dt,$$

and

$$b_k^u(\tau) = \int_0^L e^{-\frac{|t-\tau|}{d}}(t-\tau)^k u(t) \, dt$$

A calculated curve has smooth first derivatives at calculated points.

To simplify the expressions, we assign $d = 1$, and the influence of this coefficient on the power of smoothing takes into account the value τ.

$$I_i(\tau) = \int_0^L e^{-|t-\tau|}(t-\tau)^i u(t) \, dt = I_i^L(\tau) + I_i^R(\tau),$$

where

$$I_i^L(\tau) = \int_0^\tau e^{t-\tau}(t-\tau)^i u(t) \, dt, \quad I_i^R(\tau) = \int_\tau^L e^{\tau-t}(t-\tau)^i u(t) \, dt$$

$$I_i^L(\tau_0) = I_i^L(0) = 0; \quad I_i^R(\tau_n) = I_i^R(L) = 0;$$

$$I_i^L(\tau_{k+1}) = \int_{\tau_{k+1}}^L e^{t-\tau_{k+1}}(t-\tau_{k+1})^i u(t) \, dt =$$

$$= e^{\tau_k - \tau_{k+1}} \sum_{j=0}^{i} \left\{ C_i^j(\tau_k - \tau_{k+1})^{i-j} I_j^L(\tau_k) \right\} +$$

$$\int_{\tau_k}^{\tau_{k+1}} e^{t-\tau_{k+1}}(t-\tau_{k+1})^i u(t) \, dt$$

(1)

$$I_i^R(\tau_{k-1}) = \int_{\tau_{k-1}}^{L} e^{\tau_{k-1}-t}(t-\tau_{k-1})^i u(t) dt =$$

$$= \int_{\tau_{k-1}}^{\tau_k} e^{\tau_{k-1}-t}(t-\tau_{k-1})^i u(t) dt +$$

$$e^{\tau_{k-1}-\tau_k} \sum_{j=0}^{i}\left\{C_i^j(\tau_k-\tau_{k-1})^{i-j} I_j^R(\tau_k)\right\}$$

(2)

If at first you calculate the integrals $\int_{\tau_k}^{\tau_{k+1}} e^{t-\tau_{k+1}}(t-\tau_{k+1})^i u(t) dt$ and

$\int_{\tau_{k-1}}^{\tau_k} e^{\tau_{k-1}-t}(t-\tau_{k-1})^i u(t) dt$, the integrals $I_i^L(\tau_k)$ and $I_i^R(\tau_k)$, where τ_k is a parameter

corresponding to the vertex of the source polyline, can be calculated with the recursive equations (1) and (2). Omitting $u(t)$ we receive formulas for calculating $c_k(\tau)$. To calculate $u_\tau(t), \tau \neq \tau_k \forall k$ it is possible to use the formulas written above, substituting τ_{k+1} and τ_{k-1} to τ in (1) and (2), respectively. This algorithm also gives good results for smoothing centerlines.

A Scale and Rotation Parameters Estimator Application to Technical Document Interpretation

Sebastien Adam[1], Jean-Marc Ogier[2], Claude Cariou[3], and Joel Gardes[4]

[1] Laboratory PSI, University of Rouen
76 821 Mont Saint Aignan, France
Sebastien.Adam@univ-rouen.fr
[2] Laboratory L3I, University of La Rochelle
17042 La Rochelle France
Jean-Marc.Ogier@univ-lr.fr
[3] LASTI, ENSSAT Lannion
6, rue Kerampont, BP 447, 22305 Lannion, France
Claude.Cariou@enssat.fr
[4] France Telecom R&D, DMI/GRI
2, avenue Pierre Marzin - 22307 Lannion, France
Joel.Gardes@rd.francetelecom.fr

Abstract. In this paper, we consider the general problem of technical document interpretation, applied to the documents of the French Telephonic Operator, *France Telecom*. At GREC'99, we presented a new set of features, based on the Fourier-Mellin transform (FMT), allowing a good classification of multi-oriented and multi-scaled pattern in comparison with classical approach. For this GREC'01, we propose the use of this set of features for the rotation and scale parameters estimation, through the use of the shift theorem of the Fourier transform. A comparison with a parameter estimation issued from Zernike moments is also given.

1   Introduction

The aim of the interpretation of technical maps is to make the production of digital documents easier by proposing a set of stages to transform the paper map into interpreted numerical storage [1,2,3,4]. An important step of this conversion process consists in recognizing characters (sometimes grouped in strings), which often appear on technical documents in several orientations and sizes. Such a constrained pattern recognition problem has been the object of intense and thorough studies. An excellent state of the art describing different possible approaches can be found in [5]. At GREC'99, we presented [6,7] a new pattern recognition tool, developed in the context of our application, based on the Fourier Mellin Transform. In this communication, we focus our attention on an original technique which allows scale and rotation parameters estimation. The application which is considered herein is the audtomatic analysis of documents issued from the prime French telephonic operator, *France Telecom*.

D. Blostein and Y.-B. Kwon (Eds.): GREC 2002, LNCS 2390, pp. 266–272, 2002.

The paper will be organized as follows. In the second section, we briefly describe our pattern description tool, based on a set of invariants derived from the Fourier-Mellin transform. In the third section, we explain how these properties can be exploited in order to obtain an estimation of rotation and scale parameters w.r.t. a reference pattern. Indeed, the orientation estimation is an important factor in order to reconstruct strings in a further step of recognition. Then, in the fourth section, we present some results obtained by using our tool, and compare them with another known approach, i.e. Zernike moments. Finally, in the fifth section, we conclude and define some potential perspectives to this work.

2   Segmentation Specialist

The strategy that we proposed in [6,7] for multi-oriented and multi-scaled pattern recognition is based on the use of the Fourier-Mellin transform (FMT) properties. In the following, we first recall the definition of the FMT, its analytic prolongation (AFMT) and a set of complete and stable similitude invariant features. Then, we recall properties of these invariants.

2.1   The Fourier-Mellin Transform (FMT)

Let $f(r,\theta)$ be a real-valued function (the pattern) expressed in polar coordinates. The FMT of this function is defined as the Fourier transform on the group of positive similitude, as:

$$M_f(v,q) = \int_{r=0}^{+\infty} \int_{\theta=0}^{2\pi} r^{-iv} \exp(-iq\theta) f(r,\theta) \frac{dr}{r} d\theta \tag{1}$$

$$\text{with } q \in \mathbb{Z}, \quad v \in \mathbb{R}.$$

In this expression, i is the imaginary unit. It is well-known that the Fourier-Mellin integral does not converge in the general case, but only under strong conditions for $f(r,\theta)$.

2.2   Analytic Prolongation of the Fourier-Mellin Transform (AFMT)

In order to alleviate the above difficulty, Ghorbel [8] has proposed the use of the AFMT, defined as:

$$\tilde{M}_f(v,q) = \int_{r=0}^{+\infty} \int_{\theta=0}^{2\pi} r^{-iv+\sigma_0} \exp(-iq\theta) f(r,\theta) \frac{dr}{r} d\theta \quad , \tag{2}$$

$$\text{with } q \in \mathbb{Z}, \quad v \in \mathbb{R}, \text{ and } \sigma_0 \in \mathbb{R}_+^*.$$

An important property of the AFMT (as well as the FMT) relies on the application of the shift theorem for the Fourier transform. Let $g(r,\theta) = f(\alpha r, \theta + \beta)$ be a scaled and rotated version of $f(r,\theta)$, then we have the following :

$$\tilde{M}_g(v,q) = \alpha^{-\sigma_0 + iv} \exp(iq\beta)\, \tilde{M}_f(v,q) \tag{3}$$

Taking the modulus of both terms in Eq. (3) yields features which are invariant under any rotation of the pattern but not under scaling. Moreover, invariants of this type do not have the completeness property, i.e. there is no bijection between the dual representations of a single pattern, since the phase information is dropped. In [8], the following set of rotation and scale invariant features was proposed :

$$I_f(v,q) = \tilde{M}_f(v,q) \left[\tilde{M}_f(0,0)\right]^{-1+i\frac{v}{\sigma_0}} \left[\tilde{M}_f(0,1)\right]^{-q} \left|\tilde{M}_f(0,1)\right|^q \tag{4}$$

Then, if $g(r,\theta) = f(\alpha r, \theta + \beta)$, it can be easily shown that $I_f(v,q) = I_g(v,q)$, thus showing the invariance of the set of FMT descriptors under change of scaling or rotation. Recognition results obtained through the use of such invariants are not the main subject of this paper. Nevertheless, they are precisely described in [6].

3   Exploitation of Properties for Scale and Rotation Parameters Estimation

In this part we introduce some theoretical aspects and give results concerning the estimation of rotation and scale parameters with the AFMT. Indeed, an important feature of the FMT is the possibility to determine orientation and scale of shapes from the set of descriptors extracted, through a comparison with extracted descriptors from reference shapes. Indeed, from Eq. 3 and through the knowledge of a reference shape, it is possible to obtain values of α and β (α being the scale factor and β the angle between the unknown shape and a corresponding reference shape). In order to compute these similitude parameters (also called *movement parameters*), we compute the Euclidean distance between the two objects in the Fourier Mellin space. If f and g are belong to $L^1(G, d\mu_G) \cap L^2(G, d\mu_G)$, the distance between them in the FM space can be defined as :

$$d_2(f,g) = \sqrt{\sum_{k\in\mathbb{Z}} \int_{-\infty}^{+\infty} \left|\tilde{M}_g(v,q) - \tilde{M}_f(v,q)\right|^2 dv} \tag{5}$$

Squaring Eq. (5) yields the square error between f and g in the FM space, which for identical patterns is a function of movement parameters :

$$E_{fg}(\alpha,\beta) = \sum_{k\in\mathbb{Z}} \int_{-\infty}^{+\infty} \left|\tilde{M}_g(v,q) - \alpha^{-\sigma_0+iv}\exp(iq\beta)\,\tilde{M}_f(v,q)\right|^2 dv \tag{6}$$

Thus movement parameters can be extracted by minimizing E_{fg}, by computing :

$$(\alpha^*,\beta^*) = \underset{(\alpha,\beta)}{Arg\ min}\, E_{fg}\ ;\ \alpha\in\mathbb{R}_+^*,\ \beta\in[0,2\pi[\tag{7}$$

Nevertheless, the calculation of the AFMT for a finite set of (v,q) values renders the above criterion non convex: several *local minima* exist. So an optimization method where gradient information are not necessary is to be used. We have chosen Nelder-Mead's algorithm [9] in order to perform this optimization, and we have also compared its performance with a genetic algorithm (GA) approach [10]. Note that the number of local minima of the criterion generally equals q_{max}, the maximum order of q which is chosen in the computation of the AFMT. In Fig. 1, the global minimum of the function E_{fg} is obtained for correct values of (α, β). In this example, the criterion is computed between a reference shape (letter "A") and another "A" with the same size, but rotated by a 100 degrees angle. Indeed, the estimation of movement parameters using 33 invariant features(explanations concerning this value are given in [6]) yields $\alpha^* \approx 1$ and $\beta^* \approx 100°$ which are correct results. Also, in this example, we have chosen $q_{max} = 3$, and yet 3 local minima can be observed for E_{fg} in Fig. 1.

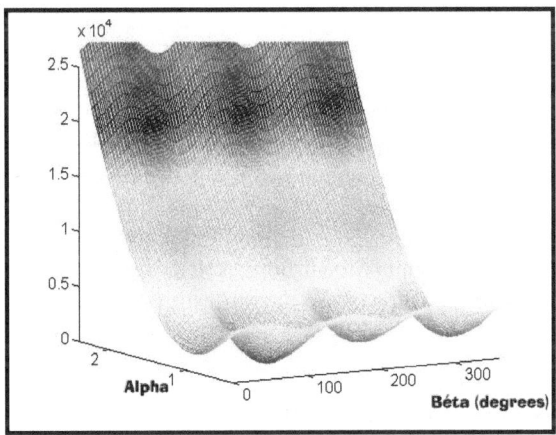

Fig. 1. Square error function calculated between a reference pattern (letter "A") and another "A" with the same

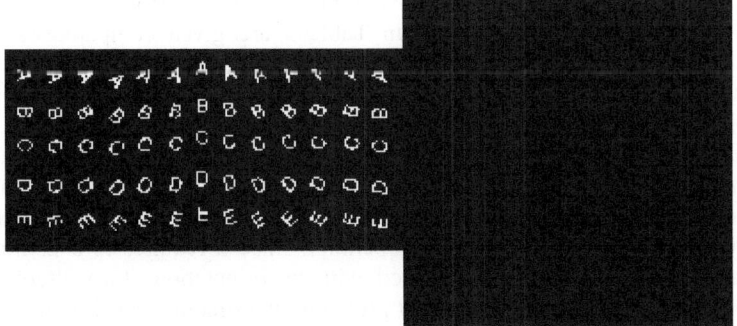

Fig. 2. Left Part : Set of shapes used for the orientation estimation. Characters of the first column are used as reference shapes. Right Part : Estimation of the rotation angle in radians of the shapes from column 2 to 13 in comparison with the reference shapes of the first column

In the right part of Fig. 2, we show the results obtained with this technique for the estimation of the orientation on the different patterns presented on the left part , by using the patterns in the leftmost column as the reference ones. 12 clusters of points can be visually identified, each of them corresponding to the common orientations of the 5 patterns, while actually each one is present in 13 different orientations. Clusters existing in the 3rd and 4th quadrants are due to recognition errors.

Table 1. Comparison of two optimization procedures for scale and orientation estimation (α/β) on a small database

Character	True orientation	Nelder-Mead	GA
E	0	1 / 9.8 E-6	0.99 / 359
A	0	1 / -1.26 E-5	0.95 / 1.47
C	0	1 / -2.77 E-7	1.00 / 2.72
E	15	0.49 / 12.15	0.49 / 12.72
A	45	1.01 / 46.56	1.01 / 47.23
A	60	0.99 / 61.40	1.02 / 63.83
C	90	0.98 / 93.74	0.98 / 93.32
A	90	1.01 / 90.74	1.01 / 90.78

Table 1 gives a comparison between the two tested optimization methods on a sample of analyzed patterns. As can be seen, results are quite the same in both cases. However, the choice of the Nelder-Mead algorithm is to be preferred, because the GA approach implies a high computational burden. It is important to note that identical tests have been performed for scale factor estimation and that the results are essentially the same.

4   Comparison of Two Rotation Estimators

Kim and Kim have proposed, in [11], a method providing an orientation estimator based on Zernike moments (combination of regular moments). In order to validate our approach, we have compared this method with our approach based on the AFMT. The results obtained with this method are also quite good since error on the estimation of the orientation never exceeds 10%. In Table 2 are given compared estimates of orientation between the two different approaches for the analysis of a character "A". Actually, results are quite similar in most cases, although a slightly better mean error is obtained with the FM approach. Our final objective with this estimation being to reconstruct strings of characters, errors (which are always less than 10 %) are small enough regarding our constraints. Obviously, further experiments must absolutely be led in order to validate the approach. The validation of this kind of technique on a consequent data base is a very heavy operation since it requires to constitute a data base for which each sample is labeled with its orientation. Two alternatives are possible when dealing with this kind of problem : the first one consists in constituting a synthetic data base, for which the data are rotated automatically. This approach is being led in our laboratory. The second consists in working on a real data set and in labeling each sample manually. This approach is the most reliable technique for the validation of the approach, but raises the problem of the assignment of a precise orientation to each shape in the labeling stage. As far as we know, on a real data set, there is no reference dealing with this kind of problem. In this domain, our current works deal with an estimation based on the fusion of different contextual information.

Table 2. Comparison of Orientation Estimators

True orientation	FM estimate	Zernike estimate
0	0	0
15	16	16
30	30	23
45	47	48
60	61	61
75	78	76
90	91	88
105	105	107
120	118	122
135	136	137
150	148	153
165	164	167
180	175	175
Mean error	1,46	2,38

5   Conclusion

In this submission, we have proposed the continuation of our studies concerning the recognition of multi-oriented and multi-scaled patterns. Since recognition of isolated patterns using this methodology have already been presented at the preceding GREC, we have focused here on a particular innovating point concerning the estimation of character orientation. Indeed, on technical documents, "consistent" strings must have the same orientation and the same size. Since results obtained through the use of the presented tools are quite good, we are currently working at the development of a system approach in order to validate the obtained estimation of orientation. Indeed, as an example, a set of aligned characters provides some information concerning a string of characters, and as a consequence, can validate the final result. Furthermore, all these aspects concerning string and symbol recognition constitute an interesting alternative to the growing needs concerning document indexing and content based information retrieval. Indeed, textual and symbolic information are among the most important cues for document indexing.

References

1. L. Boatto . et al., An interpretation system for land register maps, IEEE Computer Magazine Vol. 25, pp. 25-33, 1992.
2. S.H. Joseph, P. Pridmore, Knowledge-directed interpretation of line drawing images, *IEEE Trans. on PAMI* **14**(9), pp. 928-940, 1992.
3. J.-.M. Ogier, R. Mullot, J. Labiche, Y. Lecourtier, Multilevel approach and distributed consistency for technical map interpretation: application to cadastral maps, *Computer Vision and Image Understanding (CVIU)* **70**, pp. 438-451, 1998.

4.   P. Vaxivière, K. Tombre, CELESTIN: CAD conversion of mechanical drawings, *IEEE Computer Magazine* **25**, pp. 46-54, 1992.
5.   O.D. Trier, A. K. Jain, T. Taxt, Features extraction methods for character recognition – a survey, *Pattern Recognition* **29**, pp. 641-662, 1996.
6.   S. Adam, J.M.Ogier, C. Cariou, R. Mullot, J. Gardes, Y. Lecourtier, Combination of Invariant Pattern Recognition primitives on Technical Documents, *Proceedings of GREC'99*, Bangalore, India, pp 203-210, 1999.
7.   S. Adam, R.Mullot, J.M.Ogier, C. Cariou, J. Gardes, Y. Lecourtier, Processing of the Connected Shapes in Raster-to-Vector Conversion Process, *Proceedings of GREC'99*, Bangalore, India, pp 38-45, 1999.
8.   F. Ghorbel, A complete invariant description for gray level images by the harmonic analysis approach, *Pattern Recognition Letters* **15**, pp. 1043-1051, 1994.
9.   J. A. Nelder and R. Mead, A simplex method for function minimization, *Computer Journal* **7**, pp. 308-313, 1965.
10. D. Zongker and A. Jain, Algorithms for Feature Selection : An Evaluation, Proceedings of International Conference on Pattern Recognition, **Vol II**, pp 18-22, 1996.
11. W.-Y. Kim and Y.-S. Kim, Robust rotation angle estimator, *IEEE Trans. on PAMI* **21**(8), pp. 768-773, 1999.

Improving the Accuracy of Skeleton-Based Vectorization

Xavier Hilaire[1,2] and Karl Tombre[1]

[1] LORIA
B.P. 239, 54506 Vandœuvre-lès-Nancy, France
{hilaire,tombre}@loria.fr
[2] FS2i
8 impasse de Toulouse, 78000 Versailles, France

Abstract. In this paper, we present a method for correcting a skeleton-based vectorization. The method robustly segments the skeleton of an image into basic features, and uses these features to reconstruct analytically all the junctions. It corrects some of the topological errors usually brought by polygonal approximation methods, and improves the precision of the junction points detection.

We first give some reminders on vectorization and explain what a good vectorization is supposed to be. We also explain the advantages and drawbacks of using skeletons. We then explain in detail our correction method, and show results on cases known to be problematic.

1 Introduction

Vectorization is the process which automatically converts a raster image to a set of graphical primitives. These primitives are assumed to be those which made up the drawing when it was drafted; therefore, vectorization can be considered as some kind of "reverse engineering" process.

An "ideal" vectorization system should therefore be able to yield vectorial data as close to the "original" as possible, in number as well as in position. Unfortunately, this is not always the case, for three main reasons:

1. the images are often disturbed (printing defects, folds, smears, etc.) and noisy (scanning as well as binarization noise), and have sometimes visible distortions due to the mechanical scanning process, or to skew;
2. there is not necessarily a unique vectorial solution for the input data: two different sets of vectors may generate the same set of pixels;
3. all known vectorization processes have some imperfections of their own, which means that errors are introduced by the process.

Different kinds of documents may be drawn using different kinds of primitives, from straight line segments to β-splines. Our purpose is to design a correct vectorization method for architectural drawings. Those are usually drawn using straight line segments and circular arcs, therefore we will limit ourselves to these two primitives in the rest of the paper.

D. Blostein and Y.-B. Kwon (Eds.): GREC 2002, LNCS 2390, pp. 273–288, 2002.

At GREC'99, we discussed some qualitative elements which should be taken into account when choosing the different steps of one's vectorization method [ ]. We emphasized the problem of being able to position both the line segments and the junctions between them with sufficient precision. We pointed out that despite its weaknesses, we still consider *skeletonization* of a binary image to be the best compromise; indeed, it is a widely used technique for vectorization.

For the topology, the number of detected junction points in such a case may be too low (Fig. 1,b1) or too high (Fig. 1,b2).

There may also be numerical distortions in the areas where the geometrical features cover each other; one consequence is that the detected junction points are not located where they are expected to be. In the two most common cases, those of L and T junctions, the distortion is parabolic, and it is easy to demon-

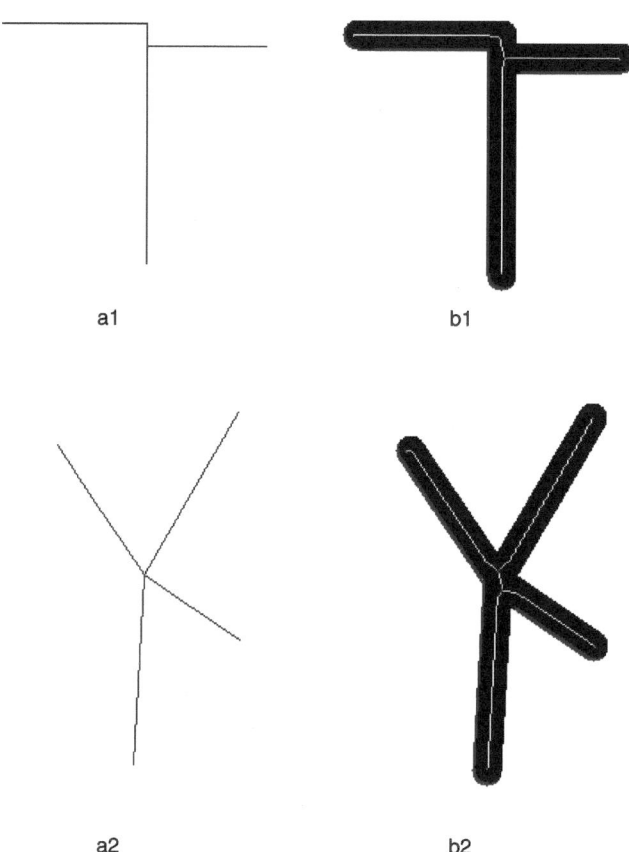

a1   b1

a2   b2

Fig. 1. Topological errors of the skeletonization. a1, a2: original vectors (without their thickness); b1, b2: skeletons of bitmaps obtained from these vectors (with their thickness)

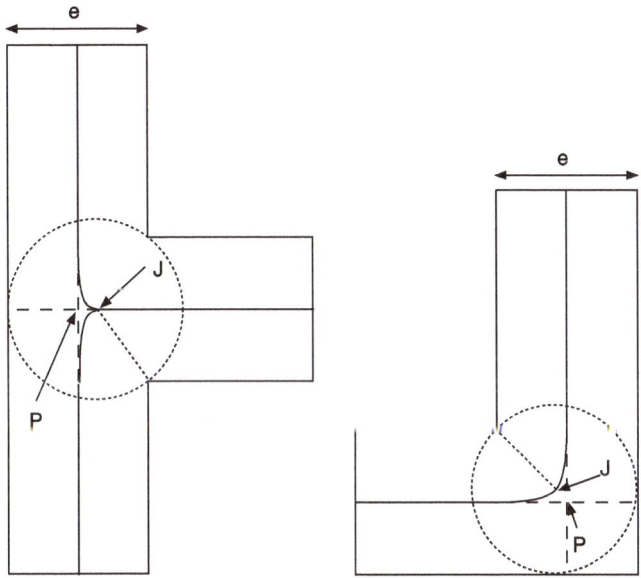

Fig. 2. Distortions introduced by skeletonization. Expected points: P. Points yielded by skeletonization: J

strate (Fig. 2) that the distance between the actual position and the expected position increases with the thickness e of the line: $d(P, J) = \frac{1}{8}e$ for a T junction, and $d(P, J) = \left(\frac{3\sqrt{2}}{2} - 2 \right) e$ for a L junction.

2   Our Method

The points raised in the previous section have led us to propose a vectorization method which is still based on the skeleton, but which is able to correct the errors introduced by polygonal approximation and to improve the precision of the junction points. In summary, our method has the following steps:

1. *Layer separation*—Separate the graphical drawing into homogeneous thickness layers.
2. *Skeletonization*—Compute the skeleton of the binary image.
3. *Segmentation*—Segment the skeleton into straight segments and circular arcs, and eliminate the geometric primitives which are shorter than the thickness e of the processed layer. Build a graph $\mathcal{G}$ whose edges are the primitives and whose nodes e_i are the connection points between them.
4. *Instantiation*—Define the support equation for each primitive, through linear regression, with the associated uncertainty domain.
5. *Topological and numerical corrections*—For all nodes e_i of $\mathcal{G}$ do
   (a) Build the set E_0 of primitives stemming from e_i

(b) If there is another node than e_i in $\mathcal{G}$
- Let e_j be the node closest to e_i (Euclidean distance)
- Add to E_0 all the primitives stemming from e_j

(c) Build the successive subsets $E_1, E_2, ..., E_{n-1}$, using the following algorithm, starting with $u = 0$:
  i. Construct F as the biggest subset of E_u for which all primitives do intersect. Let e'_u be the intersection node.
  ii. If $F \neq E_u$ then
    - Set E_{u+1} to $E_u \smallsetminus F$
    - Set E_u to F
    - Increment u and loop at 5(c)i

(d) For $u = 0, \ldots, n - 1$, replace the extremity of each primitive of E_u stemming from e_i or e_j by e'_u

(e) Connect all the e'_u

6. Repeat step 5 as long as the global topology is modified by the process.

The basic point is that the skeleton is a good descriptor of the median axis for elongated and isolated shapes, but is a poor descriptor for their intersections. The idea is therefore not to take into consideration those parts of the skeleton which are included in the overlapping between two or more shapes. Once these parts are eliminated, we perform an *analytical* reconstruction of each junction from the remaining parts of the skeleton.

We will now describe each of these steps in detail.

2.1   Layer Separation

This is an important step for the method we propose, as we need to work on image layers with lines of homogeneous thickness. Simple mathematical morphology methods can be used to get these different layers, as we have explained in previous papers [6].

2.2   Skeletonization

As we explained in detail at the last GREC, we use Sanniti di Baja's skeletonization algorithm, based on the 3–4 distance transform, to compute the skeleton [1]. The extracted skeleton is chained using the algorithm described in [7].

2.3   Segmentation

The first difference with the method we have used until now in our group [7] is that we do not perform a polygonal approximation immediately after the skeletonization. The various polygonal approximation methods can of course convert a chain of pixels to a curve in a very efficient way, but they disturb the analytical junction reconstruction algorithm we propose. The method proposed by Rosin and West [5] yields an excellent approximation of the original curve, but is very sensitive to noise and to the choice of starting points, as illustrated

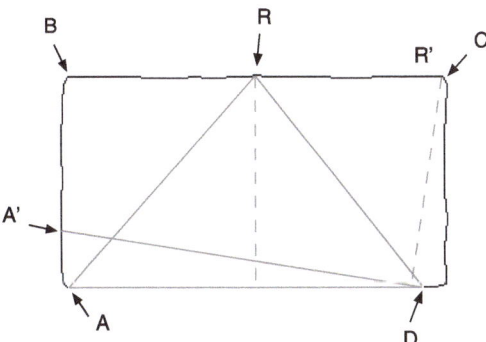

Fig. 3. Sensitivity to noise of the cutting point in the Rosin and West approximation

by Fig. 3: if the starting points are A and D, the splitting phase introduces a cutting point R which leads to splitting $[BC]$ into two segments. If A is displaced towards A', the cutting point R becomes close to a point R' which is neighbor of C, and the decomposition of $[BC]$ is thus much better, but a new problem may be introduced on $[CD]$.

The method proposed by Wall and Danielsson [9] does not have this drawback, but it is not very scalable, which makes circular arcs detection, among others, difficult: for a circle, the angle α from the center of the first cutting point, with respect to the starting point, is dependent on the radius R of the circle, as the polygonal approximation uses a fixed threshold K, and we have the relation $R(1 - \frac{\sin \alpha}{2\alpha}) = K$. Thus, a small circle will be approximated by fewer segments than a bigger circle, and vector-based circle detection methods, such as the one developed in our group [2], will be less successful.

We therefore segment the skeleton using another method. First, we note that the anchor points[1] are good cutting points for separating primitives, as they correspond to their meeting points. These anchor points are located in the zones where the skeleton is "distorted", and not in the "useful" parts of the skeleton, which define the primitives. We can therefore make an initial segmentation of the skeleton at these anchor points, and thereafter only process one branch at a time.

Each branch is then segmented using a robust sampling method close to RANSAC [3]. The segmentation algorithm has the following steps:

1. Select the best candidate segment $\mathcal{S}^\star$
2. Select the best candidate circular arc $\mathcal{C}^\star$
3. Extract the best of these two primitives $\mathcal{S}^\star$ and $\mathcal{C}^\star$ from the chain of points
4. If the remaining length of the chain is smaller than e, end the process, else go back to step 1

Let us now give some details for these steps.

---

[1] Meeting points for at least three branches of the skeleton.

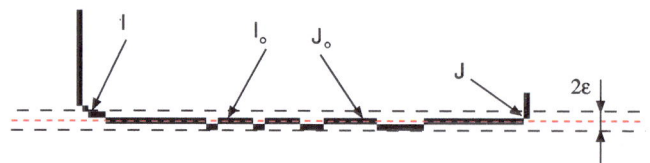

Fig. 4. Segmentation through sampling

Selection of the Best Candidate Segment We first choose two random points I_0 and J_0 from the chain to be processed (Fig. 4), and we instantiate a line model $\mathcal{S}$ passing through these points. We then enlarge the domain $[I_0, J_0]$ by "decrementing" I_0 and "incrementing" J_0 as long as these points verify $d(I_0, \mathcal{S}) \leq \varepsilon$ and $d(J_0, \mathcal{S}) \leq \varepsilon$, where ε is a tolerance constant on outliers that we set close to 1. This gives a new segment $[I, J]$, which is the longest segment containing in the initial $[I_0, J_0]$ segment without having more outliers.

This step is repeated k times, and we keep the model $\mathcal{S}^\star$ for which the number of points contained in $[I, J]$ is maximum. The number of trials k can be estimated so that the probability p that at least one of the couples of initial points (I_0, J_0) is good, is greater than a given value τ. A couple of initial points (I_0, J_0) is considered as good when I_0 and J_0 are simultaneously inlier points and belong to the same primitive. If L is the length (in number of points) of the chain to be segmented, e is the length (in number of points) of the smallest primitive which can be extracted from the chain, and α is the ratio total number of inlier points / L, the worst case (when all primitives are of length e), is:

$$p = 1 - \left(1 - \alpha^2 \frac{e}{L}\right)^k$$

We can therefore choose k such that $p \geq \tau$, which gives

$$k \geq \frac{\log(1 - \tau)}{\log\left(1 - \alpha^2 \frac{e}{L}\right)} \tag{1}$$

Selection of the Best Candidate Circular Arc The method is similar to the previous one, but we must now choose three points instead of two, and the expressions for p and k change:

$$p = 1 - \left(1 - \alpha^3 \left(\frac{e}{L}\right)^2\right)^k$$

$$k \geq \frac{\log(1 - \tau)}{\log\left(1 - \alpha^3 \left(\frac{e}{L}\right)^2\right)} \tag{2}$$

Segmentation Graph At the end of this segmentation step, we build a topological representation of the segmentation, with a structure close to a planar map. We build a graph $\mathcal{G}$ whose nodes are the connection points of the primitives, and whose edges are the primitives themselves. Each edge contains two pointers to the nodes forming its extremities, and a pointer to the set of points associated with the primitive. Each node contains pointers to the list of primitives to which they are connected, and the label of the connected component in which it is to be found.

All this information is used in the junction reconstruction phase, as explained in the next section. A first transformation of graph $\mathcal{G}$ is performed to eliminate all chain cycles whose total length is smaller than e, as suggested in [1]. When an edge of $\mathcal{G}$ disappears, the graph is updated accordingly.

This correction may still not be sufficient to represent the drawing, as the initial topology is computed from the skeleton, which is not necessarily correct, as we have seen in Fig. 1. Therefore, additional corrections are needed. We must also compute the actual position of each node. This is the objective of the following steps.

2.4   Instantiation of the Primitives

This step aims at defining the support equations of the primitives, and their associated uncertainty domains, and from there, to compute the initial positions to give to each node. The instantiation consists in finding the support equation $F(x,y) = 0$ for each primitive. In our case, this may be line equations $F_d(x,y) = x\cos\theta - y\sin\theta + C = 0$ or circle equations $F_c(x,y) = (x-\alpha)^2 + (y-\beta)^2 - r^2 = 0$. These equations can be obtained:

- From the I_0, J_0 pair of points of the best model found during the sampling step,
- or by another robust technique, or through linear regression (e.g. algebraic least squares).

Even if it has the potential to yield a good final precision, linear regression must be handled with care: unknown shapes, such as quadrics, will usually be vectorized as chains of small segments, and linear regression may introduce large distortions in the intersections of all these segments. When a recognition step for such shapes is foreseen in a later phase, linear regression must therefore be avoided.

To each primitive p_i with its equation $F_i(x,y) = 0$, we also associate an uncertainty domain $\Delta(p_i)$, which is the space delineated by the two curves distant of ε from $F_i(x,y) = 0$, on both sides. This domain is used to solve possible topological ambiguities which may appear during the last correction step, which we will now describe.

2.5   Topological and Numerical Correction

This last step consists in changing the topology and the estimated positions of the junction points. These two changes are performed simultaneously. We will

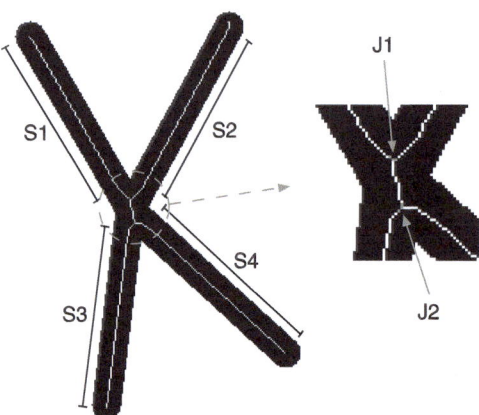

Fig. 5. Intersection of $N \geq 3$ primitives

describe the method by distinguishing the general case, where $N \geq 3$ primitives intersect, and the specific case of the intersection of $N = 2$ primitives.

General Case: $N \geq 3$ We illustrate this with Fig. 5, which corresponds to $N \geq 3$ primitives. Here, the skeletonization yields two distinct junction points J_1 and J_2, which are represented in the graph by nodes e_1 and e_2. This is not correct, as the 4 primitives have been constructed to intersect at a unique point. The correction we propose relies on two remarks. First, if the position of junction points J_1 and J_2 are obtained by skeletonization with a poor precision, the *analytical* computation of their position, through the segment pairs to which they are connected, will yield two points J_1' and J_2' with a better precision. Then, if J_1' and J_2' are close enough, we can suppose that the primitives they connect do also intersect, whereas they must be considered as non intersecting if $d(J_1', J_2') > e$. The reader should note that this is only a hypothesis–we have no formal proof that this is always true, but practice has shown it is reasonable to assume so.

These two observations lead us to the following correction method. We first look for the two closest nodes e_i and e_j of $\mathcal{G}$, belonging to the same connected component, and we build the set E_0 of primitives connected to these nodes, except the primitive (e_i, e_j)[2]. We include the condition that they must belong to the same connected component of the image, so that we avoid connecting primitives that were disconnected in the original drawing. The set E_0 simply represents a list of primitives which *may* intersect at one or more points. We must then find these points.

---

[2] This can be extended without loss of generality to the case when $\mathcal{G}$ contains only one node e_i. The primitives stemming from e_i are grouped in the same way and we just write $e_j = e_i$.

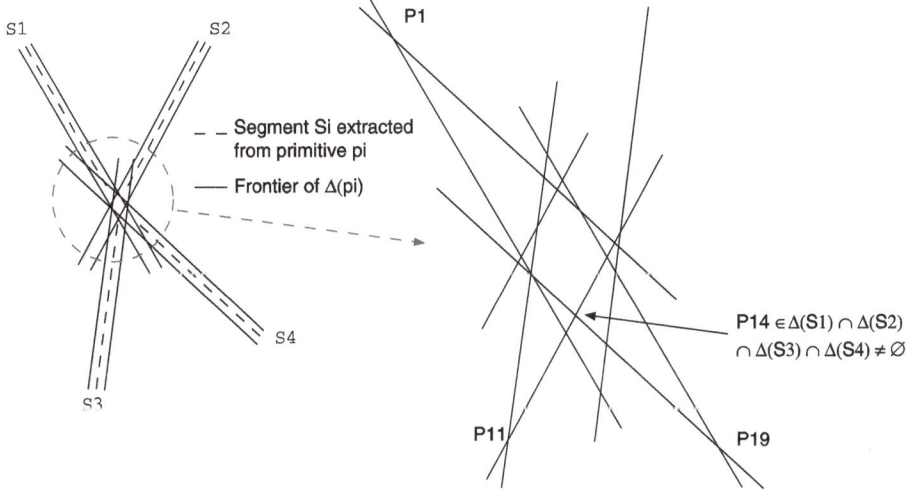

Fig. 6. Intersection of uncertainty domains

We therefore construct, from E_0, the set P_0 of intersection points for the borders of the uncertainty domains $\Delta(p)$ associated with each primitive p of E_0, as illustrated by Fig. 6.

Then, we check for each intersection point $X \in P_0$ that it is inside the uncertainty domain $\Delta(p)$ of each primitive $p \in E_0$. This check can be easily performed by evaluating the sign of $F_{sup}(X).F_{inf}(X)$, where F_{sup} and F_{inf} are the equations of the borders of $\Delta(p)$. When X belongs to a given domain $\Delta(p)$, we consider X as being also a point of p. Then, to each candidate point $X \in P_0$, we associate the set $\mathcal{U}$ of primitives to which p does belong, and we keep the point M which has the largest associated set $\mathcal{U}$. The fact that M exists implies that the primitives of $\mathcal{U}$ do intersect, so we can define a new set $E_1 = p \in E_O : \Delta(p) \notin \mathcal{U}$ and change the value of E_0: $E_0 - E_0 \smallsetminus E_1$. The procedure is then repeated with E_1 and so on, until there are no more primitives to group.

At the end of this step, we have n groups of primitives $E_i, i = 0, ..., n-1$, each group containing a variable number of intersecting primitives. We then consider separately the cases $|E_i| = 1$ and $|E_i| \geq 2$ for the following steps.

Case a: $|E_i| = 1$. This is a special case, where a single isolated primitive does not intersect with any other, despite the estimation which had been implied by the skeleton information. Although it is exceptional, this may happen, as illustrated by Fig. 7, where the primitive pairs (P_1, P_2) and (P_3, P_4) cannot intersect *by construction*. As the pair (P_3, P_4) has no intersection either, these two primitives are isolated. From a topological point of view, these isolated primitives are left unchanged. This means that their common node (e_i or e_j in the present case) is kept as it was. In the example given by Fig. 7, the two extremities of P_3 and P_4 are merged and represent the same node. The resulting

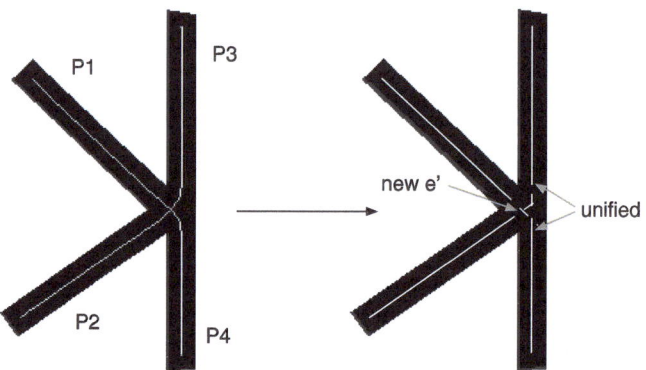

Fig. 7. Isolated primitives after computing the sets E_i

junction is formed by $n = 2$ primitives, which constitutes a particular case described in § 2.

Case b: $|E_i| \geq 2$. This is the general case, where the intersection of the uncertainty domains is non empty, so that there is a junction point M inside this intersection. We then propose to compute the position of M through the weighted orthogonal least squares method. For each set E_i of primitives $p_1, ...,$ we compute the point M which minimizes the sum

$$S = \sum_{j=1}^{|E_i|} \omega_j d^2(M, p_j)$$

where $d(M, p_j)$ is the distance from M to primitive p_j, and where each ω_j is set to be the length—in number of pixels—of primitive p_j. This weighting gives a better confidence to the final estimation, as the least squares error induced by a displacement of M, even a small one, is proportional to the length of the reference primitive.

This minimization problem is linear as long as the p_j are lines; thus, its solution can be computed analytically. The linearity is lost as soon as there is a circular arc, but the solution remains analytical as long as E_i contains only two primitives. With three or more primitives, of which one at least is a circular arc, we must use a numerical method, as there is no general analytical solution anymore. We do not yet take into account the case where $|E_i| \geq 3$, as a numerical method may end up finding a local minimum, and there are probably many such minima inside the intersection domain. Also, we have until now never met this case in the architectural drawings we have processed, as circular arcs usually appear in the thin-line layer, where they are associated with simple segments to represent doors, windows, etc. But in a more general framework, we must probably take this problem into account...

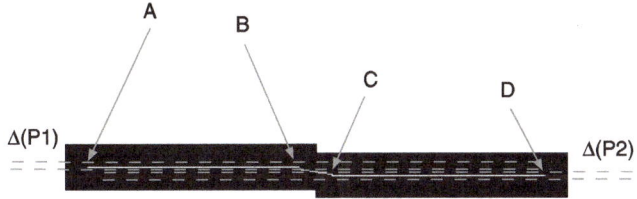

Fig. 8. Primitives without apparent intersection

Once the intersection point M has been determined, the node to which it corresponds is given M as its new geometrical position. The supports of the primitives stemming from this node are then re-instantiated using the following rules:

1. the position of the other extremity remains unchanged,
2. the supports of the segments are re-instantiated by the line passing through the two extremities,
3. the supports of the circular arcs are re-instantiated by the circle passing through the two extremities, with the same radius, and with the center located in the same half-plane as in the previous model[3].

Special Case $N = 2$ This case may generate difficult ambiguities, if the supports of the primitives are parallel, depending on the kind of primitives. We process separately three types of possible junctions: segment–segment, segment–arc, arc–arc.

Case a: segment–segment. The difficult case appears when the two supports are strictly parallel, while being topologically connected. This is illustrated by Fig. 8, where the borders of $\Delta(P_1)$ and $\Delta(P_2)$ have no point in common, whereas the primitives they define are topologically connected (the BC chain is removed as its length is smaller than the line thickness). The problem is to find a way to vectorize this part of the graphics.

There may also be a numerical problem, when the value of the angle formed by two segments is sufficiently close from π to put the intersection points of the borders outside of the acceptable zone (Fig. 9).

We propose the following solution to this problem: using arbitrary-precision computation, we determine if $[AB]$ and $[CD]$ define a computable intersection point J. If this is the case, the validity of J is given by the sign of $(\boldsymbol{AB}.\boldsymbol{BJ}).(\boldsymbol{DC}.\boldsymbol{CJ})$, which must be equal or greater than zero. If J is not computable, we disconnect the segments, and leave them as they are.

---

[3] We are aware that this way or re-instantiating circular arcs supports is somewhat arbitrary. In fact, we assume that the displacement induced by the new M remains small. It would certainly be more general and robust to determine the position of M by trying to minimize the error of a model with N branches stemming from M.

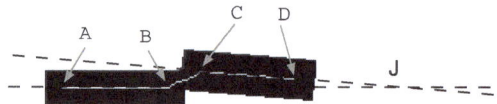

Fig. 9. Combined effects of shift and small angle

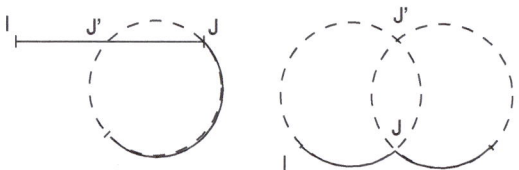

Fig. 10. False junction points

Case b: segment–arc and arc–arc. There are no special robustness problems with these two cases, except the fact that the intersection of the primitive supports may create a false junction point J', in addition to J, as illustrated by Fig. 10. A simple comparison of the lengths of the arcs IJ and IJ' is actually sufficient to eliminate this false point.

2.6   Connecting the e'_u Nodes

This is probably the most critical part of the process, as we only have two pieces of information on the nodes e'_u which have been created. First, we know that they are located in the intersection zone of a set $\mathcal{E}$ of primitives which are topologically connected, according to the skeleton. Secondly, we know that they correspond to intersections of subsets of $\mathcal{E}$, according to the uncertainty computation. We have no other *a priori* information which may help us in deciding whether these points should be connected or not.

As a first try, we propose to connect these points pairwise, starting with the closest pair, and ending with the pair of points whose distance is the largest.

3   Results

3.1   Usual Cases (Real Data)

In Fig. 11, we report typical results obtained on fragments of architectural drawings that have been scanned and binarized. The method easily corrects the usual distortions on L and T junctions. The extraction of the arc and the computation of its intersection with the segment is also correct in case c). Case d), on the contrary, is more difficult. First, the result of vectorization depends on the value given to the constant ε during processing. With $\varepsilon = 1.2$, detail A shows that

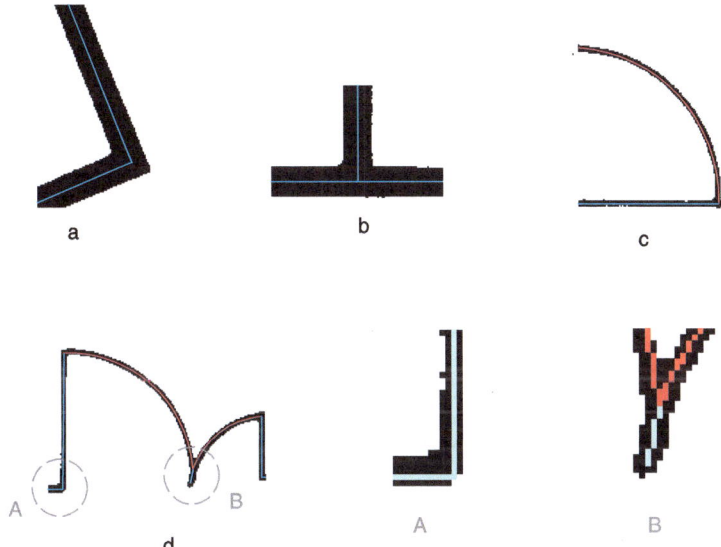

Fig. 11. Corrections made in the most frequent cases

2 segments are extracted. When ε comes closer to 1, a circular arc is inserted between the two segments; actually, the binary image may be interpreted in both ways. Secondly, there is still a problem at the intersection between the two arcs. Here, the left arc is intercepted by the right arc, and is thus truncated. The skeleton contains a local anchor point connecting the two chains which correspond to the arcs, and a third chain which prolongates the left arc. The latter chain is interpreted as being a segment, and cannot be removed, as its length is larger than the line thickness. This defect is not corrected by our method.

3.2   Difficult Cases (Synthetic Data)

Fig. 12 shows other results, obtained on synthetic data. The first group shows the initial construction of junctions. The second group shows the lines thickened to a width of 15 pixels, and skeletonized. The last group shows the extracted vectors. The X junction is processed in two steps. First, the upper and lower pairs are grouped into two junction points J_1 and J_2, as suggested by the skeleton. It is only when step 5 is iterated that J_1 and J_2 are merged.

The Y junction is processed differently, and has a defect. As for the X junction, the upper segments are grouped to yield an intersection point J. But the lower branch remains disconnected. Initially, the chain connecting its upper extremity J' to the anchor point is removed (length $<$ 15), so there is no valid intersection with one or the other of the two upper branches. When topology is checked in the last step of the algorithm, J and J' are linked, but the final result is not exactly the expected one. This emphasizes a default of the method: if the

Fig. 12. Corrections made in more difficult cases

two left segments had been grouped first, the final resulting topology would have been different.

There is a similar problem with the last case: the left and right pairs of branches are correctly grouped, but the intermediate segments are not; their lengths are shortened by the cutting points computed during segmentation.

4   Conclusion and Perspectives

The method we have presented yields good results on "classical" drawings, and has the advantage to be simple and easily extendible. As an example, it is possible to introduce more complicated curves, such as conics and splines, typically found in CAD drawings. However, the method still suffers from some drawbacks, and we think that it is necessary to make several improvements in future work.

First, the random sampling method used to segment the skeleton is not time efficient at all. Among the methods used to achieve the segmentation of a 2D curve, the Hough transform is known to be time efficient [4], but also space inefficient. Indeed, the required amount of memory space increases exponentially with the dimension of the parameter space of the features to extract. For this reason, Hough or Hough-like methods are not appropriate for our purposes.

From equations 1 and 2, it is easy to see that the number of trials to be performed reaches huge values when the value of e decreases. This problem actually happens because the method always assumes that the worst case occurs. As noted in [8], an adaptive version of the random method may be considered. Such a method should be able to reduce the number of trials, as well as to focus on a promising region by taking into account the results computed at each step.

Next, the method is also not able to extract interrupted patterns. This means, for example, that the vectorization of two primitives, one intercecting the other, will always consist of at best four primitives, not two. Actually, this problem requires interpretation in many cases: are we talking about two segments locally forming an isolated X junction, for example, or about a very long line intercepting a small arc that is a part of a door? But both situations actually weaken the precision of the estimated position of the junction point–if, of course, the existence of this point makes sense.

A still more complicated problem is probably the way we connect the computed junction points e'_u in the last stage of the method. In fact, there are many possible connections of those points, but the rules used to connect them should certainly not come exclusively from the vectorization.

References

1. G. Sanniti di Baja. Well-Shaped, Stable, and Reversible Skeletons from the (3,4)-Distance Transform. *Journal of Visual Communication and Image Representation*, 5(1):107–115, 1994. 276, 279

2. Ph. Dosch, G. Masini, and K. Tombre. Improving Arc Detection in Graphics Recognition. In *Proceedings of 15th International Conference on Pattern Recognition, Barcelona (Spain)*, volume 2, pages 243–246, September 2000. 277

3. Martin A. Fischler and Robert C. Bolles. Random Sample Consensus : A Paradigm Model Fitting with Applications to Image Analysis and Automated Cartography. *Communications of the ACM*, 24(6):381–395, 1981. 277

4. J. Illingworth and J. Kittler. A survey of the Hough transform. *Computer Vision, Graphics and Image Processing*, 44:87–116, 1988. 287

5. P. L. Rosin and G. A. West. Segmentation of Edges into Lines and Arcs. *Image and Vision Computing*, 7(2):109–114, May 1989. 276

6. K. Tombre, C. Ah-Soon, Ph. Dosch, A. Habed, and G. Masini. Stable, Robust and Off-the-Shelf Methods for Graphics Recognition. In *Proceedings of 14th International Conference on Pattern Recognition, Brisbane, Australia*, pages 406–408, August 1998. 276

7. K. Tombre, Ch. Ah-Soon, Ph. Dosch, G. Masini, and S. Tabbone. Stable and Robust Vectorization: How to Make the Right Choices. In A.K. Chhabra and D. Dori, editors, *Graphics Recognition – Recent Advances*, volume 1941 of *Lecture Notes in Computer Science*, pages 3–18. Springer Verlag, 2000. 274, 276

8. Aimo Törn and Antanas Žilinskas. *Global Optimization*, volume 350 of *Lecture Notes in Computer Science*. Springer-Verlag, Berlin, 1989. 287

9. K. Wall and P. Danielsson. A Fast Sequential Method for Polygonal Approximation of Digiti zed Curves. *Computer Vision, Graphics and Image Processing*, 28:220–227, 1984. 277

Structural Rectification of Non-planar Document Images: Application to Graphics Recognition

Gady Agam and Changhua Wu

Department of Computer Science, Illinois Institute of Technology
Chicago, IL 60616
{agam;wuchang}@iit.edu

Abstract. Document analysis and graphics recognition algorithms are normally applied to the processing of images of 2D documents scanned when flattened against a planar surface. Technological advancements in recent years have led to a situation in which digital cameras with high resolution are widely available. Consequently, traditional graphics recognition tasks may be updated to accommodate document images captured through a camera in an uncontrolled environment. In this paper the problem of document image rectification is discussed. The rectification targets the correction of perspective and geometric distortions of document images taken from uncalibrated cameras, by synthesizing new views which are better suited for existing graphics recognition and document analysis techniques. The proposed approach targets cases in which the document is not necessarily flat, without relaying on specific modeling assumptions, and by utilizing one or more overlapping views of the document. Document image rectification results are provided for several cases.

Keywords: perspective correction, joint triangulation, view synthesis, document pre-processing, graphics recognition, document analysis, image processing.

1   Introduction

Document analysis and graphics recognition algorithms are normally applied to the processing of images of 2D documents scanned when flattened against a planar surface. Distortions to the document image in such cases may include planar rotations and additional degradations characteristic of the imaging system [1]. Skewing is by far the most common geometric distortion in such cases, and have been treated extensively [2].

Technological advancements in recent years have led to a situation in which digital cameras with high resolution are widely available. Consequently, traditional graphics recognition tasks may be updated to accommodate document images captured through a camera in an uncontrolled environment. Examples of such tasks include analysis of documents captured by a digital camera, OCR in images of books on bookshelves [3], analysis of images of outdoor signs [4],

D. Blostein and Y.-B. Kwon (Eds.): GREC 2002, LNCS 2390, pp. 289–298, 2002.

license plate recognition [5], and text identification and recognition in image sequences for video indexing [6]. Consequently, distortions characteristic of such situations should be addressed.

Capturing a document image by a camera involves perspective distortions due to the camera optical system, and may include geometric distortions due to the fact that the document is not necessarily flat. Rectifying the document in order to enable its processing by existing generic graphics recognition algorithms require the cancellation of perspective and geometric structural distortions.

A treatment of projective distortions in a scanned document image have been proposed [7] for the specific case of scanner optics and a thick bound book modeled by a two parameter geometric model. Extending this approach to more general cases requires the generation of parametric models for specific cases, and the development of specific parameter estimation techniques. Conversely, in the proposed approach modeling specific cases is avoided, hence facilitating treatment of various scenes.

In general, geometric distortions introduced to a planar document do not include non-uniform stretching of document parts and so the introduced distortions maintain distances between points when measured along the document's surface. Hence, if the distorted planar document does not contain self occlusions, it is possible to flatten it without distortions. Nevertheless, since the interpretation of the geometric structure of the document may inevitably be inaccurate, some distortions during the structural rectification may appear.

In this paper we describe a novel view synthesis [8] approach to perspective and geometric correction of images of planar documents. The proposed approach is then demonstrated in several different cases.

The following sections discuss the proposed approach in greater detail. Section 2 provides a problem statement and outlines the proposed approach. Section 3 describes the proposed approach to document image rectification. Section 4 provides demonstration of the proposed approach. Section 5 concludes the paper.

2   Problem Statement

Let $F \equiv \{f_p\}_{p \in \mathbb{Z}_m \times \mathbb{Z}_n}$ be an $m \times n$ image in which the value of each pixel is represented by an s-dimensional color vector $f_p \in \mathbb{Z}^s$ where in this expression $\mathbb{Z}_m$ represents the set of non-negative integers $\{0, \ldots, m-1\}$. Given two images of the same scene F_1, F_2, taken from different partially overlapping viewpoints, it is possible to determine a set of matching points between the two images:

$$M_{F_1, F_2} \equiv \{(p, q) \mid f_p \in F_1, \quad f_q \in F_2, \quad \mathcal{C}(\mathcal{N}_{F_1}(p), \mathcal{N}_{F_2}(q)) > \tau\} \quad (1)$$

where $p, q \in \mathbb{Z}_m \times \mathbb{Z}_n$, the notation $\mathcal{C}(a, b)$ represents a correlation measure between a and b, the notation $\mathcal{N}_F(p)$ represents a neighborhood of p in F, and τ represents a preset threshold.

Given the point set $P \equiv \{p_i\}_{i \in K}$, a triangulation of that point set is defined as an irreflexive symmetric binary relation over P, denoted by $\mathcal{T}(P) \equiv \{E_i\}_{i \in L}$.

It should be noted that each pair of points $E_i \equiv (p_i, q_i) \in \mathcal{T}(P)$ corresponds to a triangle edge. In general, several different triangulation relations may exist over the same point set P.

Given the point set P and a triangulation relation over it $\mathcal{T}(P)$, the set of triangles induced by the triangulation is defined as a set $\mathcal{M}(P) \equiv \{T_i\}_{i \in M} \subset P^3$ such that:

(a) Each point in a triangle is related to exactly two other points:

$$\forall (p_0, p_1, p_2) \in \mathcal{M}(P) \quad : \quad (p_i, p_{(i \mid 1) \bmod 3}) \in \mathcal{T}(P) \;\; \forall i \in [0, 2] \qquad (2)$$

(b) Each three points define only one triangle between them:

$$\forall T_i, T_j \in \mathcal{M}(P) \quad : \quad \#(\mathcal{V}(T_i) \cap \mathcal{V}(T_j)) < 3 \qquad (3)$$

where $T_i \equiv (p_i, q_i, r_i)$ and $\mathcal{V}(T_i) \equiv \{p_i, q_i, r_i\}$.

In the case of a set of matching points between two images, a joint triangulation is required in order to generate consistent triangles in the two images. Given the set of matches M_{F_1, F_2}, a joint triangulation of it is defined by:

$$\mathcal{J}(M_{F_1, F_2}) \equiv \{(T_i, T_i') \mid T_i = (p_i, q_i, r_i) \in \mathcal{M}(P_1) \wedge T_i' = (p_i', q_i', r_i')\} \qquad (4)$$

where $P_1 \equiv \{p_i \mid (p_i, p_i') \in M_{F_1, F_2}\}$, and $(p_i, p_i'), (q_i, q_i'), (r_i, r_i') \in M_{F_1, F_2}$. As mentioned earlier the triangulation of a point set is not unique, and so there may be several options for a joint triangulation of the matches M_{F_1, F_2}. A physically correct joint triangulation $\mathcal{J}(M_{F_1, F_2})$ is such that the correlation between the triangles in each triangle pair (T_i, T_i'), when rectified to a common view plane, is high. Such a triangulation is generated when the triangles in the images belong to a single planar surface in the scene. Given an initial, arbitrary, joint triangulation, it is shown in Section 3 how this triangulation may be modified in order to maximize this correlation measure.

After correcting the joint triangulation $\mathcal{J}(M_{F_1, F_2})$, the corresponding 2D triangles may be used to generate a 3D triangular mesh $\mathcal{T}(M_{F_1, F_2})$ by using the stereo disparity of each pair of matching points to approximate the depth of the corresponding 3D mesh vertices. Assuming that the images F_1 and F_2 are of a planar document, perspective distortions may be corrected by projecting the triangular mesh $\mathcal{T}(M_{F_1, F_2})$ onto an image plane parallel to the reconstructed plane surface. If, however, the reconstructed 3D structure of the document, as represented by the triangular mesh, is non-planar, a rectification stage is required prior to projecting the triangular mesh. Such a stage is necessary in order to flatten $\mathcal{T}(M_{F_1, F_2})$ onto a plane so as to correct geometric distortions. Examples to cases in which geometric distortions should be corrected include paper documents which are folded or bent, pages in a thick book, labels on a round object, and labels on boxes. Finally, the rectified mesh, and the original images are used to synthesize a rectified view of the document.

3   Document Image Rectification

Arbitrary triangulation may not necessarily produce physically valid joint triangulation. A physically valid joint triangulation is such that each triangle in it completely belongs to a planar surface in the scene. In such a triangulation, the structure representation of the scene through a triangular mesh is more accurate, and hence new views from new viewpoints synthesized based on it are expected to produce smaller distortions. Distortions in new views is in fact a key feature that may be used in order to modify an existing triangulation so as to become physically correct, as defined here.

Given a set of matching points M_{F_1,F_2} and a joint triangulation of it $\mathcal{J}(M_{F_1,F_2})$, as defined earlier, adjustment of the joint triangulation is possible through modification of the triangulation relation over the matched point set $\mathcal{T}(P_1)$ and/or through modification of the matched point set M_{F_1,F_2} itself by adding or removing matched points to it. In both cases, an essential feature required in order to execute the modification is the ability to assess the correlation between two triangles when rectified to a common viewpoint.

Given a pair of joint triangles $(T_i, T_i') \in \mathcal{J}(M_{F_1,F_2})$ where $T_i \equiv (p_i, q_i, r_i)$ and $T_i' \equiv (p_i', q_i', r_i')$, the triangle T_i may be warped onto T_i' so as to represent viewing of T_i and T_i' from a common viewpoint. When the triangle vertices are represented in homogeneous coordinates, perspective warping of T_i onto T_i' is defined up to a scale factor by a matrix $H \in \mathbb{R}^{3 \times 3}$ such that $p' = Hp$. This transformation may be approximated [9] by affine warping in which the transformation matrix H is given by:

$$H = [p_i' \ q_i' \ r_i'][p_i \ q_i \ r_i]^{-1} \tag{5}$$

The correlation between the triangles is then computed by a normalized, zero-mean, cross correlation operation which is modulated by the joint variation in the triangles so as to eliminate the correlation strength between uniform triangles. Based on the ability to assess the correctness of a triangle through the evaluation of the texture similarity between the triangles, incorrect triangles may be detected and modified.

Given a triangular mesh $\mathcal{T}(M_{F_1,F_2})$ obtained as described in Section 2, its rectification may be obtained by first aligning all the triangles with a common plane, and then aligning them with each other. Let $T_i \equiv (p_0, p_1, p_2)$ be a triangle in $\mathcal{T}(M_{F_1,F_2})$. As the plane with which all the triangles should be aligned with can be chosen arbitrarily, we choose the plane $z = 0$. When representing the vertices of the triangles in homogeneous coordinates, the triangle T_i may be aligned with $z = 0$ by pre-multiplying its vertices by a compound 4×4 translation-rotation transformation matrix given by:

$$RT = \begin{bmatrix} u_x^T & 0 \\ u_y^T & 0 \\ u_z^T & 0 \\ \mathbf{0} & 1 \end{bmatrix} \begin{bmatrix} I & [0 \ 0 \ -(p_0)_z]^T \\ 0 & 1 \end{bmatrix} = \begin{bmatrix} u_x^T & -(u_x)_z(p_0)_z \\ u_y^T & -(u_y)_z(p_0)_z \\ u_z^T & -(u_z)_z(p_0)_z \\ \mathbf{0} & 1 \end{bmatrix} \tag{6}$$

where $u_x, u_y, u_z \in \mathbb{R}^3$ are given by:

$$u_z = ((p_1 - p_0) \times (p_2 - p_0))/|(p_1 - p_0) \times (p_2 - p_0)| \tag{7}$$
$$u_y = (p_2 - p_0)/|(p_2 - p_0)| \tag{8}$$
$$u_x = u_y \times u_z \tag{9}$$

and $(u)_x$ represents the x component of u. In this description it is assumed that the vertices of each triangle are arranged in a counterclockwise order so that the area of each triangle $\mathcal{A}(T_i)$ is non-negative. If the area is negative the vertices p_1 and p_2 of the triangle are swapped. Given two triangles in the plane $z = 0$: $T_i \equiv (p_i, q_i, r_i)$ and $T_j \equiv (p_j, q_j, r_j)$, where without loss of generality it is assumed that the common edge between them is (p_i, q_i) in T_i and (p_j, q_j) in T_j, triangle T_j is aligned with T_i by pre-multiplying its vertices by the following 4×4 transformation matrix:

$$M = T(p_i)R_z(\angle(v_j, v_i))T(-p_j) \tag{10}$$

where $v_i \equiv q_i - p_i$, $v_j \equiv q_j - p_j$, the matrix $T(p) \in \mathbb{R}^{4 \times 4}$ represents translation by p, the matrix $R_z(\alpha) \in \mathbb{R}^{4 \times 4}$ represents rotation about z by α, and $\angle(v_j, v_i)$ is the angle between the vectors v_j and v_i. The angle $\angle(v_1, v_2)$ is given by:

$$\angle(v_1, v_2) \equiv \cos^{-1}(\hat{v}_1 \cdot \hat{v}_2) \, \text{sgn}((\hat{v}_1 \times \hat{v}_2)_z) \tag{11}$$

where $\hat{v}_i \equiv v_i/|v_i|$ and $\text{sgn}(x) = 1$ when x is non-negative or -1 otherwise.

An alternative approach to rectifying the triangular mesh may be obtained by an iterative algorithm. The iterative algorithm considers the edges of the triangles to be springs which when made shorter or longer apply a force proportional to the length modification on their vertices. In this approach the triangles are aligned with the $z = 0$ plane by projecting their vertices onto that plane and then minimizing the potential energy within the spring mesh in an iterative process. Let p_i and q_i be two vertices. The directed edge between p_i and q_i may be expressed by the vector $e_i - q_i - p_i$. Given the original and modified edges e_i and e'_i the potential energy E_i of the edge e_i is defined by the distortion of its length:

$$E_i \equiv (|e_i| - |e'_i|)^2 \tag{12}$$

and the total mesh energy to be minimized is then given by: $E = \sum_i E_i$.

Given a vertex p^t and the set of directed edges that connect to that vertex $\{e_j\}$, the position of p^t in the next iteration is given by:

$$p^{t+1} = p^t + \sum_j s * |e_j| * E_j \tag{13}$$

where s is a constant. The position of the vertices that minimize the energy may be refined by reducing the constant s. Consequently, the constant s is decreased whenever the previous iteration did not manage to reduce the total energy E. While the complexity of the iterative approach is higher than that of the direct approach described earlier, it is better suited to handle cases where the initial mesh contains errors.

The triangular mesh rectification described so far is a general approach that does not depend on the content of the document image. Nevertheless, the rectification results may be improved when imposing constraints on the input image. Particularly, in many cases it is possible to assume that the document image is composed of text lines and/or tables. In such a case the rectification may be obtained from a single image by detecting the ends of text lines and imposing a constraint requiring that text lines should be straight. In such a case the rectification may be done by projecting text lines onto the straight lines connecting their ends. The results of this approach are demonstrated in the following section.

4   Results

This section provides three examples to document image rectification. Document image rectification by direct mesh manipulation is presented in Figure 1. The corrected triangular mesh of a cylindrical scene obtained as described in Section 3 is presented in Figure 1-(a) whereas the rectified triangular mesh is presented in Figure 1-(b). As explained earlier, in Section 3, the rectification of the triangular mesh is based on a rough reconstruction of the scene. This rough reconstruction for the box scene is presented in Figure 1-(c). As can be observed, the reconstruction contains inaccuracies due to numerical errors, the fact that the original views were not calibrated, and the fact that the reconstruction is based on a sparse set of matching points. These structural errors are reflected in the rectified mesh by distortions around certain vertices. Figure 1-(d) presents the error in the rectified mesh, where light color represents error. The error is measured as the cumulative distance between common vertices of triangles after rectification. Finally, the synthesized views of the box scene before and after rectification are presented in Figures 1-(e) and 1-(f) respectively. Figure 1-(e) is generated by mapping the image of each triangle in one of the views onto the reconstructed triangular mesh, whereas Figure 1-(f) is generated by mapping the image onto the rectified triangular mesh. It should be noted that the synthesized results may be further improved by using more than one input view for the synthesis.

Results of document image rectification by iterative energy minimization are presented in Figure 2. Figure 2-(a) presents a view of a low-resolution approximated cylindrical surface. The texture on the surface demonstrates perspective and geometric distortions to the surface. Figure 2-(b) shows another view of the same surface where the curved nature of the surface is clearer. Figure 2-(c) displays the distorted surface with document texture. Finally, Figures 2-(d) – 2-(f) show three stages in the rectification process, whereas Figures 2-(g) – 2-(i) display the same three stages with document texture. As can be observed, the rectification process results in reduced perspective and geometric distortions in the input image.

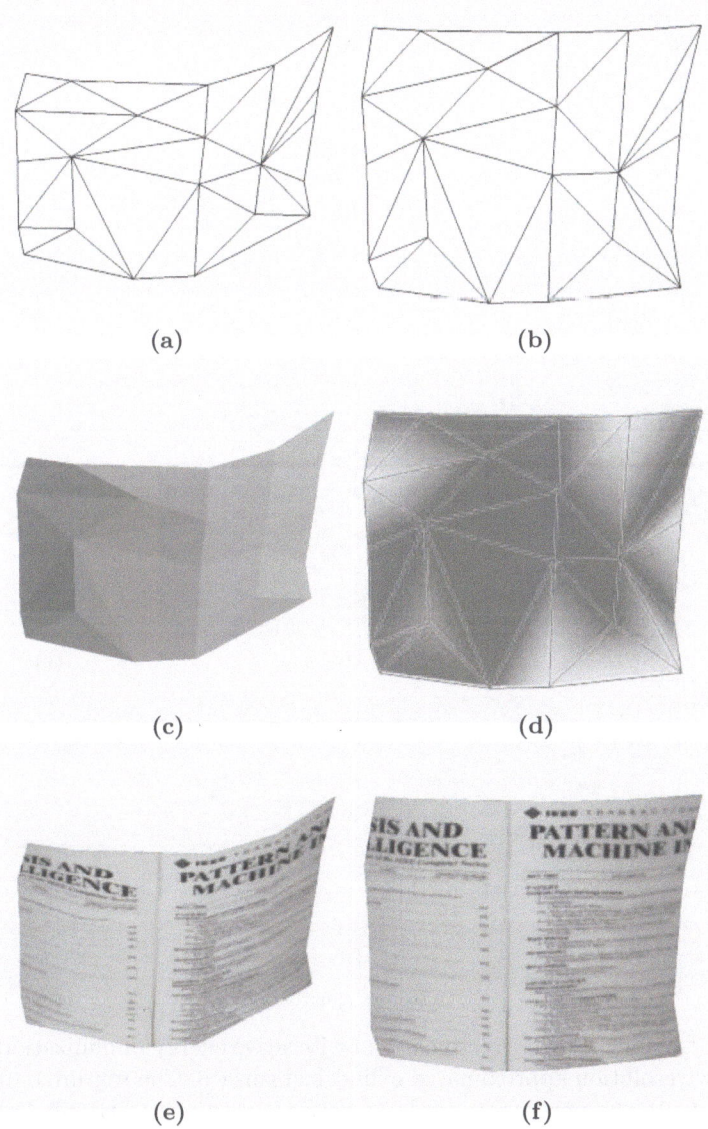

(a)   (b)

(c)   (d)

(e)   (f)

Fig. 1. Document image rectification by direct mesh manipulation. **(a)**–**(b)** The triangular mesh before and after rectification. **(c)** rough 3D reconstruction of the scene. **(d)** Error introduced by the rectification. **(e)**–**(f)** The synthesized surface image before and after rectification

Document image rectification by line straightening, as described earlier, is presented in Figure 3. Figure 3-(a) presents the input book image in which there are perspective and geometric distortions. Figure 3-(b) shows the correction transformation applied for rectification. The rectangle texture represents

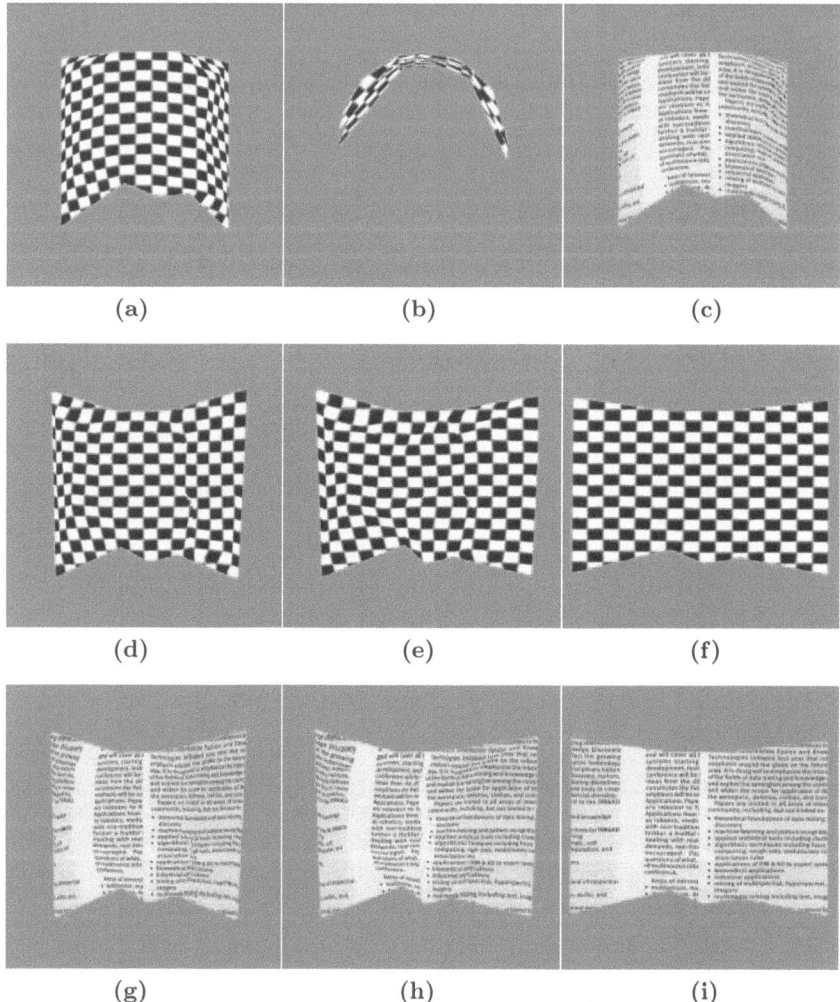

Fig. 2. Document image rectification by iterative energy minimization. (a) View of a low-resolution approximated cylindrical surface. The texture on the surface demonstrate the perspective and geometric distortions to the surface. (b) Another view of the same surface. (c) The distorted surface with document texture. (d)–(f) Three stages in the rectification process. (g)–(i) The same three stages with document texture. As can be observed, the rectification process results in reduced perspective and geometric distortions in the input image

the opposite distortion needed for the rectification. As explained earlier this transformation is computed based on the assumption that the document image is composed of straight text lines and so the text lines should be rectified.

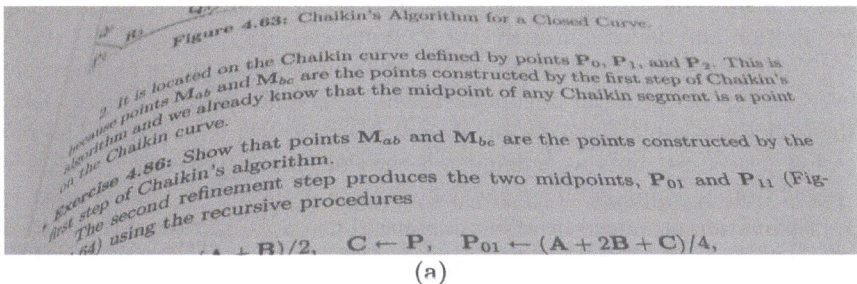

(a)

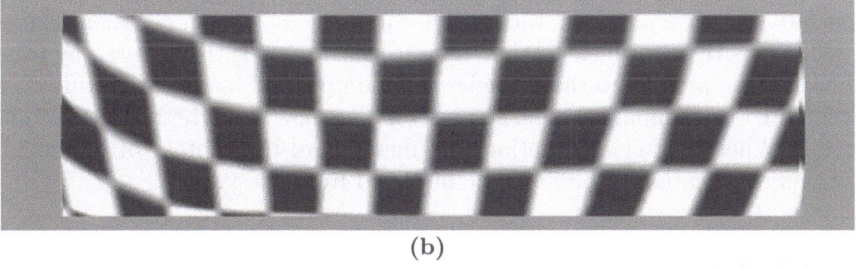

(b)

Figure 4.63: Chaikin's Algorithm for a Closed Curve.

2. It is located on the Chaikin curve defined by points P_0, P_1, and P_2. This is because points M_{ab} and M_{bc} are the points constructed by the first step of Chaikin's algorithm and we already know that the midpoint of any Chaikin segment is a point on the Chaikin curve.

Exercise 4.86: Show that points M_{ab} and M_{bc} are the points constructed by the first step of Chaikin's algorithm.

The second refinement step produces the two midpoints, P_{01} and P_{11} (Fig. 4.64) using the recursive procedures

(c)

Fig. 3. Document image rectification by line straightening. (a) Input book image with perspective and geometric distortions. (b) The correction transformation applied for rectification. (c) The rectified image after correction. As can be observed, the curved lines in the original image are straightened in the rectified image

Figure 3-(c) shows the rectified image after correction. As can be observed, the curved lines in the original image are straightened in the rectified image.

5 Conclusion

In this paper a method is presented for perspective and geometric correction of document images. The images are assumed to be taken from uncalibrated cameras, and it is assumed that the document is not necessarily flattened against a planar surface. The purpose of the proposed approach is to produce a rectified image in which distortions are minimized so as to facilitate conventional graphics recognition and document analysis techniques.

The proposed method relies on the availability of one or more, uncalibrated, overlapping views of the document taken from different viewpoints. Based on these views a joint triangulation is generated and then modified. The modification procedure aims at maximizing the correlation between triangles when viewed from a common viewpoint, where the correlation evaluation relies on an approximation produced by affine warping. The maximization of the correlation criterion, aims at generating a physically valid triangulation in which each triangle belongs to a planar surface in the scene. Based on the produced triangulation, a 3D mesh representation of the scene is generated and then rectified. The rectification process aligns all the triangles in the mesh with a common plane, and then aligns them with each other. The rectified mesh is then viewed by using an image plane parallel to the triangles common plane. Finally the rectified view is generated by mapping the image pixels in one view onto the rectified triangular mesh. When further assumptions are made as to the content of the document image, rectification results may be obtained from one view.

References

1. H. Baird, "Document image defect models", In Proc. SSPR'90, pp. 38–46, 1990. 289
2. A. Amin, S. Fischer, A. F. Parkinson, R. Shiu, "Comparative study of skew algorithms", Journal of Electronic Imaging, Vol. 5, Iss. 4, pp. 443–451, 1996. 289
3. M. Sawaki, H. Murase, and N. Hagita, "Character recognition in bookshelf images by automatic template selection," Proc. ICPR'98, pp.1117–1120, Aug. 1998. 289
4. H. Fujisawa, H. Sako, Y. Okada, and S. Lee, "Information capturing camera and developmental issues", In Proc. ICDAR'99, pp. 205–208, 1999. 289
5. M. Shridhar, J. W. V. Miller, G. Houle, and L. Bijnagte, "Recognition of license plate images: issues and perspectives", In Proc. ICDAR'99, pp. 17–20, 1999. 290
6. H. Li, D. Doermann, and O. Kia, "Automatic text detection and tracking in digital video", IEEE Trans. Image Processing, Vol. 9, No. 1, pp. 147–156, 2000. 290
7. T. Kanungo, R. Haralick, and I. Philips, "Global and local document degradation models", In Proc. ICDAR'93, pp. 730–734, 1993. 290
8. G. Agam, G. Michaud, J. S. Perrier, J. L. Houle, and P. Cohen, "A survey of image based view synthesis approaches for interactive 3D sensing", Technical Report GRPR-RT-9901, The Perception and Robotics Laboratory, Ecole Polytechnique, Montreal, Canada, April, 1999. 290
9. J. S. Perrier, G. Agam, and P. Cohen, "Image-based view synthesis for enhanced perception in teleoperation", in *Enhanced and Synthetic Vision 2000*, J. G. Verly, ed., Proc. SPIE 4023, pp. 213–224, 2000. 292

An Effective Vector Extraction Method on Architectural Imaging Using Drawing Characteristics

Young-Jun Park and Young-Bin Kwon

Department of Computer Engineering, Chung-Ang University
221 Heuksukdong, Dongjakku, Seoul, 156-756, Korea
ybkwon@visionnet.cse.cau.ac.kr

Abstract. In this paper, vectorization is achieved by the recognition of architectural drawing images. To obtain vector components, an architectural drawing recognizer (ADR) is developed. The ADR recognizes line components such as main wall, dimension line and points that indicate the dimension of scanned architectural drawing images. A merging process from separately recognized line components is accomplished in order to increase the accuracy of result vectors. The recognition ratio and the vectorization of ADR is 98.3% over 9 real scanned apartment building images.

1 Introduction

The field of automatic vector generation from printed or hand printed architectural drawing is a part of pattern recognition. This process differs slightly from the vectorization of engineering drawings in the characteristics of drawing information. Thus, the vector extraction process of scanned architectural drawings may cause difficulty because of its generation characteristics during their generation stage.

Architectural drawings consist of lines, symbols, characters, and numbers. Now CAD tools do most of the drawing. Traditionally, lettering devices were manually used for this process. Architectural drawings are published as blueprints for storage. This type of documents is inherently low in its quality and degrades even further over time. This type of documents are now scanned and stored in an optical filing system. However, optical filing system is only a scanned version of blue printed image and has difficulty in the process of image retrieval and modification. For the modification of scanned image, it must first be converted into vector form manually using CAD software. This process is time consuming and routine work. Thus, an automatic process for vectorization is strongly required [1].

For this purpose, Fletcher and Kasturi propose the connected component method on the engineering circuits [2]. Tombre et al. uses the connected component and morphological operation in order to obtain recognition of the main wall [3]. For the vectorization, orthogonal zig-zag method is proposed by Dori[4]. This method is useful with the thickness. The architectural drawings used in this paper are of apartment building. These drawings have different characteristics. That is, they are

D. Blostein and Y.-B. Kwon (Eds.): GREC 2002, LNCS 2390, pp. 299-309, 2002.

composed of both horizontal and vertical walls. Thus, we propose another method for the vectorization. Main walls and dimension line are major vector components. A series of main wall component vectors constructs a closed loop, and main wall has its own dimension line around its closed loop. Thus, a hierarchical approach to detect main wall and dimension lines is proposed in order to produce vectors efficiently.

Following introduction, section 2 explains the properties of architectural drawings. Section3 shows the recognition process and vector extraction, simulation results and conclusions are followed.

2   Architectural Drawing Analysis

Figure 1 illustrates an architectural drawing used in this research. We use images with similar structure with the scale of 1:150.

These images are printed using plotter and are scanned with 300dpi with 256 gray levels. In principle, the drawings used are the basement floor of apartment buildings. The figures contain lines, dashed lines, symbols, characters and numbers. Many of the dimension lines and dimension size are located outside the images. The end of dimension is generally marked by a dot, called the end point. The size of the dimension, which is expressed by a number, is located between the end points. Main wall and symbols are located in the center part of the drawings. Main wall is the part of the building that must not be torn down in order to maintain the structure of the building, even when remodeling or restructuring. Main wall components make a closed loop, forming the outbound area of the structure. The thickness of main wall has a tendency to be thicker than the dimension line. In the drawing used, main walls are detected in the outbound rectangular area in the center part of figure 1. The horizontal and vertical lines indicate the dimension lines with a numerical figure representing the dimension. In general, the dimension extension line crosses over in the middle of the main wall. The detection of the main wall and dimension lines is accomplished using the characteristics of the architectural drawings.

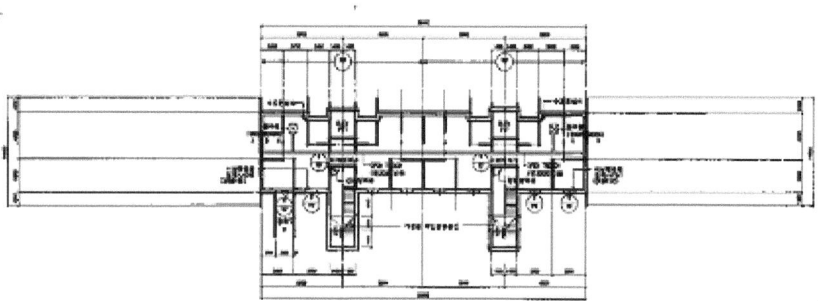

Fig. 1. Architectural drawing: Basement floor of apartment

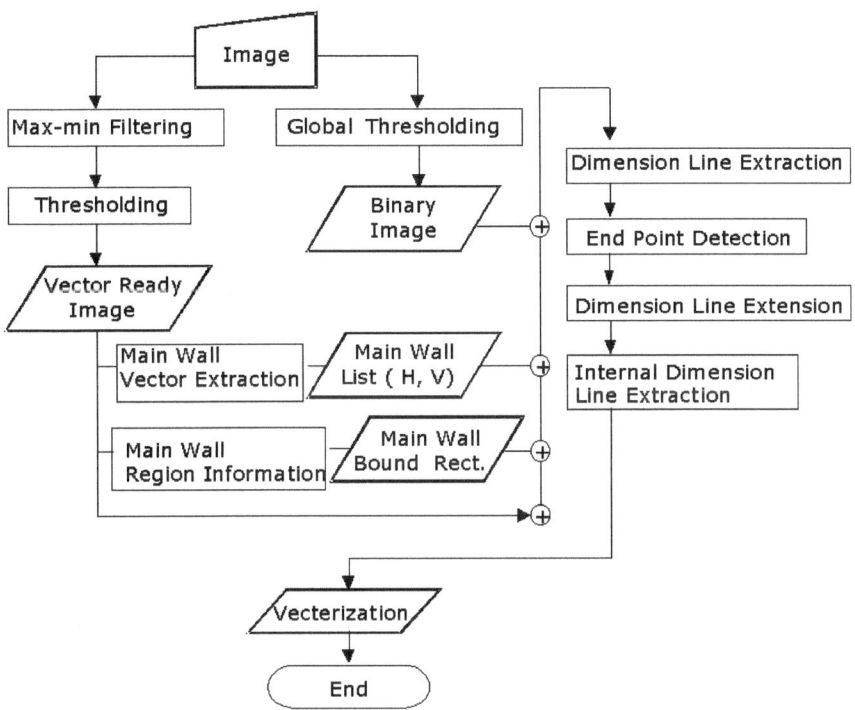

Fig. 2. Architectural Drawing Recognizer(ADR) and Vectorization Process

3   Architectural Drawing Recognition and Vectorization

In order to extract the necessary information, several different methods are used to find out reliable vector parts. Figure 2 shows a schematic diagram of the architectural drawing recognizer (ADR) and vectorization process.

Input image produces intermediate results through two different preprocessing. Max-min filtering is used to detect the main wall and its out-bounding region. Global thresholding is used for binarization. After the binarization process, the average run length of both the horizontal and vertical direction and toggling number are calculated in order to detect the peak points of the average run length. The vectors that are located near the peak points are used to indicate the vectors, which are independent of the symbols such as the characters and graphic parts. Dimension lines and end points end are detected from the combination of these two different images and related information such as the main wall list and main wall boundary. Vectorization on the main wall and dimension line is accomplished using the two preprocessed images. The internal dimension line is detected using the information of the extended dimension line and information of the end-points detected through the max-min filtering process.

3.1   Max-min Filtering

The major purpose of max-min filtering is on line separation. Since architectural drawings used in this paper contain an attribute for the thickness of lines. Size of 3*3 template is used for the filtering. For max filtering, a maximum value of 8 is selected from the surrounding template values. This value replaces the center point value. The effect of main filter is erosion. It means that max filter shrinks the thin lines such as the indication line and the dimension line. The thick line, such as the main wall, keeps its boundary line even after the max filtering, whereas most of the thin lines are eliminated. Figure 3(b) shows the max-filtering result after 2 repetitions. On the other hand, min filtering which takes a minimum value of 8, selected from the surrounding template values, demonstrates dilation effects. It expands its boundary width. Figure 3(c) shows the effect of min filtering after 2 repetitions. After the max-min filtering process, contrast adjustment on histogram is applied to detect the main wall effectively. Figure 4(d) shows the result. Using this result, main wall vectors are obtained.

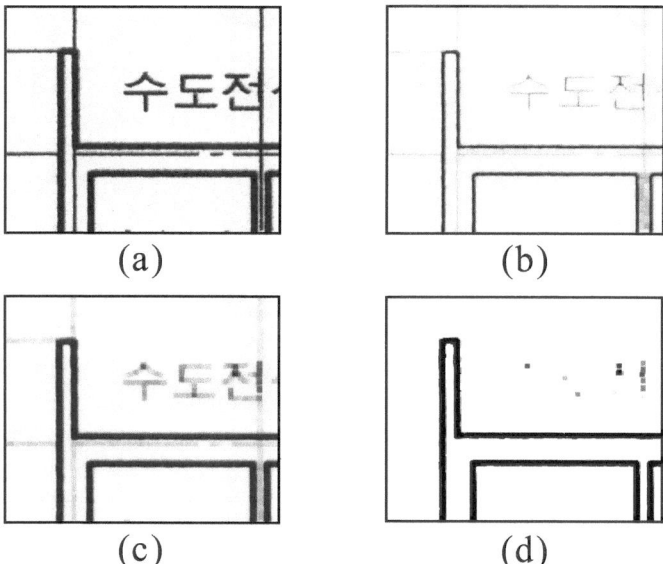

Fig. 3. Preprocessing by Max-min Filtering   (a) original image (b) Max filtering (c) Min filtering (d) contrast adjustment

3.2   Vectorization

Most of the vector elements from the image of the basement of an apartment composed of vertical lines and horizontal lines. A line in a drawing consists of several elementary lines with the width of one pixel in image but elementary lines must be recognized as a single line. The elementary lines are called vector elements, and become a parameter to the vector merging process that produces the merged main wall component vector with its thickness indicated. The merging process is an expansion on both horizontal and vertical direction. The list of elementary vector of current point has a relation with the vector list that is detected or merged by previous point.

In reality, cutting of lines during max-min filtering and skew of dimension line may cause a separate elementary vector. Thus, a newly obtained elementary vector must be examined for the possibility of merging. Width expansion of previous vector process may also be required.

Figure 4 shows the vector-merging process. A threshold value is used to eliminate the line element, which is attached to the character or symbols.

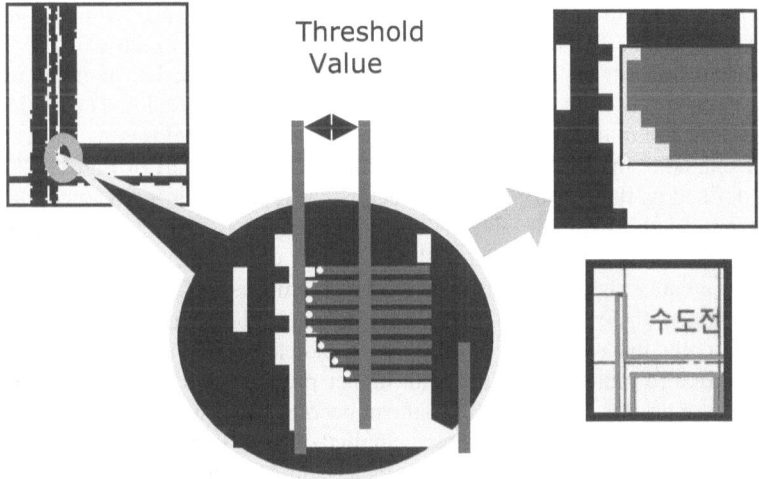

Fig. 4. Vector merge process

The data structure of line element, extraction and the merging algorithm is as follows:

```
Struct{
        Cpoint Sp;
        Cpoint Ep;
        bool SpC;
        bool EpC;
        }
```

Where, CPoint is a class provided by MFC and abstraction of coordinate points. Sp and Ep are the starting point and end point, respectively. SpC and EpC are Boolean parameters, which indicate the existence of vector connected to the starting point and the existence of vector connected to the end point, respectively. Initial value of SpC and EpC is zero. Using this Boolean parameter, small elementary vectors that are isolated during the filtering process are merged.

3.2.1   Main Wall Recognition

In the drawing, main wall is sketched with thick lines, as explained in Figure 1 in section 2. Figure 5 is a highlight of the upper left part of Figure 1. The walls above the "ELEV. PIT" sign, indicated with the thick lines is the main wall in this figure. Another main wall is detected on the right side of the staircase with windows on it. Only the main wall itself, excluding the windows, is detected when the vectorization process has been applied to the figure. The line crossing the center of the figure, indicated as "OPEN TRENCH", is not determined to be a main wall and thus has been drawn using thin lines. Dashed lines, thin lines and characters must be filtered out in order to perform the main wall detection. Thus, preprocessing with max-min filtering gives rise to an effect of thick line detection, as indicated in Figure 3-(d).

The main wall consists of main wall component vectors. A series of main wall component vectors are used to construct a closed loop, forming a region as explained in Section 2. After the max-min filtering, some parts were left unclosed as shown in Figure 6. For example, there are 4 unclosed region in the upper part of Figure 6. An end point is defined as the end of each line vector that is left unclosed. Hence, there are 8 end points in the example above. The line element of the closed loop has connection properties at each end point with the nearest end point. Hence, main wall is selected on the data structure when the Boolean parameters SpC and EpC are both true. In the first step, the candidate for main wall is detected like figure 6. Blue color lines and black color lines are used in Figure 6 to indicate different conditions. The black lines are used to indicate the lines with higher probability of becoming a main wall. Blue lines are used to indicate lines that can become main wall candidates when the end-point connection process has been applied. There may be three possible combinations of the color to indicate the characteristic of each line. The end point connecting algorithm is applied for the actual detection of the main wall candidate vector. The main wall candidate vectors become a set of wall components through the end point connection process. The main wall vector set is decided through this process.

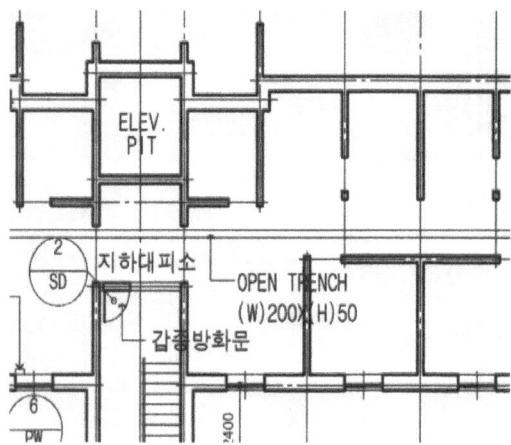

Fig. 5. Upper-left part of figure 1

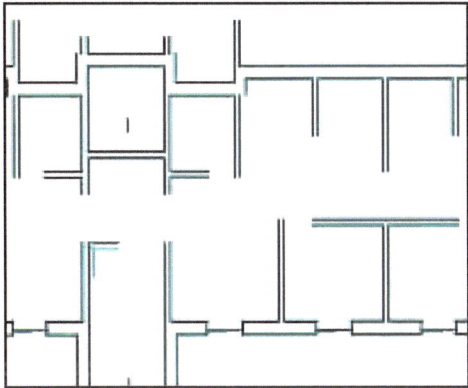

Fig. 6. Main wall candidates before the end-point connection

A connection comparison is made for every candidate of the main wall vector set. Figure 7 shows the intermediate state during the process of end-point connection applied to check the status of the connection comparison. Connection comparison is used to connect the set of end points for connection by sorting the points with common end point. For example, the upper left part of Figure 7 shows six different vector that are connected with a common end point to form a main wall vector. Thus, the main wall extension is accomplished through the connection comparison process. The pink lines are used to indicate line vectors that are still being processed.

Figure 8 is a part of the final result of the main wall detection obtained from Figure 1. The upper part has become a single area through the end-point connection and the connection comparison process. The lower part of the figure forms 6 different areas as a result of the above process. The lines detected as the final result are vectorized along lines with maximum straight length. This set becomes a main wall vector list in Figure 2.

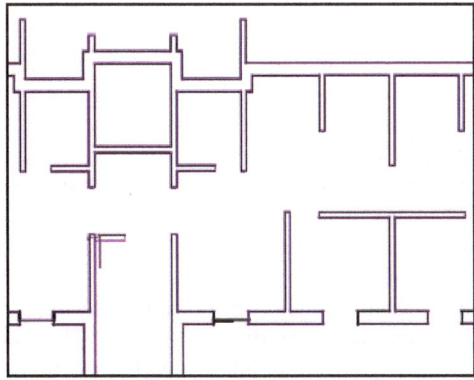

Fig. 7. Intermediate result during the connection comparison process

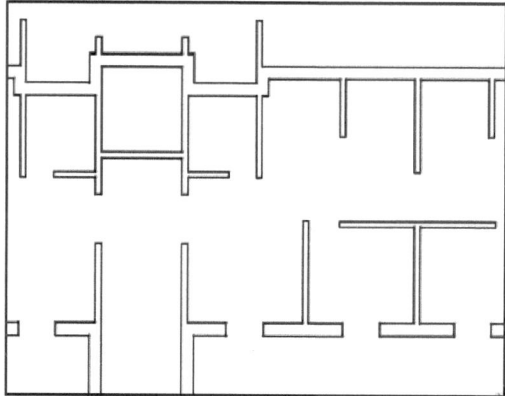

Fig. 8. Final result of main wall detection

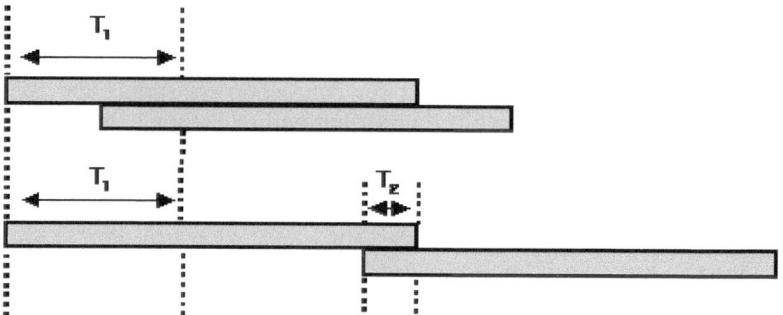

Fig. 8. Merge condition for dimension line

3.2.2   Extraction of Dimension Line and Its End Point

After detecting the main wall, the maximum out-bounding rectangle ratio on the main wall is calculated. The area outside this rectangle is composed of dimension lines and end points. From the global threshold, indicated in Figure 2, the binary image value obtained using the Get function. Line elements can be extracted efficiently through this process. Skew orientation is a sensitive factor in detecting the dimension line. A line drawing with A0 size is scanned with 300dpi. Black run of the dimension line lies between 50 to 4,000 pixels. Thus, the variation of 5 per 1000 pixel may cause a disconnection of dimension line vectors. A merge process to the candidate vector is proposed, as shown in Figure 8.

There are 2 different criteria for merging the candidate line vectors. The upper case illustrates the merging process for the dimension lines. This condition is to detect the tightly coupled lines. If the two end points are within the threshold range of T1, then the merge condition only arise when the difference between the y coordinates value is 1. The lower case shows the other merging process of when one end point is larger than T1 and the line is connected to the other line with a y pixel difference of 1. This case may occur when a long dimension line is scanned. This is the process for merging loosely coupled lines. Merging is done when the difference between the y pixel coordinate is 1. To detect the end points of the dimension line, the max filtering results are also applied. Max filter shrinks the point information, but the core of each end point is not eliminated. Dimension line and indication line have the same width. Dimension line is connected to one or two end point. Thus, end point connection is a key factor in distinguishing the dimension line and the indication line.

After detecting the dimension line, line vector components extend to the interior of the main wall of the out bounding rectangle. Because the dimension line is across the middle or on the side of the main wall vector, it needs to be extended. However, dashed lines are generally used to express the internal dimension line. The other type of internal dimension line is also detected by wall width information. In order to detect the internal dimension line, difference image which is obtained by taking the difference between the out-bounding rectangle area of original image and the main wall area. The internal dimension line has 3 different dashed line elements, as shown in Figure 9.

In Figure 9, type 1 is detected by an extension of the already detected dimension line. This is easily detected through a process of simple expansion of the already detected dimension line. Type 2 is a dimension line that is not related to the external dimension line. This type of line is located in the internal part of the main wall. This type of line is detected using the region width information of the main wall. Type 3 lines are neither related to externally detected lines nor the main wall. This type may exist in theory but have not occurred in our research.

3.2.3   Vectorization of Detected Lines

In this system, 3 kinds of vectorization results are obtained. First of all, the main wall vector generation is accomplished after the main wall detection using max-min filtering. The second is the dimension line detection stage. In this stage, we use binary image obtained by global thresholding. The dimension line and its extension make up a long vector. The last stage is accomplished through detecting the internal dimension line in the interior of the main wall.

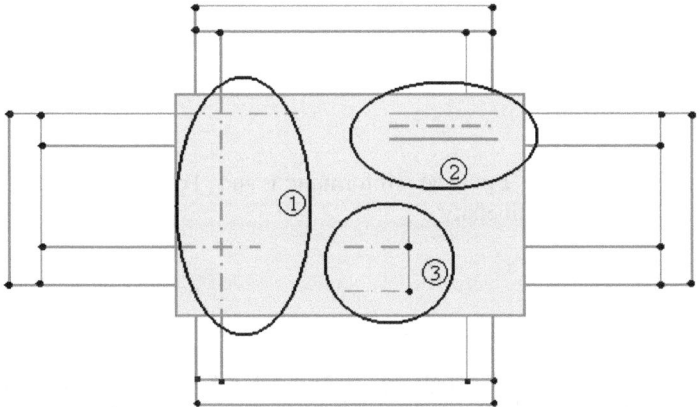

Fig. 9. Types of internal dimension lines

4   Simulation Results

The efficiency of vectorization depends on the recognition ratio of the line drawing images. From the drawings, the elements of recognition are main wall, dimension line, and end points. For information, we calculate the number of line segments of horizontal and vertical components and points. 9 line drawing images are tested. The recognition results are listed in Table 1. Overall recognition ratio indicates 98.3%. Unrecognized objects are occurred by rectangular main wall parts of small threshold value. In general, these parts are isolated small area that maybe considered as noise. False recognition is occurred by door symbol. Door symbol that is close to main wall with indication line satisfies main wall vector decision conditions.

Table 1. Recognition results

Item	Number of Objects	Correct Recognition	Unrecognized Objects	False Recognition	Recognition Ratio (%)
Main Wall	556	542	8	6	97.5
Dimension Line	474	466	8	0	98.3
End Points	458	455	3	0	99.3
Total	1,488	1,463	19	6	98.3

5   Conclusion

In this paper, a hierarchical approach of vectorization on the architectural drawing is proposed. We propose a 3-stage vectorization using two different images of max-min filtered image and global threshold binary image. Vectorization is accomplished by sequential order: detection of main wall vector component, dimension line and its end point detection, and internal dimension line. A vector-merging method is also proposed to obtain the vectors more correctly.

For further research, distinction of unrecognized objects with noise and blue print processing with noise filtering and restoration are considered.

Acknowledgement

This work is supported by ITRI (Information and Telecommunication Research Institute), Chung-Ang University.

References

1.   Rid D. T. Janssen and Albert M. Vossepoel, "Adaptive Vectorization of Line Drawing Images", Computer Vision and Image Understanding, Vol. 65, No. 1, pp. 38-56, 1997.

2.   L. A. Fletcher and R. Kasturi, "A Robust Algorithm for Text String Separation from Mixed Text/Graphics Images", IEEE Transactions on PAMI, Vol. 11, 4, pp. 411-414, 1988.
3.   Karl Tombre, Christian Ah-Soon, Philippe Dosch, Adlane Habed, and Gerald Masini, "Stable, Robust and Off-the-Shelf Methods for Graphics Recognition", ICPR, Vol. 1, pp. 406-408, 1998.
4.   Dov Dori, "Orthogonal Zig-Zag: an algorithm for vectorizing engineering drawings compared with Hough Transform", Advances in Engineering Software, Vol. 28, pp. 11-24, 1997.

Thresholding Images of Line Drawings with Hysteresis

Tony P. Pridmore

Image Processing & Interpretation Research Group,
School of Computer Science and Information Technology University of Nottingham,
Nottingham, UK
{tpp@cs.nott.ac.uk}

Abstract. John Canny's two-level thresholding with hysteresis is now a de facto standard in edge detection. The method consistently outperforms single threshold techniques and is simple to use, but relies on edge detection operators' ability to produce thin input data. To date, thresholding with hysteresis has only been applicable to thick data such as line drawings by top-down systems using a priori knowledge of image content to specify the pixel tracks to be considered. We present, and discuss within the context of line drawing interpretation, a morphological implementation of thresholding with hysteresis that requires only simple thresholding and idempotent dilation and which is applicable to thick data. Initial experiments with the technique are described. A more complete evaluation and formal comparison of the performance of the proposed algorithm with alternative line drawing binarisation methods is underway and will be the subject of a future report.

1   Introduction

Binarisation of a grey level image via some form of thresholding is a key stage in line drawing interpretation. Many thresholding schemes have been proposed [1-6], Static, adaptive thresholding methods are the most commonly used in line drawing interpretation. A threshold value is defined as a function of (only) the grey level distribution of a local image region. The image is typically divided into usually fixed, sometimes overlapping, regions, a threshold is computed independently for each region and applied only within that region, e.g. [7,8,9].

In local, dynamic, adaptive methods other regions are considered. Specifically, the threshold applied to a given region is a function of the position of that region relative to other regions having particular local properties. Sauvola and Pietikainen [9], for example, use a dynamic technique to select binarisation methods. Perhaps the best-known dynamic thresholding technique, however, is that proposed for use in edge detection by John Canny [10].

Canny employed the calculus of variations to find the operator which, assuming a perfect step edge in noise, is optimal with respect to good detection, good localisation and minimal response. The result may be closely approximated by seeking local, significant maxima in the first derivative of a Gaussian smoothed image. A key prob-

D. Blostein and Y.-B. Kwon (Eds.): GREC 2002, LNCS 2390, pp. 310-319, 2002.
© Springer-Verlag Berlin Heidelberg 2002

lem, however, is how to determine which maxima should be considered significant; as in line drawing interpretation, threshold selection is problematic. The widespread acceptance of Canny's algorithm is due in large part to its use of thresholding with hysteresis, which may be summarised thus:

1. Select a starting pixel whose first derivative is locally maximal in some direction (recall that step edges are elongated features that form ridges when differentiated) and above an upper threshold u. Mark that pixel as having been visited.
2. Select and move to an adjacent pixel whose first derivative is also a local maximum in some direction. Mark that pixel as having been visited.
3. Repeat 2 until the value of the first derivative at the selected pixel falls below a lower threshold l.
4. Repeat 1 until all maxima above u have been marked as visited.

Thresholding with hysteresis consistently outperforms single-value edge thresholding techniques and is simple to use. In Canny's method u and l are set by the user; though selecting upper and lower limits is often easier than choosing a single critical value. Hancock and Kittler [11] have shown that the method can be formulated as a discrete relaxation process within the framework of bayesian contextual decision theory. This allows hysteresis thresholds to be related to the parameters of an image model that can in turn be derived from image statistics. The resulting algorithm is therefore adaptive. Hancock and Kittler also introduce a third threshold that incorporates information about the connectivity of non-edge configurations. Poggio and Voorhees [12] have also modelled hysteresis thresholds, but neglected to explicitly consider connectivity information.

Thresholding with hysteresis is now a de facto standard in edge detection. Maxima in the output of the Canny operator, however, generally form thin tracks. Edges can therefore be followed without any prior knowledge of their shape; it is sufficient to assume that connectivity is non-accidental. To date, thresholding with hysteresis has only been applicable to thick data such as line drawings by top-down systems using a priori knowledge of image content to specify the pixel tracks to be considered. The line tracking system of Joseph [13] provides a good example. Its basic operation is to test a linear sequence of pixels and, while they satisfy some criterion of blackness, extend that sequence. During tracking two thresholds are employed. Meeting a single pixel with a value below a fatal threshold of around 1/3 to 1/4 of the line's greyness stops tracking immediately. A second, blacker threshold (around 1/2 of the line's greyness) is also used to mark provisional line ends. After marking a provisional end, tracking, however, continues and if the pixel track becomes darker again the provisional end is removed. This mechanism allows small light sections to be jumped as long as further sufficiently dark pixels follow on the same linear path, and is effectively an application-specific implementation of thresholding with hysteresis. There is evidence [13] that the use of the dual threshold method noticeably improved the output of Joseph's system.

In what follows we present, and discuss within the context of line drawing interpretation, a morphological implementation of thresholding with hysteresis that requires only simple thresholding and idempotent dilation and which is applicable to thick data. The algorithm is first described in Section 2. Section 3 considers the automatic determination of upper and lower thresholds. The results of initial experiments are presented in Section 4 and before future work is outlined and conclusions drawn in Section 5. A more complete evaluation and formal comparison of the performance of the proposed algorithm with alternative methods is underway and will be the subject of a future report.

2   Thresholding with Hysteresis Using Mathematical Morphology

It is useful to restate the thresholding of edge data with hysteresis thus:

1.   Discard all local maxima whose values are below l
2.   String together any remaining maximal values that are connected
3.   Discard all strings which do not contain at least one value above u

When applied to the binarisation of general grey level images this becomes:

1.   Discard all pixels whose blackness is below l
2.   Group together any remaining connected pixels
3.   Discard all connected components that do not contain at least one blackness value above u

The extraction of connected components and test for blackness values above u may be combined. Identical results are achieved by identifying all pixels whose blackness is above u, then using their image locations as seed points for region growing operations which are allowed to extend beyond this initial coordinate set, but not into regions marked as below l. Regions above u and below l may be identified by simple thresholding, The required region growing may be achieved via a variant of the familiar morphological dilation operator which is both idempotent and employs initial and transformable sets (Appendix 1 and [14]).

Initial sets define the image coordinates to a given operator may be applied; image locations outside the initial set are therefore prevented from influencing the outcome of the transformation. Transformable sets also restrict operations to regions of interest, but allow areas of the image outside the transformable set to influence the outcome of the transformation. A morphological operation is said to be idempotent if its subsequent application produces no further change in the image. Efficient algorithms implementing idempotent dilation (and other morphological operations) with initial and transformable sets are available [14]. If B is the input grey level image and I, T, O are binary images, the above algorithm may be implemented as follows:

1.   Set $I = \text{thresh}(B, u)$, where 'thresh' is the threshold operation and u the upper threshold. The coordinates of non-zero pixels of I define the initial set.

2.   Set $T = \text{thresh}(B, l)$, where l is the lower threshold. The coordinates of non-zero pixels of T define the transformable set.

3.   Set $O = B \oplus_{Coord(I),Coord(T)} \check{V}^{\infty}$, where Coord(<image>) denotes the coordinate set of an image and V is a flat unit square structuring element.

Coord(O) then gives the locations of pixels in *B* with values greater than *l* and which are connected to at least one pixel with a value above *u*.

3   Threshold Selection for Line Drawing Interpretation

Choice of threshold selection method is the central step in the design of any adaptive thresholding algorithm, and many approaches have been put forward as suitable for use in the binarisation of line drawing images [1-9]. Most work from a grey level histogram of some image area, assuming the grey level distribution to comprise two normally distributed components; one centred on plain (white) paper and the other on inked (black) regions. In single threshold methods the problem is to identify a grey level value which best (according to some criteria) separates these two overlapping distributions. Two problems complicate this task. First, in line drawings at least, the black peak is frequently diffuse and ill defined. It is often impossible to clearly identify even the modal 'black' value. Second, changing the threshold value by even a small amount in either direction can have a significant effect on the resulting binary image. Thresholding with hysteresis' need to specify an intensity band, rather than single cut-off, eases the latter problem somewhat. The low threshold can be a little too low and the high a little too high without serious effects. The problem of poorly defined 'black' pixel distributions, however, remains. Two previous approaches to threshold selection have had notable success in dealing with this problem.

Minimum error thresholding [7] assumes that the grey-level histogram is a reasonable estimate of the probability density function *p(g)* of a mixture population containing object and background pixels. It is further assumed that the two components of the mixed population $p(g \mid i)$, where $i = 1, 2$, are normally distributed with means μ_i, standard deviations σ_i and a priori probabilities P_i. The aim of the algorithm is to identify the threshold value t for which

$$P_1.p(g \mid 1) < P_2.p(g \mid 2) \quad \text{if } g <= t$$

and

$$P_1.p(g \mid 1) > P_2.p(g \mid 2) \quad \text{if } g > t$$

This value is the Bayes minimum error threshold; thresholding the original image at t will minimise the number of incorrectly classified pixels. Kittler and Illingworth [7] propose an iterative algorithm in which an initial threshold $t = T$ is chosen arbitrarily. The parameters μ_i, σ_i, P_i are then estimated for each of the two resulting sections of the distribution without explicit fitting of Gaussian profiles. These two models may be used to estimate the conditional probability *e(g, T)* of grey level g being correctly classified after thresholding at *T*. *e(g, T)* then forms the basis of an index of classification performance which reflects the amount of overlap between the two (assumed Gaussian) populations obtained by splitting the histogram at *T*. Note that the method cannot provide an exact solution, as the use of a threshold to separate the two modes of the histogram will of necessity truncate both components to some degree, causing errors in their estimated parameters. The possibility of selecting *u* and *l* to straddle an estimate of the Bayes minimum error threshold is, however, attractive, and will be the subject of a future report.

Dunn and Joseph [8] ignore the 'black' pixel distribution completely, basing their threshold selection method upon simple measurements of the location and width of the white peak alone; dark areas are effectively defined as those with a low probability of being light. The proposed method is a local one. Grey level histograms are first computed over 64 x 64 pixel sub-images, then smoothed slightly by averaging over three grey levels. After smoothing, Dunn and Joseph report experiments that suggest that the peak can be located to +/- one grey level by simply seeking the modal grey level g_m. Once g_m has been identified, the noise level associated with white paper is estimated by measuring the half-width of the white peak at half its height (Fig. 1). This value is approximately 1.2σ for a normally distributed peak of standard deviation σ. Experiments have shown [8] that although the half-width may vary somewhat from sub-image to sub-image this variation does not seem to relate to visually obvious variations in background noise. The mean half- width across all sub-images is therefore used as a background noise measure. This global measure is, however, used in conjunction with each g_m to generate local threshold values; thresholds were set at g_m -3.average half-widths, i.e. g_m - 3.6σ.

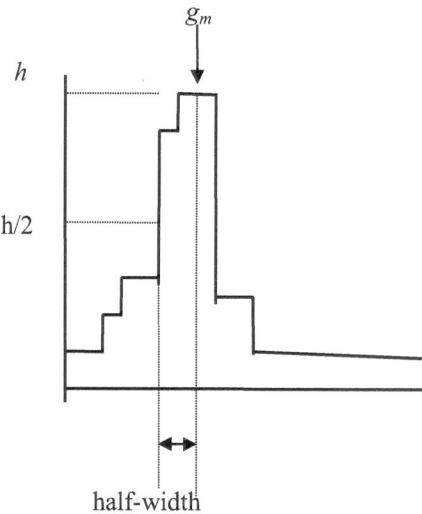

Fig. 1. Dunn and Joseph's [8] measure of noise in the grey level histogram of clean paper

Dunn and Joseph's method has several strengths. First, making the threshold a function of the standard deviation of the white peak means that, assuming a normal distribution of white pixels, the noise level expected from a given threshold may be estimated. The method also relies only upon the most reliable section (the top half) of the most reliable histogram peak. Comprehensive experimental evaluation [8] further shows the technique to deal well with the systematic noise that so often arises in dyeline and electrostatic copies of line drawings. The method therefore provides a simple, principled, well-tested benchmark which can easily be generalised to provide the two thresholds required by thresholding with hysteresis; values of u and l can be set at g_m - $N\sigma$, for different values of N.

More recently Rosin [15] has proposed a technique similar in spirit to [8] in that it considers only the largest peak in the histogram, which is assumed to occur at one end of the intensity range. Rosin's algorithm takes a straight line from the top of the histogram peak to the first zero-valued bin following the last filled bin. The threshold value is the index of the histogram bin whose value provides the minimum perpendicular distance to this line. Though it is less easy to see how a threshold pair can be generated to straddle this point in a principled fashion the method's performance on unimodal histograms makes it worthy of further consideration.

4   Initial Results

It will be noted that the algorithm outlined in Section 2 assumes that the goal is, as in edge detection, to identify significantly bright pixels, while the algorithm of Dunn and Joseph is expressed in terms of a search for significantly dark pixels. To resolve this, in the initial experiments described here we first invert the input image, making the goal the identification of bright lines on a dark background. The upper and lower thresholds u and l referred to in Sections 1 and 2 are then given by

$$l = g_m + N_l \sigma$$

and

$$u = g_m + N_u \sigma$$

respectively, where g_m captures the half-width of the now black background peak and $N_u > N_l$. The resulting binary image is then inverted to produce the more standard view of black ink on white paper. It should also be noted that although the aim of this work is to provide a local, dynamic, adaptive thresholding method based on thresholding with hysteresis, at time of writing only a global implementation is complete. The application of the current technique to local image regions is, of course, straightforward.

Fig. 2a shows a section of the poor quality drawing employed by Dunn and Joseph [8]. As the original drawing is no longer in existence the image used here was obtained from a hardcopy of [8] using an HP Scanjet. After creation of a smoothed grey level histogram, the modal grey level of the inverted image as identified as 80, with the half-width being 28 grey levels. Figs. 2b and c show the result of simple thresholding of the (inverted) fig. 2a at values of $g_m + 2.5$ and $g_m + 3.5$ halfwidths respectively. Thinking in terms of the 'blackness' of the original image, the former is a lower threshold value than the latter, being closer to the modal value of the background.

As one might expect, fig. 2c displays less residual noise than fig. 2b, but some lines (e.g. the diagonal alongside "DEEP") have significantly reduced thickness. The higher threshold also noticeably rounds the letter E in 'HOLES'. Fig. 2d shows the result of applying morphological thresholding with hysteresis to the image of fig. 2a. The threshold values employed were those used to create figs. 2b and c. Noise is reduced to a level similar to that in fig. 2c, while line thickness, and therefore representational accuracy, is similar to that seen in fig. 2b. This is to be expected from a method that takes connectivity to high confidence (i.e. very dark) pixels into account when considering lower confidence (lighter) regions. Note however, that no prior knowledge of the geometry of the regions concerned has been employed here; as long as a suitable threshold selection method can be determined, this method may be applied to any grey level image.

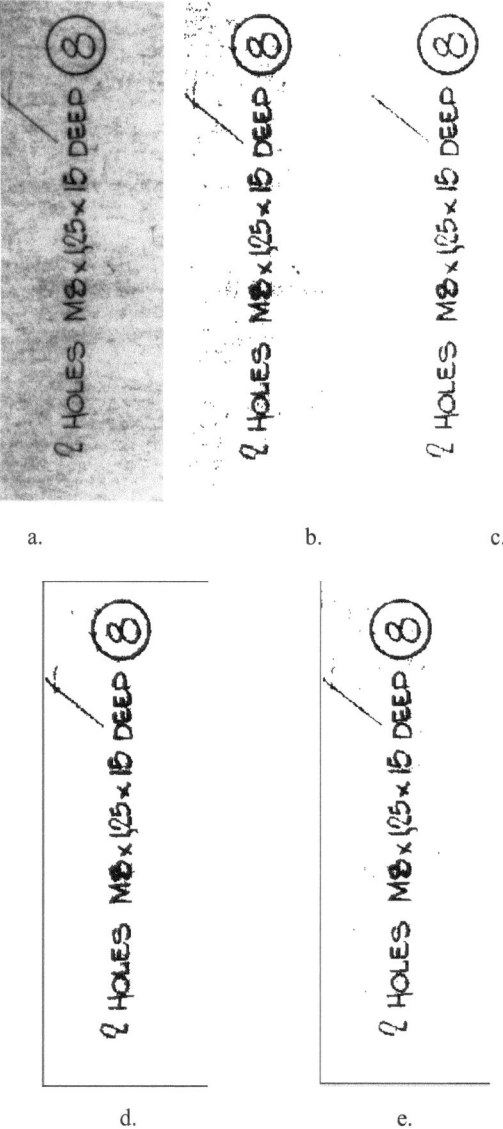

Fig. 2. a) a section of poor quality drawing employed by Dunn and Joseph [8], b) simple thresholding of the (inverted) image of fig. 2a at $t = g_m + 2.5$ halfwidths, c) simple thresholding of the (inverted) image of fig. 2a at $t = g_m + 3.5$ halfwidths, d) the result of applying morphological thresholding with hysteresis to the image of fig. 2a using the thresholds employed in the generation of figs 2b and c, e) simple thresholding of the (inverted) image of fig. 2a at $t = g_m + 3.0$ halfwidths (following Dunn and Joseph [8])

The threshold values used here were chosen to straddle the value recommended by Dunn and Joseph. Fig. 2e shows the result of simple thresholding at $g_m + 3$ half-widths (3.6σ), for comparison. The hysteresis method again shows less noise, though the edges of the characters are cleaner in fig. 2e. Further work is required to determine the optimal threshold values and spacing for different situations.

5   Conclusion and Future Work

Canny's [10] thresholding with hysteresis is now a de facto standard method of thresholding the output of gradient-based edge detection operators, but has previously only been applicable to line drawing images by systems which use prior knowledge to decide to which pixel tracks the method should be applied. We have described and presented initial results generated by a morphological implementation of thresholding with hysteresis which may be applied to thick data without prior knowledge of image content. In the current implementation, threshold values are generated following Dunn and Joseph's half-width method. Several questions remain to be addressed:

How can stable and reliable threshold values be determined from the input image? Attention will focus initially on histogram-based methods [3], beginning with the half-width and minimum error approaches outlined above, and going on to consider approaches based on Rosin's [15] algorithm.

What effect does varying the parameters have on the performance of the method? One would expect both thresholds to affect the level of noise that can be accommodated, while the lower value plays a greater role in determining the shape of the regions recovered. We shall follow Abak et al [6] in measuring pixel classification errors between artificially-generated ideal and thresholded images and using Hausdorf distances to measure shape changes under various levels of additive noise.

How does the method perform against other algorithms? The evaluation protocol outlined above will also be used to compare performance on artificial images against, in the first instance, the systems described in [7], [8], [9] and [15]. A collection of paper line drawings dating back to the beginning of the 20th century has also been assembled and, after ground-truthing by a human operator, will be used to assess the performance of the algorithm given real thresholding problems.

Development and evaluation of the technique continues.

Appendix 1: Idempotent Dilation, Initial and Transformable Sets

Employing the Image Algebra [16] notation, let $\Im$ be a value set such that $-\infty \in \Im$ and let X be a co-ordinate set. We define an $\Im$-valued image A on X as the graph of the function $A: \Re^n \rightarrow \Im$, that is:-

$$A = \{(x, A(x)): x \in X\} \cup \{(y, -\infty): y \in \Re^n \setminus X\}$$

This simplifies the analysis of operations defined on a neighbourhood of a pixel with co-ordinate $x \in X$, since any member of the neighbourhood which does not have co-ordinates in X has, by definition, pixel value $-\infty$.

Let A, V be $\mathfrak{I}$-valued images on X, Y respectively, where V is a structuring element. Let $I \subset Y$. We define the dilation of the image A by the structuring element V with *initial set I*, denoted $A \oplus_{iI} \check{V}$, thus:-

$$A \oplus_{iI} \check{V} = \left\{ (x, C(x)) : C(x) = \max \left\{ A(x+y) + V(y) : x + y \in I, y \in Y \right\}, x \in X \right\}$$

where $\check{V}$ denotes the transpose of V about its origin. This is equivalent to setting the pixel value of each co-ordinate outside the initial set to $-\infty$, reducing the co-ordinate set of the image to I.

Let A, V be $\mathfrak{I}$-valued images on X, Y respectively, where V is a structuring element. Let $T \subset X$. We define the dilation of the image A by the structuring element V with *transformable set T*, denoted $A \oplus_{tT} \check{V}$, by :-

$$A \oplus_{tT} \check{V} = \left\{ (x, C(x)) : C(x) = \begin{cases} \max \left\{ A(x+y) + V(y) : y \in Y \right\}, x \in T \\ A(x), x \in X \setminus T \end{cases} \right\}$$

Dilation with an initial set and dilation with a transformable set can be combined, an operation denoted by $A \oplus_{I,T} \check{V}$.

Let A, V be $\mathfrak{I}$-valued images on X, Y respectively, where V is a structuring element. Let $I, T \subset X$. Then we define *idempotent* dilation of the image A with initial set I and transformable set T, denoted $A \oplus_{I,T} \check{V}^\infty$, by :-

$$A \oplus_{I,T} \check{V}^\infty = A \oplus_{I,T} \check{V}^k$$

where k is the smallest integer such that further dilation by the structuring element V produces no further change in the image.

References

1. Fu K.S. and J.K. Mui. A survey on image segmentation. Pattern Recognition 1981; 13: 3- 16.
2. Sahoo P.K. et al. A survey of thresholding techniques. Computer Vision, Graphics and Image Processing 1988: 41; 233-260.
3. Glasby E. An analysis of histogram-based thresholding algorithms. Graphical Models and Image Processing 1993: 55; 6.
4. Trier O. and T. Taxt. Evaluation of binarisation methods for document images. IEE Transactions on Pattern Analysis and Machine Intelligence 1995: 17; 3; 312-315.

5.   Den Hartog T., T. ten Kate and J. Gerbrands. Knowledge-based segmentation for automatic map interpretation. Lecture Notes in Computer Science 1996: 1072; 159-178.
6.   Abak, A., U. Barns, B.Sankur. The performance evaluation of thresholding algorithms for optical character recognition. In: Proceedings of the 4th Int. Conf. on Document Analysis and Recognition 1997; 697-700.
7.   Kittler J. and J. Illingworth. Minimum error thresholding. Pattern Recognition 1986; 19: 41-47.
8.   Dunn M.E. and S.H. Joseph. Processing poor quality line drawings by local estimation of noise. In: Proceedings of the 4th International Conference on Pattern Recognition 1988, 153-162.
9.   Sauvola J. and M, Pietikainen. Adaptive document image binarisation, Pattern Recognition 2000: 33: 225-2366.
10. Canny J. A computational approach to edge detection. IEEE Transactions on Pattern Analysis and Machine Intelligence 1986; 8: 679-698.
11. Hancock E.R. and J. Kittler. Adaptive estimation of hysteresis thresholds. Proceedings IEEE Computer Vision and Pattern Recognition Conference, IEEE Computer Society Press, 1991; 196-201.
12. Voorhees H. and T. Poggio. Detecting textons and texture boundaries in natural images. Proceedings of the 1st International Conference on Computer Vision; 1987: 250-258.
13. Joseph S.H. Processing of line drawings for automatic input to CAD. Pattern Recognition 1989; 22: 1-11.
14. Bleau, A., J. de Gruise and A.R. Leblanc. A new set of fast algorithms for mathematical morphology. CVGIP: Image Understanding 1992; 56: 2: 178-209.
15. Rosin P. Unimodal thresholding. Pattern Recognition 2001: 34; 2083-2096
16. Ritter G.X., J.N. Wilson and J.L. Davidson. Image algebra: an overview. Computer Vision, Graphics and Image Processing 1989; 49: 297-331.

A Recognition Method of Matrices
by Using Variable Block Pattern Elements
Generating Rectangular Area

Kanahori Toshihiro and Suzuki Masakazu

Graduate School of Mathematics, Kyushu University
36, Fukuoka, 812–8581 Japan `kanahori@math.kyushu-u.ac.jp`

Abstract. In this paper, we propose our new method to recognize matrices including *repeat symbols* and *area symbols*. The method consists of 4 parts; detection of matrices, segmentation of elements, construction of networks and analysis of the matrix structure. In the construction of networks, we regard a matrix as a network of elements connected each other by links representing their relative relations, and consider its horizontally projected network and vertically projected one. In the analysis, we obtain the areas of variable block pattern elements generating the minimum rectangular area of the matrix by solving the simultaneous system of equations given by the two projected networks. We also propose a format to represent the structure of matrices to output the result of the matrix recognition.

1 Introduction

The technology of OCR is very efficient to digitize printed documents. However, current OCR systems can not recognize mathematical formulae which are very important in scientific documents. Several algorithms for recognizing mathematical formulae have been reported in literature ([1]-[3]). Some of them can be applied to very simple matrices, such as gridironed matrices. However, no method to recognize matrices including abbreviation symbols, which are used in mathematics, is reported. Besides, there is no standard format to represent the structure of complicated matrices. So, we can not keep the result of matrix recognition.

In this paper, we present a method to recognize matrices including *repeat symbols* or *area symbols*, which appear in scientific documents, and a format to represent their structure to output the result of our recognition, and report the experimental results of this method. Matrices which we are going to recognize consist of *formula elements*, *area symbols* and *repeat symbols* (Fig. 1). In the recognition method, we assume that lines are distinguised and characters are correctly recognized by a character recognition, matrices have at least two rows, and elements of matrices can be gridironed. When the segmentation and extraction of formula elements, repeat symbols and area symbols are exactly completed, the structure analysis, is *always exactly* done. The *decorators*, which

D. Blostein and Y.-B. Kwon (Eds.): GREC 2002, LNCS 2390, pp. 320–329, 2002.
© Springer-Verlag Berlin Heidelberg 2002

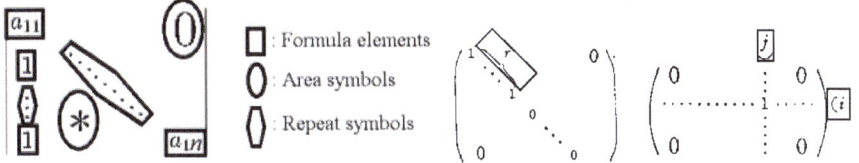

Fig. 1. The components of a matrix (the left) and Decorators (enclosed in boxes on the right)

represents detail numbers of elements, positions of elements, etc., are excluded at present (Fig. 1).

2   Representation of Matrix

We classify the components of a matrix into the following 3 classes;

1. Formula element
   - It is a component of a matrix.
   - It has only one grid as its own area.
   - It can connect to other elements in the 8 directions.
2. Area symbol
   - It has several grids as its own area.
   - Its area has a free boundary.
   - Common area symbols are $O, 0, 1, *$, etc. , and a space is also an area symbol.
3. Repeat symbol
   - It means that formula elements are continuously aligned on the straight line in its direction; $\downarrow, \rightarrow, \searrow$ or $\nearrow$.
   - It can connect to formula elements and other repeat symbols with different directions.
   - We assume that it consists of 3 points or more and they are put on straight line.

   We represent an area of a matrix element by the set of couples of indices representing the row and column on the matrix. In Fig. 2, the formula element 'a_{11}' has (1,1) as its area, 'a_{nn}' has (4,4), and the area symbol '0' has (1,2), (1,3), (1,4), (2,3), (2,4) and (3,4) as its area, and '*' has (3,2), (4,2), (4,3).

   The format to represent the structure of matrices is summarized in Table 1. The results of our matrix recognition are output in it. For example, the matrix in Fig. 2 is represented by Table 2, where we omitted the coordinates of the elements' bounding rectangles.

Table 1. The rule of matrix representation

Symbol Names	Informations
MATRIX	Coordinate of its own bounding rectangle on the image, Parentheses on right and left, Numbers of the row and column List of ELEMENT, List of CONNECTION
ELEMENT	Coordinate of its bounding rectangle, Results of the recognition Set of its areas
CONNECTION	Couple of the positions of the repeat symbol's end points

Table 2. Example of matrix representation

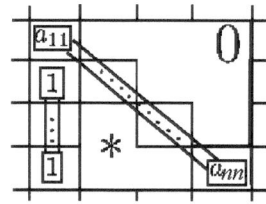

Fig. 2. Example of positions of matrix elements

MATRIX				
Parentheses	), (		Row, Column	4, 4

ELEMENT LIST				
Formula	Areas		Formula	Areas
a_{11}	(1,1)		1	(2,1)
1	(4,1)		a_{nn}	(4,4)
0	(1,2),(1,3),(1,4),(2,3),(2,4),(3,4)			
*	(3,2),(4,2),(4,3)			

CONNECTION LIST			
End Points	(2,1),(4,1)	End Points	(1,1),(4,4)

3   Matrix Recognition

For our matrix recognition of a page image, we assume that its lines are distinguished, the characters are recognized, the coordinate of the bounding rectangle of each character is obtained.

The method consists of 4 parts;

1. Detection of matrices in a page image, and extraction of characters in each area of the matrices.
2. Segmentation of the characters into elements for each matrix.
3. Construction of the network where formula elements are connected by repeat symbols or adjacent relations.
4. Structure analysis of the matrix.

In the 1st step, the algorithm of the detection of matrices is very simple at present. Its outline proceeds by finding couples of parentheses in the given character sequences.

In the 4th step, we let the minimum length of repeat symbols in a matrix be 2 on the network. Then, we set up equations for the height and width of the matrix from its vertically projected network and horizontally one. By solving the equations, we obtain the areas of variable block pattern elements generating the

minimum rectangular area of the matrix, and decide the minimum numbers of its rows and columns. After these steps, if a detected matrix has only one row, it is not recognized as a matrix.

In the followings, we assume that the results of the character recognition are always correct.

3.1 Segmentation of Elements

By the detection of matrices, each detected matrix has the set of characters in its own area. It is necessary to group them into matrix elements. In this section, we explain the method of the segmentation.

We let $L = \{C_1, \cdots, C_n\}$ be the set of the characters in the matrix. We define the distance $d(C, D)$ between C and $D \in L$ by

$$d(C, D) := \alpha_x d_x(C, D) + \alpha_y d_y(C, D),$$

where d_x (or d_y) is the distance between the intervals projected to x-axis (resp. y-axis), but we let d_x (resp. d_x) be 0 if the intersection of the intervals is not empty. The coefficients, α_x and α_y, also depend on C and D. We let the coefficient α_x for the horizontal distance be smaller than α_y for the vertical distance, so that the horizontal connections are tighter than the vertical ones. If C or D is a binary operator, we set α_x smaller value than ordinary, because binary operators have bigger spaces on their both sides than other characters in many cases. The operator, '$-$' (minus), is also used as a sign at head of elements. Therefore, if the left space of a character '$-$' (minus) is longer than the right, we consider it as a sign, and cut the connection to its left-hand side. Big symbols, $\sum$, $\prod$, etc. , and fractional lines often have vertical connections. If they have formulae above themselves or below (upper limit formulae, lower one, numerators or denominators), the formulae are closer to them. So, we let α_y for big symbols and fractional lines be bigger to prevent them from connecting to formulae which are not their limit formulae, numerators or denominators.

For two characters, C and D, we say that they are *clearly unadjacent*, when there is some other character whose bounding rectangle is at least on one of lines from 4 corners (upper-left, upper-right, lower-left and lower-right) of C's rectangle to the same corners of D's. We use the distance between clearly unadjacent characters to evaluate of the thresholds. We define the threshold $t(C)$ with respect to C as the average of the distances to the two closest unadjacent characters. We let $G(L)$ be the directed graph derived from the adjacency matrix $A(L) := (a_{ij})_{i,j=1,\cdots,n}$. Then, we can obtain the elements of the matrix as the connected components of the graph $G(L)$. The elements a_{ij} of $A(L)$ where characters, $C_i, C_j \in L$, are not clearly unadjacent are defined by

$$a_{ij} := \begin{cases} 1 \ (d(C_i, C_j) < t(C_i)) \\ 0 \ (d(C_i, C_j) \geq t(C_i)) \end{cases} (C_i, C_j \in L),$$

and for clearly unadjacent characters, C_i and C_j, a_{ij} are always 0.

Table 3. Example for the flow of construction of networks

The special process for the class of dots is done after the above segmentation. First, we take a dot which have very close elements on its both sides for a comma, and combine them into one element. Next, we extract repeat symbols from the dots. We classify repeat symbols into 4 types $(\downarrow, \rightarrow, \searrow, \nearrow)$ according to their directions. The extraction of repeat symbols proceeds by tracing the dots.

3.2   Construction of Networks and Equations

This section describes the algorithm to construct *the horizontally projected network* and *the vertically projected network* using the following simple examples in Table 3-(0).

1. Connection by repeat symbols (Table 3-(1))
   For each repeat symbol, we connect its origin to its terminal. If the origin and the terminal of a repeat symbol are other repeat symbols, we divide the repeat symbols by virtual elements, we call them *pivots*. We consider the lengths of these connections as variable. If a repeat symbol is divided by a

pivot, let the minimum value of the variable corresponding to the symbol be 1. Otherwise, let the minimum values of repeat symbols be 2.

2. Segmentation into lines (Table 3-(2))

First, we set each pair of the horizontally connected elements on the same line. Second, we segment the elements into lines by using the lengths of overlapping of their bounding rectangles on their horizontal projection, their sizes and baselines. If there are bigger $O, 0, 1, *$ than the average of characters' sizes in the document, or elements laying on several lines, we let them be area symbols.

3. Vertical connection of elements (Table 3-(3))

We connect each pair of vertically adjacent elements by a vertical 1-length path, its vertical length is 1, and its horizontal length is 0.

4. Horizontally connection of elements (Table 3-(4))

We connect each pair of horizontally adjacent elements by a horizontal 1-length path, its vertical length is 0, and its horizontal length is 1.

5. Diagonal connection of elements (Table 3-(5))

For each element which is not connected to others by 1-length paths, we connect it to its diagonally adjacent element by a diagonal 1-length path, its vertical length is 1, and its horizontal length is 1.

6. Horizontal projection of the network (Table 3-(6))

By identifying elements on each line, we horizontally project the network constructed on the matrix by the above connections. We also identify elements on the upper end (or the lower end) of the matrix. If the projected area covers some area symbols, we do not project them. For area symbols uncovered by the projected area, we project them and connect to other close nodes to them.

7. Vertical projection of the network (Table 3-(7))

Similarly, we vertically project the network by identifying elements which are vertically connected each other. We also identify elements on the left end (or the right end) of the matrix.

8. Identification of nodes and paths (Table 3-(8))

On each projected network, we identify nodes having a common node at 1-distance from both of them including the directions. Moreover, we identify paths whose origin and terminal are same. We store the information of these identifications.

9. Complement of paths

If there is a node which no path enters, or if there is a node which no path leaves, we connect the node to the nearest node by a virtual path. We let the minimum length of a virtual path be 1.

Thus, we obtain two projected networks. All the lengths of total paths from the upper end (or the left end) to the lower end (resp. the right end) must be equal to the number of rows (resp. columns) of the matrix. We let v be the number of rows and h be the number of columns, and assign an variable to each arc of the path. Then, we can set up the simultaneous system of equations by the lengths from end to end and the information of the paths' identification. In

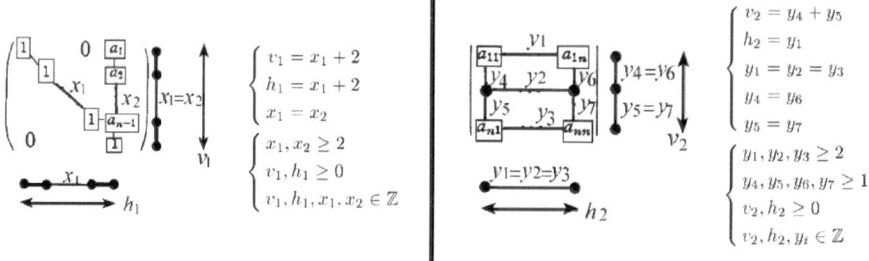

Fig. 3. Example of the networks and the simultaneous systems

the examples, we can obtain the following simultaneous systems and conditions (Fig. 3).

3.3   Structure Analysis

By solving a simultaneous system introduced from each projected network so that v and h are minimum, we can obtain the minimum numbers of rows and columns and the relative positions between the connected elements. We gridiron the matrix and put the elements on the grids by obtained values. The area symbols have the connected components separated by paths as their own areas.

To solve the simultaneous system of equations, we do not use an existing algorithm for linear programming, because solutions of the simultaneous system are limited to natural numbers and an algorithm for linear programming does not always guarantee integer solutions generally. We use our original algorithm to solve the simultaneous system. This algorithm directly solves the simultaneous system and possesses higher speed.

Briefly, the algorithm to the simultaneous system of equations proceeds as follows:

1. Evaluate the temporary values by substituting minimum values
   We substitute minimum values of the variable. Then, we let the maximum value for v (or h) be a temporary value of v (resp. h), and let the minimum values of the variables included in the equations attaining the maximum value of v or h be their own temporary values.
2. Substitution of the temporary values
   We substitute the temporary values of v, h and variables obtained in the step 1. We solve the monomial equations changed by this substitution, and let the solutions be the temporary values. Using these temporary values, we repeat this step until new temporary values are not obtained. If there are different temporary values of a variable, we let the maximum value among them be the minimum value of the variable, and try from the first step again.

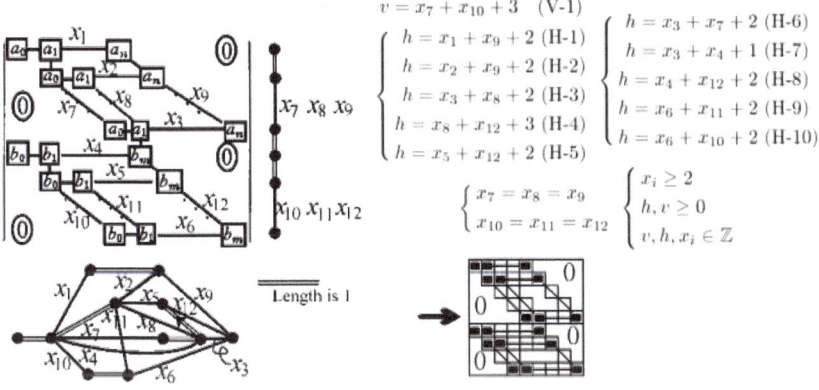

$$v = x_7 + x_{10} + 3 \quad \text{(V-1)}$$

$$\begin{cases} h = x_1 + x_9 + 2 & \text{(H-1)} \\ h = x_2 + x_9 + 2 & \text{(H-2)} \\ h = x_3 + x_8 + 2 & \text{(H-3)} \\ h = x_8 + x_{12} + 3 & \text{(H-4)} \\ h = x_5 + x_{12} + 2 & \text{(H-5)} \end{cases}$$

$$\begin{cases} h = x_3 + x_7 + 2 & \text{(H-6)} \\ h = x_3 + x_4 + 1 & \text{(H-7)} \\ h = x_4 + x_{12} + 2 & \text{(H-8)} \\ h = x_6 + x_{11} + 2 & \text{(H-9)} \\ h = x_6 + x_{10} + 2 & \text{(H-10)} \end{cases}$$

$$\begin{cases} x_7 = x_8 = x_9 \\ x_{10} = x_{11} = x_{12} \end{cases} \qquad \begin{cases} x_i \geq 2 \\ h, v \geq 0 \\ v, h, x_i \in \mathbb{Z} \end{cases}$$

Fig. 4. Example for the result of the structure analysis

3. Substitution of the common temporary values

   In a manner similar to the step 2, we substitute the common temporary values for remaining equations, solve monomial equations. If there are different solutions of a variable, we change its minimum value and try from the first step again. We repeat this step until new temporary values are not obtained.

4. Solving remaining equations

   After those steps, if some equations are remaining, we obtain the values of the remaining variables by using the elementary transformation of a matrix. From the following simultaneous system (in this case, $m < n$), its *coefficient matrix* is introduced, and then we can deform it into 4-parted matrix by the elementary transformation, where we let r be the rank of the matrix, I_{rr} be a $r \times r$ identify matrix, and O_{pq} be a $p \times q$ zero matrix.

$$\begin{cases} a_{11}x_1 + \cdots + a_{1n}x_n = a_1 \\ a_{21}x_1 + \cdots + a_{2n}x_n = a_2 \\ \cdots \\ a_{m1}x_1 + \cdots + a_{mn}x_n = a_m \end{cases} \Rightarrow \begin{pmatrix} a_{11} & \cdots & a_{1n} \\ a_{21} & \cdots & a_{2n} \\ \cdots\cdots\cdots \\ a_{m1} & \cdots & a_{mn} \end{pmatrix} \Rightarrow \left(\begin{array}{c|c} I_{rr} & O_{r,n-r} \\ \hline O_{m-r,r} & O_{m-r,n-r} \end{array} \right)$$

When the values of the variables corresponding to columns whose elements are zero are given, the others are determined (remarking the reshuffle of the columns). Then, we give the minimum values of $x_{r+1}, \ldots, x_n$, and solve the remaining equations in a manner the similar to the step 2. Hence, the solutions depends on the order of naming of variables and the elementary transformation.

For almost all matrices which we used to evaluate our methods in Section 4, their solutions are obtained in 1 or 2 steps. For example, the solutions of the matrix in Figure 4 are obtained in 2 steps, and the result of the structure analysis is represented on the grids.

Table 4. Experimental results of segmentation and extraction

Data	English	Japanese	Total
Total matrices	200	229	429
Detected matrices	198	221	419
Detection rates	99.0	96.5	97.7
Detection errors	2	4	6
Segmentation errors of elements	15	6	21
Segmentation rates of elements for detected matrices	89.4	85.5	87.4
Proper extraction of symbols	145	178	323
Proper extraction of symbols for total matrices	72.5	77.7	75.3
Extraction rates of symbols for detected matrices	81.9	94.2	88.3

4   Experimental Results

In order to evaluate our methods, we implemented them into our original OCR System ([3]). In our method, when the segmentation and extraction of formula elements, repeat symbols and area symbols are exactly completed, the structure analysis, is *always exactly* done (its success rate is *100%*!). Therefore, we evaluated the segmentation and extraction of elements and symbols, and the detection of matrices, using Infty.

We used two English textbooks and two Japanese ones of mathematics including many matrices. For 50 page images of each text (total 200 pages) scanned in 600 dpi, we counted the numbers of matrices where errors were made with respect to the 3 parts. The 50 page images included about 10 pages where matrices did not appear but big parentheses did in order to evaluate the detection. We show the experimental results below.

In Table 4, "Detection errors" means the numbers of detected non-matrix parts as matrices.

Table 4 shows that the detection of matrices has some measure of high accuracy. The misdetection are mainly caused by errors of line segmentation and broken big parentheses by bad conditions of the prints. It was found that the segmentation of elements had a tendency to segment an element including long subscripts into several elements. In the experiment, the recognition rates depend on textbooks. One of the main reasons for the dependence is that all textbooks have their own distinctive notations of matrices. Interestingly, error frequency of the judgment of area symbols on simple matrices was higher than on complicated one, because the information of character sizes on simple matrices was less than on complicated one. Comparing the construction rates of networks for total matrices with the construction rates for detected matrices proves that if the segmentation of elements succeeds, the next analysis will be successful at a remarkable rate.

5   Conclusion

We proposed a practical method to recognize matrices containing abbreviation symbols and a format to represent their structure to output the recognition result. We defined the domain, allowing matrices to contain formula elements, area symbols, and repeat symbols in the matrix coordinate.

The method consists of 4 independent parts; detection of matrices, segmentation of elements, construction of networks and analysis of the matrix structure. In the detection of matrices, we use very simple algorithm using correspondence of big parentheses, and the experimental results prove its high accuracy. To segment characters in a matrix into elements, we use distances between the characters and some characters' features in the mathematical structure. However, the features are not enough to segment, so there are many errors in the experiment. In the construction of networks, we project the connections on a matrix vertically or horizontally. The projected networks are robust against mis-connection between elements on a matrix, namely if there are some lost connections, we can obtain the element positions on a matrix. The construction of networks has some measure of accuracy in the experimental results, but there are some errors in irregular notations. In the analysis of the matrix structure, we let the length of repeat symbols in a matrix be variable. Then, we set up equations for the height and width of the matrix from its vertically projected network and horizontally one. Using the minimum values and ranges of solutions (all solutions are positive integers) instead of the linear programming, we solve the equations and obtain the width and height of a matrix and element positions. When the segmentation and extraction of formula elements, repeat symbols and area symbols are exactly completed, the structure analysis, is *always exactly* done.

For further improvement, we will try the following problems: (1) To use more mathematical information for improvement of the segmentation of elements. (2) To recognize the decorators of matrices which we excluded in this paper.

References

1. D. Blostein and A. Grbavic, *Recognition of Mathematical Notation*, Handbook of Character Recognition and Document Analysis, Eds. H. Buke, and P. Wang, Word Scientific, 1997. 320
2. M. Okamoto and H. Twaakyondo, *Structure analysis and recognition of mathematical expressions*, Proc. 3rd Int. Conf. on Doc. Anal. and Recog., Wontreal, pp. 430-437, 1995.
3. Y. Eto, M. Sasai and M. Suzuki, *Mathematical formula recognition using virtual link network*, Proc. 6th Int. Conf. on Doc. Anal. and Recog., Seattle, pp. 430-437, 2001. 320, 328

Music Manuscript Tracing

Kia Ng

Interdisciplinary Centre for Scientific Research in Music
School of Music & School of Computing
University of Leeds, Leeds LS2 9JT, UK
omr@kcng.org
www.kcng.org/omr

Abstract. This paper presents an ongoing project working on an optical handwritten music manuscript recognition system. A brief background of Optical Music Recognition (OMR) is presented, together with a discussion on some of the main obstacles in this domain. An earlier OMR prototype for printed music scores is described, with illustrations of the low-level pre-processing and segmentation routines, followed by a discussion on its limitations for handwritten manuscripts processing, which led to the development of a stroke-based segmentation approach using mathematical morphology. The pre-processing sub-systems consist of a list of automated processes, including thresholding, de-skewing, basic layout analysis and general normalization parameters such as the stave line thickness and spacing. High-level domain knowledge enhancements, output format and future directions are outlined.

1 Introduction

Nowadays, the computer has increasingly become an important device in many aspects of Music. Whether in multimedia processing (e.g. digital audio processing or synthesis), or in performing time consuming tasks, such as part extraction or transposition, the computer offers speed and accuracy. In order to carry out any machine processing of a music score, it has to be represented in a machine readable format, and hence input methods became an important factor. Current input methods, such as using an electronic keyboard (e.g. MIDI) or notation software, are time and labour intensive, and require manual human intervention.

Optical Music Recognition was first attempted over thirty years ago [21], and the potential benefits of such a system have been widely recognized. OMR research has been especially active over the last decade [2,3,4,5,14,15,17,19,23,27], and there are currently a number of commercially available packages [13,20,25]. Reviews and background of various OMR systems can be found in [7] and [26].

With an effective and robust OMR system, it can provide an automated and time-saving input method to transform paper-based music scores into a machine readable representation, for a wide range of music software, in the same way as Optical Character Recognition (OCR) is useful for text processing applications. Besides direct

D. Blostein and Y.-B. Kwon (Eds.): GREC 2002, LNCS 2390, pp. 330–342, 2002.
© Springer-Verlag Berlin Heidelberg 2002

applications, such as playback, musical analysis, re-printing, and archiving, such an application would enable efficient translations, for example, to Braille notations or other non-western musical notations.

2   Obstacles

Musical symbols and score layout are highly interconnected. Western Common Music Notation (CMN) uses a grid system with horizontal stave lines to guide the relative pitch and the stave lines introduce interconnectivity to most of the musical symbols on the scores. Additionally, musical symbols may connect horizontally (e.g. beams), vertically (e.g. chords) or be overlaid (e.g. slurs or phrase markings cutting through stems or bar lines). When symbols are grouped (horizontally and/or vertically), they vary in shape, size and visual appearances. For example, consider the shape of four isolated semi-quavers, and the many possible appearances of a four semi-quaver beamed groups, hence standard approaches, including template matching and deformable model methods, are difficult to apply at this level.

Besides the inherent complexities of music notation as mentioned earlier, handwritten manuscripts present another layer of uncertainties, including the inconsistencies in writing styles, their typically slanted or curved line segments, fussy and uneven fillings (e.g. solid or hollow note-head) and many others. As with other forms of optical document analysis, imperfections introduced during the printing and digitising process, for example, noise, decolourisation and uneven (non-planner) paper surface that are normally tolerable to the human eye can often introduce complications to the recognition process.

3   Printed Music Score Recognition

This section outlines the approaches of an earlier OMR prototype designed and developed for printed music scores. The pre-processing stages of the prototype are illustrated in Fig. 1 and 2.

3.1   Pre-processing

The prototype takes grey level bitmap from a scanner or read directly from image files as input. Generally the input bitmap is 300dpi in 256 grey-level. Firstly, the input grey image is converted into a binary image using the iterative threshold selection method [22,28]. Typical residues or noises are simply some isolated and small points, which can be identified and removed easily (see Fig. 1(a)).

In practice, it is difficult to position the music score with the staves perfectly horizontal, during digitisation. To simplify further processing, the skew of the input image is detected and corrected by rotation, and binarisation is re-applied to the corrected image (see Fig. 1(b)).

Initial stave line positions are located by horizontal projection, and a measure of horizontality at a range of possible skew angles with a fixed step (e.g. -5° to +5°, at

0.1° intervals) is taken with the four most dominant estimated stave line position (assuming a global skew).

The middle column of the image, which is the most likely to cut through any stave line, is scanned from the top to the bottom. When a foreground pixel is found, a digital line template at each possible skew is overlaid onto the image, and the number of foreground pixels, which fall on the template line, is counted. The count for each angle is then accumulated for each row, after which the maximum of the pixel-count versus angle plot yields the skew angle.

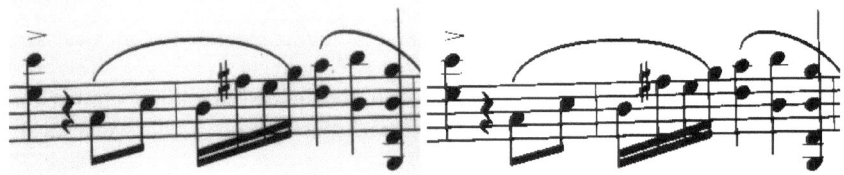

(a) Iterative thresholding: Input (left) and binarised output (right).

(b) Deskewing (left) and binarisation (right)

Fig. 1. Low-level pre-processing

Stave line is the fundamental element of the CMN. It forms a grid system for musical symbols, of which most are related to the geometry of the staves; for example, the height of a note head must approximate the distance between two stave lines plus the thickness of the one or two stave lines. Hence, it is important to detect their positions accurately in order to extract certain parameters of other related symbols, which is represented by the relative position of the symbols with respect to the stave line. For example, the pitch information of a note-head. However, stave lines interconnected most of the symbols that are printed on or around them. Hence, once the stave lines have been located, a line tracing algorithm with a local vertical projection window is used to detect the average stave line thickness, and electively mark the pixels that belong to each stave line, and thereby removing the stave lines to isolate the musical features for further processing [14].

Stave lines are typically not completely straight or of even thickness, even on printed manuscript papers. Hence, the left and right most coordinates and the minimum or maximum tuning point of each stave line are recorded. [3] noted that

> *"Music recognition would be simple if music images had straight staff lines, clear printing ..."*

At this stage, the stave line position, average stave line thickness (α_t) and the space between stave lines (α_s) are detected. When the stave lines are removed, the image is left with blocks of connected foreground (see Fig. 2), which may be isolated or grouped musical symbols (composite objects). α_t and α_s are used to provide the fundamental unit ($\alpha_f = \alpha_t + \alpha_s$) for normalization, since the stave lines form the grid system which constrains the relative sizes of musical symbols.

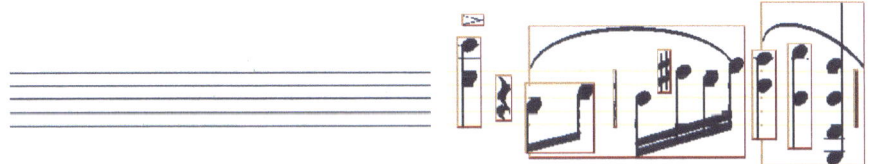

Fig. 2. Stave lines removal: Stave line positions (left), stave pixels detection and initial segmentation (right)

3.2   Level Graphical Primitive

As mentioned above, musical symbols can be grouped to form composite objects, which complicate the recognition process. The primitives with which the composer deals (e.g. quavers or eighth-notes, and others) are often not as simple for an automatic system to classify as their component parts. A sub-segmentation algorithm is designed to disassemble musical symbols into lower level graphical primitives, which sometimes, but not always, do not correspond to the musical symbols. For example, a stem is regarded as a primitive to be classified independently of its note-head, and beam is traced and recognized independently of the connected stems. Examples of primitives include note-heads, vertical lines, horizontal lines, curves, isolated rest signs, isolated accidental signs and others.

3.3   Classification

After initial segmentation, the first iteration of the classification module recognises isolated primitive musical symbols (e.g. dot and rest), and symbols that are normally located at certain positions with respect to their staff (e.g. clef and time signature).

Recognition of other primitives is performed by interplay between the classification and the sub-segmentation modules. The classification module relies on the sub-segmentation module to extract the primitive features for recognition, and the sub-segmentation module relies on the classification module to satisfy one of its termination criteria.

The simplest model explaining the observation is the best representation [24]. The classification module only uses simple features such as the aspect ratio and normalised width and height of a feature's bounding box, based on a pre-sampled training set, using a k-Nearest-Neighbour (kNN) classifier [15].

The prototype has been trained and tested on live data. The classifier was built to recognise 12 different (sub-)symbols. The training set consists of 15 samples for each group of features, captured from six different scores.

With the chosen feature set, the clusters are well separated. Feature with overlapping clusters, for example the quaver-rest and accidental, are merged into a multi-class (despite dissimilar semantics), to be resolved with syntactical conventions at a later stage. This simple classifier demonstrates over 95% reliability. Misclassification can usually be detected and corrected with high-level knowledge at the reconstruction stage (see Section 5).

3.4 Sub-segmentation

This module takes each block of connected composite feature in turn and passes it to the classification module. If the object is not classified and is too large as a primitive feature (typically in low density and with its normalised width and height > 1.5 α_f), it is sub-segmented into two or more objects.

The first iteration of the sub-segmented module is specially targeted on long and thin features, such as slurs and beams. Instead of breaking the features into two or more smaller sub-parts, any long and thin features are traced and recognised.

Three sub-segmentation approaches are applied:

1. *horizontal tracing (typically for separating long and thin features, such as slurs and beams, as discussed earlier)*
2. *vertical cut (example usage include separating note-head from its stem)*
3. *horizontal cut (example usage include separating vertically connected note-heads)*

Criteria for the appropriate choice of sub-segmentation method are based on the normalised width and height, and density of the object.

Within a composite object, a sudden change in vertical projection histogram usually suggests a possible junction point of two separate features. Hence, the horizontal position (x) with maximum of the measure

$$(V_{x-1} - 2V_x + V_{x+1}) / V_x \tag{1}$$

across the vertical projection V_x is used to indicate the best break point [12]. Fig. 3 illustrates the separation of detected slurs and beams of input from Fig. 2, using the horizontal tracing sub-segmentation algorithm.

Fig. 4 shows a completely sub-segmented input, using the divide and conquer technique, which over-segmented the musical symbols into lower level graphical primitives before recognition and reconstruction, and a basic reconstruction of the recognised features.

4   Handwritten Manuscript

This section discusses the limitation of the earlier prototype (designed for printed music scores) for the processing of handwritten manuscripts, and outlines the development of a stroke-based segmentation approach using mathematical morphology.

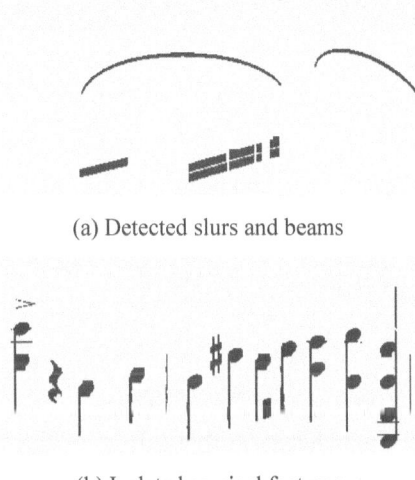

(a) Detected slurs and beams

(b) Isolated musical features

Fig. 3. (a) Long and thin segment tracing and (b) separated features

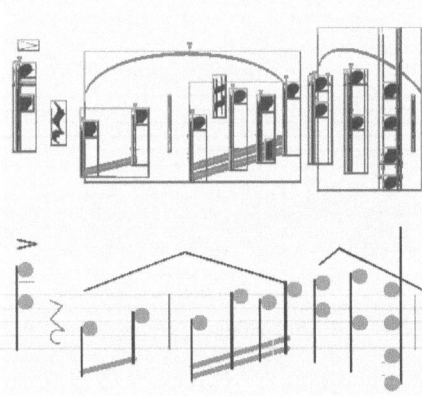

Fig. 4. Complete sub-segmentation (top) and a basic reconstruction (bottom)

4.1 Pre-processing

The pre-processing processes discussed above, for printed music, are reusable for handwritten manuscript analysis. However, the sub-segmentation approach relies on the vertical orientation of the musical features and performs unsatisfactorily for handwritten music due to the characteristically slant and curve line segments. Following general methods used in the domain of cursive script analysis and recognition, a mathematical morphology approach [1] is adopted.

Using the binarised input image (see Fig. 5), a skeletonisation process is applied to reduce features to their skeleton representations (see Fig. 6).

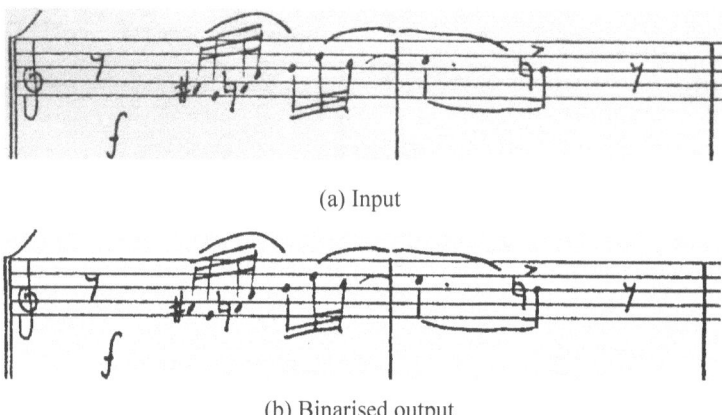

(a) Input

(b) Binarised output

Fig. 5. Pre-processing

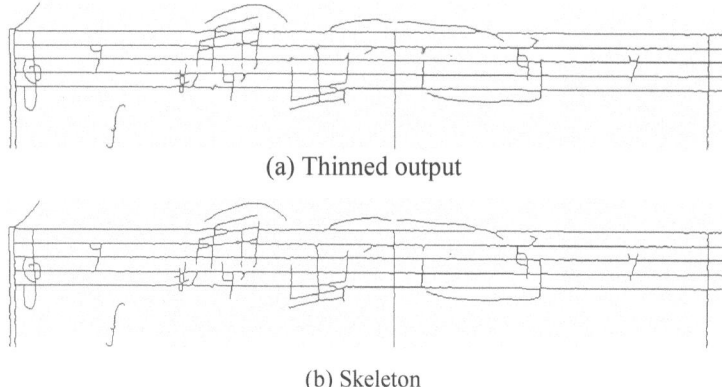

(a) Thinned output

(b) Skeleton

Fig. 6. Thinning process

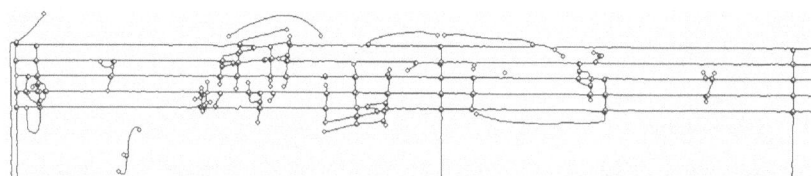

Fig. 7. Detected junction- and termination-points (illustrated by circles)

At this stage, musical features of handwritten manuscripts may be viewed as a set of interconnected curved lines. From the skeleton representations, junction- and termination-points are extracted. Junction points are pixels on the skeleton with more than two connected foreground pixels and termination points are skeleton pixels that have one (or zero, i.e. a dot) connected foreground pixels (see Fig. 7).

4.2 Stave-Line

Similar to printed OMR approaches, the stave-line pixels are detected and marked at this stage to segment the musical writing that has been interconnected by the staves. For the process, all horizontal line segments are parsed to determine if they belong to part of a stave using basic notational syntax and an estimated stave-line height.

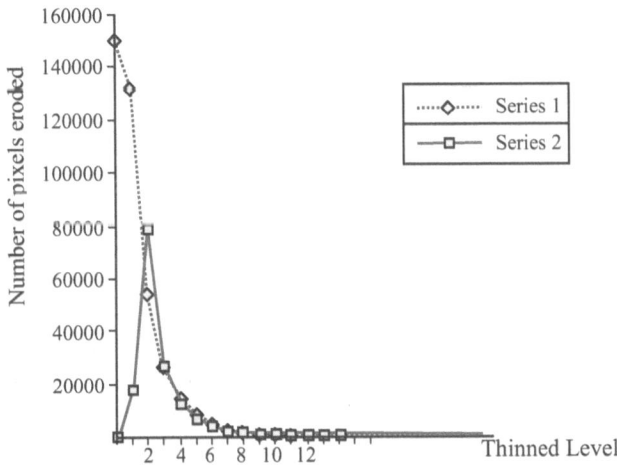

Fig. 8. A distribution of eroded pixels per thinning iteration

With the mathematical morphology approach, the estimated stave-line height can be obtained from the histogram of the number of pixels eroded during each skeletonisation iteration. Fig. 8 shows a typical distribution of the histogram, and the estimated height is taken from the largest change in the number of pixels eroded. Series 1 illustrates the number of pixels eroded at each iteration, and Series 2 is a second differential of Series 1, denoting the estimated stave-line height (maximum).

4.3   Sub-segmentation

Initial segmentation technique using junction- and termination-points manage to separate some of these connections, but does not detect joints (points of discontinuity between segments) that are smooth and form continuous curve segments (see Fig. 9).

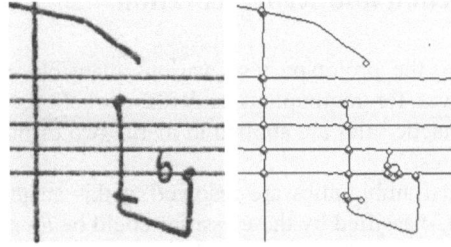

Fig. 9. Input (left). Thinned output with detected junction- and termination-points (right)

Hence an additional process using a combination of edges, curvature and variations in relative thickness and stroke direction to perform further sub-segmentation, segregating the writings into lower-level graphical primitives, such as horizontal and vertical line segments, curves and ellipses, before passing to the classifier.

4.4   Classification

As the sub-segmentation approach for handwritten manuscripts is designed for the same graphical primitives as the printed version, a similar underlying methodology can be used for both. Handwritten primitives are classified using a kNN classifier, with appropriate training dataset, and additional feature vectors to make use of information extracted during the skeletonisation process, and the junction point distributions.

4.5   Stroke Continuity

From the sub-segmented short line segment, each terminal points are parsed to search for any other nearby (α_f x α_f search window) terminal points which are collinear with the current segment or following a polynomial (using a least square estimate) extrapolation from the terminal points of the current segment. The algorithm search for vectors that might form part of the current segment. All segments within the search area are examined and the parameters in consideration include the distance and angle between the two termination points. There are two types of segment assembly:

1.   *tracing through a gap (e.g. connecting parts of broken stem)*
2.   *tracing through a junction-point (e.g. inner junction-point of a treble clef)*

Works in hand include an improved tracing routine using a database of isolated handwritten music symbols (stroke directions and sequences) in order to provide logical stroke tracing and provide better classification with the mean shape and possible deformation and variation (learnt from the database). The dataset is captured using a pen-based graphic tablet.

Further work in the stroke tracing approach may include writing style extraction for writer identification, and to automatically learn the writing style in order to reproduce a consistently written manuscript, using the mean shape as the handwritten font. This could result in a manuscript *beautifier* system.

5   Reconstruction and Representation

After classification, the prototype tests various plausible combinations of the sub-segmented primitives for regroupings and reconstruction. For each regrouping hypothesis, basic syntactic rules are applied to merge two or more primitives into higher level musical features.

Firstly, contextual ambiguities are resolved at this stage using relative positions. For example, a dot, classified by the classifier could be an expression sign or a duration modifier depending on its relative position with respect to a note-head nearby.

Music score contains redundant information and the reconstruction module offers an intermediate stage where extensive heuristic, musical syntax and conventions could be introduced to enhance or confirm the primitive recognition and regroupings. For example, a solid-note-head connected to a stem to form a quarter-note. The main function of this module is to analyse the 2-D localizations and the spatial relationships of the recognized graphical primitives (e.g. left, right, above or below) and connectivity (e.g. near, touch, or intersect). The current prototype uses a rule-based system to process the primitives in three levels:

1. *local: at a note by note level, for example, a dot to the right of a note-head implies dotted note;*
2. *intermediate: at a bar level, with a database of frequently found rhythmical patterns and beam groupings; and*
3. *global: within an estimated sections with similar tonality (including clef and key information) and timing information to further clarify uncertainties.*

Possible future enhancements include formal approaches, such as graph-grammar processing [8,10].

The default output format is currently set to *exp*MIDI [9], which is compatible with the standard MIDI file format, and capable of storing expressive symbols such as accents, phrase markings and others. This format encompasses the compactness of the widely supported MIDI file format as well as notational details needed for the music visualisation and archiving.

There have been active developments on musical file formats, for example Notation Information File Format (NIFF), Standard Music Description Language (SMDL), GUIDO [11], Score, WEDELMUSIC [6], MusicXML and many others. It is intended to include input and output filters for some of these musical data representations.

5.1   High Level Enhancement

Global information such as the time and key signatures of music score, written in the period of common-practice tonality, and many contemporary popular pieces, influence the layout, grouping and beaming of notes, and introduce probabilistic constraints to the usage of certain note classes in accordance to the tonality of the piece or section of the music. Since machine recognition of handwritten manuscript is frequently fought with uncertainties and ambiguous results, it is beneficial to the recognition system if these global parameters are provided, or automatically detected, and the reconstructed results can be re-examined in the light of the information and provide corrections to resolve any uncertainties.

5.1.1   Tonality

From a collection of experimental data, the distribution of note classes (with known tonality) is observed. It was observed that two of the most significant note classes of any key are the tonic (1st note) and the dominant (5th note), and there is a distribution pattern which reflects the harmonic weighting of the note classes with their total duration of individual note classes. Using these empirical results a heuristic search with rules derived from high-level musical knowledge, is conducted in order to estimate the tonality of the music. Details of this approach can be found in [16].

With the detected tonality and modality (major or minor) information, the relevant section of this score can be parsed to verify all isolated accidental signs.

5.1.2   Time

Many possible time signatures can be resolved using a simple count of the linearly disposed note values between a pair of barlines (ignoring the first and last bars), and beaming information can be used to clarify simple and compound times (e.g. ambiguity of 3/4 and 6/8 times).

Once the timing information is obtained, the total duration of a bar is established, and any discrepancies between the estimated duration and the expected duration of the time signature indicates missing or mis-classified items. The relevant section of the score can then be re-parsed, with a database of commonly found rhythmical patterns for the time-signature (e.g. three quavers group in 6/8 time) to resolve ambiguous note length, note-head type and others. The conventions of grouping and rhythmical patterns offer additional analysis at the sub-bar level, and can provide plausible solutions to missing or mis-recognised duration information. Further discussions on time-signature detection and rhythmical grouping can be found in [18].

6   Conclusion and Future Direction

In this paper, low- and high-level approaches for an OMR system are presented, with discussions on heuristic and conventions employed to enhance recognition.

To provide flexible and efficient transcriptions, a graphical-user-interface with built-in basic musical syntax and some contextual intelligence to assist the transcription and output is currently under development.

Besides direct applications to convert paper-based manuscripts into a machine-readable format, the framework could be extended to process other form of notations. Initial experimentations are currently under development with Benesh Movement Notation (choreographic scores) due to the similar five-line-stave grid system, as in CMN, and the pre-processing modules are directly reusable.

Complex document analysis, with incomplete and ambiguous input, requires clues and guessing works. The system exploits redundant information available in music notation. In addition to high-level enhancements using time- and key-signatures, future development include melodic, harmonic and stylistic analysis to improve recognition.

References

1.   Ablameyko, S. and Pridmore, T.: Machine Interpretation of Line Drawing Images, Springer-Verlog (2000)
2.   Anquetil, É, Coüasnon, B. and Dambreville, F.: A Symbol Classifier Able to Reject Wrong Shapes for Document Recognition Systems. In Chhabra, A.K. and Dori, D. (eds.): Lecture Notes in Computer Science, Graphics Recognition, Recent Advances, Springer-Verlag, Vol. 1941 (1999) 209–218

3. Bainbridge, D. and Bell, T.C.: Dealing with Superimposed Objects in Optical Music Recognition. In Proceedings of the 6th International Conference on Image Processing and its Applications (1997) 756–60

4. Bainbridge, D. and Bell, T.: The Challenge of Optical Music Recognition. Computers and the Humanities, Kluwer Academic Publishers (2001) 35(2): 95–121

5. Bellini, P., Bruno, I. and Nesi, P.: Optical Music Sheet Segmentation. In Proceedings of the First International Conference on WEB Delivering of MUSIC (2001) 183–190

6. Bellini, P. and Nesi, P.: Wedelmusic Format: An XML Music Notation Format For Emerging Applications. In Proceedings of the First International Conference on WEB Delivering of MUSIC (2001) 79–86

7. Blostein, D. and Baird, H.S.: A Critical Survey of Music Image Analysis. In Baird, H.S. Bunke, H. and Yamamoto K. (eds.): Structured Document Image Analysis, Springer-Verlag (1992) 405-434

8. Coüasnon, B. and Rétif, B.: Using a Grammar for a Reliable Full Score Recognition System. In Proceedings of the International Computer Music Conference (1995) 187–194

9. Cooper, D., Ng, K.C. and Boyle, R.D.: Expressive MIDI. In Selfridge-Field, E. (ed.): Beyond MIDI: The Handbook of Musical Codes, MIT press (1997) 80–98

10. Fahmy, H. and Blostein, D.: A Graph-Rewriting Paradigm for Discrete Relaxation: Application to Sheet-Music Recognition. International Journal of Pattern Recognition and Artificial Intelligence, (1998) 12(6): 763 – 99

11. Hoos, H.H., Hamel, K.A., Renz, K. and Kilian, J.: The GUIDO Music Notation Format - A Novel Approach for Adequately Representing Score-level Music. In Proceedings of the International Computer Music Conference (1998) 451-454

12. Kahan, S., Pavlidis, T. and Baird, H.S.: On the Recognition of Printed Characters of Any Font and Size. IEEE Trans. on PAMI (1987) 9(2): 274–288

13. Midiscan: URL: http://www.musitek.com/midiscan.html

14. Ng, K.C.: Automated Computer Recognition of Music Scores, Ph.D. Thesis, School of Computer Studies, University of Leeds, UK (1995)

15. Ng, K.C. and Boyle, R.D.: Reconstruction of Music Scores from Primitives Subsegmentation, Image and Vision Computing (1996)

16. Ng, K.C., Boyle, R.D. and Cooper, D.: Automatic Detection of Tonality using Note Distribution, Journal of New Music Research, Swets & Zeitlinger Publishers (1996)

17. Ng K.C., Cooper D., Stefani E., Boyle R.D., Bailey N.: Embracing the Composer: Optical Recognition of Hand-written Manuscripts. In Proceedings of the International Computer Music Conference (ICMC'99) – Embracing Mankind, Tsinghua University, Beijing, China (1999) 500–503

18. Ng K.C. and Cooper D.: Enhancement of Optical Music Recognition using Metric Analysis. In Proceedings of the XIII CIM 2000 – Colloquium on Musical Informatics, Italy (2000)

19. Sayeed Choudhury, G., DiLauro, T., Droettboom, M., Fujinaga, I. and MacMillan, K.: Strike Up the Score: Deriving Searchable and Playable Digital Formats from Sheet Music, D-Lib Magazine 7(2) (2001)

20. PhotoScore: Neuratron PhotoScore, Sibelius, URL: http://www.sibelius.com

21. Pruslin D.H.: Automated Recognition of Sheet Music, Dissertation, MIT (1966)
22. Ridler, T.W. and Calvard, S.: Picture Thresholding using an Iterative Selection Method, IEEE Trans SMC (1978) 8(8): 630–632
23. Roach, J.W. and Tatem, J.E.: Using Domain Knowledge in Low-level Visual Processing to Interpret Handwritten Music: An Experiment. Pattern Recognition, 21(1) (1988)
24. Sclaroff, S. and Liu, L.: Deformable Shape Detection and Description via Model-Based Region Grouping, IEEE Trans. PAMI (2001) 23(5): 475–489
25. Scorscan: npc Imaging, URL: http://www.npcimaging.com
26. Selfridge-Field, E.: Optical Recognition of Music Notation: A Survey of Current Work. In Hewlett, W.B. and Selfridge-Field, E. (eds.): Computing in Musicology: An Int. Directory of Applications, CCARH, Stanford, USA, Vol. 9 (1994) 109–145
27. Todd Reed, K. and Parker, J.R.: Automatic Computer Recognition of Printed Music. In Proceedings of ICPR, IEEE (1996) 803–807
28. Venkateswarlu, N.B. and Boyle, R.D.: New Segmentation Techniques for Document Image Analysis, Image and Vision Computing (1996) 13(7): 573-583

Extended Summary of the Arc Segmentation Contest

Liu Wenyin[1], Jian Zhai[2], and Dov Dori[3]

[1] Dept of Computer Science, City University of Hong Kong
Hong Kong SAR, PR China
csliuwy@cityu.edu.hk
[2] Dept. of Computer Science and Technology, Tsinghua University
Beijing 100084, PR China
zhai98@mails.tsinghua.edu.cn
[3] Faculty of Industrial Engineering and Management
Technion, Haifa 32000, Israel
dori@ie.technion.ac.il

Abstract. The Arc Segmentation Contest, as the fourth in the series of graphics recognition contests organized by IAPR TC10, was held in association with the GREC'2001 workshop. In this paper we present the extended summary of the contest: the contest rules, performance metrics, test images and their ground truths, and the outcomes.

1 Introduction

This contest on arc segmentation held at the fourth International Workshop on Graphics Recognition (GREC'2001), Kingston, Ontario, Canada, September 7-8, 2001 is the fourth in the series of graphics recognition contests organized by the International Association for Pattern Recognition's Technical Committee on Graphics Recognition (IAPR TC10). The first contest, held at the GREC'95 workshop, focused on dashed line detection [1,2,3]. The second contest, held at the GREC'97 workshop, attempted to evaluate complete raster to vector conversion systems [4,5,6,7]. The third contest, held off-line in association with the GREC'99 workshop, also aimed to evaluate complete raster to vector conversion systems. The purpose of this series of contests is to encourage third-party independent and objective evaluation of the industrial and academic solutions to the graphics recognition problem and therefore push the research in this area.

Following the success of these three contests, in this contest we focused on testing the abilities of participating algorithms and systems to detect solid arcs from raster images. The participating systems were tested on-site with synthesized and real scanned images at different levels of quality and complexity. An overall performance metric based on a previously developed line detection performance evaluation protocol [8] was used to measure each system and determine the winners. In this paper we present the extended summary of the contest, including the contest rules, performance metrics, test images and their ground truths, the outcomes, and discussions.

D. Blostein and Y.-B. Kwon (Eds.): GREC 2002, LNCS 2390, pp. 343-349, 2002.
© Springer-Verlag Berlin Heidelberg 2002

2   General Rules

We require the participants to correctly recognize solid arcs in raster images. There-fore we measure the recognition accuracy of solid arcs as the performances of the participants. The recognition accuracy is measured in terms of the difference between the geometric parameters of the recognized solid arcs and the corresponding expected solid arcs (which are also referred to as ground truth) by the differences in the geome-try of the detected and expected (ground truth) arcs. The smaller the difference, the higher the accuracy, and hence, the better the performance. The details of the per-formance metrics are defined in Section 4.

Although we measure solid arcs only, the participating systems can compute any-thing they wish in order to improve the arc segmentation. For example, text and other type of graphics can be segmented to help reduce false alarms of arc segmentation. However, since detection of text, symbols, and other graphics may also produce false alarms, which are actually solid arcs, the impact of other detections on the overall performance is not known precisely. There is a tradeoff in this aspect. Participants were required to fix the graphics recognition processes and configurations (e.g., thresholds) of their systems during the test. That is, the participating systems were required to run on-site as black boxes, which accepted input files and yielded output files. No other human intervention was allowed.

Each participating system was required to run on the same set of testing images, which include four synthesized images and three scanned images. We present details of the test images in Section 3. All images are binary (black and white) and in com-mon TIFF format. The output file format was VEC, defined by Atul Chhabra [9]. The system's score on each test images is calculated independently according to perform-ance metrics defined in Section 4. If there were no result produced for any image due to any reason (e.g., system crash, dead loop, etc.), the score for that image is taken as zero. An overall average score for these seven images is used as the unique measure to judge the system's performance.

The contest rules are summarized below.

- Recognition accuracy of solid arcs was measured.
- System configuration was fixed during the contest.
- Seven images were tested: four synthesized and three scanned.
- The Input file format was binary TIFF and the output file format was VEC.
- An overall average score was used as the unique measure of performance.

3   Test Images and Their Ground Truths

The four synthesized images were generated by the following way. First of all, we composed a ground truth vector drawing containing arcs of various radii, central an-gles, and line widths, as shown in Fig.1. The vector drawing was then converted into four images. During the conversion processes, four types of noise: Gaussian (Level 5), high frequency (Level 3), hard pencil (Level 3), and geometry distortion (Level 3), were added to the four images separately. Parts of the four test images are shown in

Fig. 2, which also show the effects of these noise types. The modeling processes of these noises are described in detail in [10].

Fig.1. The ground truth vector drawing for the four synthesized test images

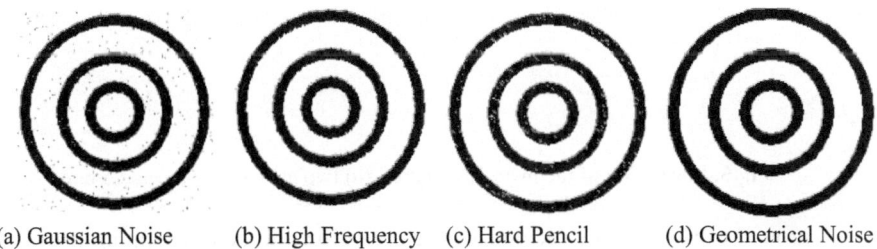

(a) Gaussian Noise   (b) High Frequency   (c) Hard Pencil   (d) Geometrical Noise

Fig.2. Parts of the four synthesized test images, each showing the effect of one type of noises

The three scanned test images were yielded by scanning two real engineering drawings from a textbook. One test image is scanned from the drawing shown in Fig. 3.The other two test images are scanned from the same drawing shown in Fig. 4 but with different resolutions. The ground truth vector files for these real images are obtained by manually measuring their geometry parameters.

All ground truth vector files were also stored in the VEC format. These test images and other pre-contest training images are now available at the contest website [11].

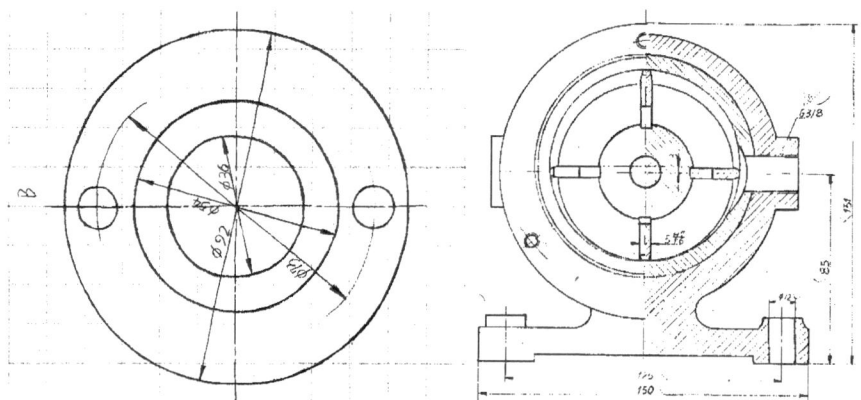

Fig.3. Once real scanned drawing   **Fig.4.** The other real scanned drawing

4   Performance Measurement

As we mentioned in Section 2, the recognition performance was measured in terms of the differences between the recognized arcs and ground truth arcs. To measure the difference between ground truth arcs and recognized arcs, correspondences between them should be established by matching ground truth arcs with the recognized ones. We define the matching degree of a pair of arcs based on the overlap area and the extent to which the endpoints match. Let $c = k \cap g$ denote the arc segment of the recognized arc k that overlaps the ground truth arc g. The method of calculating c is presented and illustrated in Fig. 5.

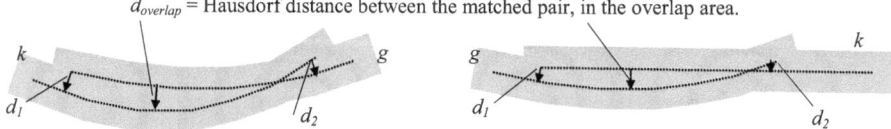

Fig.5. Illustration of the matching of arcs. (a) two arcs. (b) an arc and a line

A ground truth arc and a recognized arc are defined to be overlapping if the overlapping distance ($d_{overlap}$) is smaller than half the ground truth line width. As **Fig. 5** shows, the overlap distance is defined as the Hausdorf distance—the maximum of all minimum distances between the points on k and g in the overlap area. The overlapping distance and the distances between the two arcs at the overlapping segment's two endpoints (denoted by d_1 and d_2) are used to define the vector detection quality of the overlapping segments and to calculate the following performance metrics. In a nutshell, the vector detection quality is measured by the detection accuracy using a number of criteria, including the endpoints, the location offset, the line width, the geometry, and the line style. We compute and record the vector detection quality of these overlapping segments, as well as their lengths, for both the ground truth arcs and their matched recognized arcs. The Basic Quality, (Q_b) is then computed as the length weighted sum of the vector detection qualities of overlapping segments. Additionally, we compute the Fragmentation Quality, (Q_{fr}), which is the measure of the detection fragmentation and/or consolidation, and the Total Quality, (Q_v), which is the product of Q_b and Q_{fr}, for both the ground truth arcs and their corresponding recognized arcs. The total Vector Detection Rate, D_v, is length weighted sum of Q_v of all ground truth arcs. The total Vector False Alarm Rate, F_v, is the length weighted sum of $1-Q_v$ of all recognized arcs. Since a good algorithm/system should recognize as many arcs as possible and produce as few false recognition as possible, we define the overall recognition performance (called Vector Recovery Index) for a single test image as $VRI=(D_v+1-F_v)/2$. These concepts were presented in detail in [8].

Once we obtain the VRI for each image, we define the overall performance of a system as the average of the average performance of the four synthesized images and the average performance of the three real scanned images: $VRI_{all} = \frac{1}{2} [(VRI_1 +VRI_2 +VRI_3 +VRI_4)/4 + (VRI_5 + VRI_6 + VRI_7)/3]$, where $VRI_{1~4}$ are the VRI values of the four synthesized test images and $VRI_{5~7}$ for the three real scanned images. The performance evaluation software is also available at the contest website [11].

5   Winners and Their Scores

Initially, several groups had expressed interest in participating in the contest. However, only two systems actually made it to the contest: Dave Elliman [12] and Xavier Hilaire [13]. The scores (VRIs) of their systems are listed in Table 1. Elliman's system obtained an overall score of 0.681 and won the First Place while Hilaire's obtained an overall score of 0.630 and won the Honorable Mention Prize.

Table 1. The scores of the participants

Test	Synthesized Images (filenames)				Scanned Images (filenames)		
Images	Gau_05	Frq_03	Pen_03	Geo_03	P1	P2	P3
Dave	**0.904**	0.853	0.896	0.927	0.547	**0.482**	**0.371**
Elliman's	Average: 0.895				Average: **0.467**		
scores	Overall (VRI_{all}): **0.681**						
Xavier	0.891	**0.889**	**0.944**	**0.958**	**0.707**	0.311	0
Hilaire's	Average: **0.921**				Average: 0.339		
scores	Overall (VRI_{all}): 0.630						

6   Summary and Discussion

We have presented the rules, the test images, the performance metrics, and the outcomes of the arc segmentation contest. From the scores in Section 5 we can see that both systems were very good at synthesized drawings. Elliman's system was better on Gaussian noise and Hilaire's was better on the other three images. The average difference is only 2%. However, both exhibited almost equally insufficient performance on the three scanned drawings. While Hilaire's is better on the simple drawing p1, Elliman's is better on the other two complex drawings. An unexpected event was that Hilaire's system crashed on the scanned image P3 and therefore obtained a zero score. This made its overall score about 5% behind Elliman's.

From the contest results we can see that current solutions to real life test drawings are still far from satisfactory and this leaves the graphics recognition research community with a lot of room for improvement in arc segmentation. Another finding is that an individual approach may work better on some images but worse on others. Hence, a large number of test images should be used to obtain a reliable overall performance metric, as mentioned by Karl Tombre.

From this contest and other previous contests we can see that it is possible to set up such a third-party independent and objective performance evaluation database within this research community although this requires some effort. Actually, it has taken us more than 10 hours to manually obtain the ground truth data for the three scanned test drawings. We have also spent a lot of time on developing the noise models and pre-testing all of these test images. We plan to continue to augment the database and welcome any help from researchers in this community. Especially, we would like the feedback on this contest from the community, on all issues, including rules, test drawings, performance scores. Moreover, we suggest that researchers use this performance evaluation protocol and the dataset to test their algorithms and compare with other algorithms. In this way, more problems can be found and the performance evaluation protocol can then be improved. It is even better to publish and contribute their performance evaluation results on a public site, such that comparison among existing algorithms can be carried out efficiently, thereby advancing the state-of-the-art in this area.

Obviously, much work still needs to be done. Particularly, as Karl Tombre pointed out just after the contest, we need more real-life drawing images with ground truths as test images. We also need more experiments to improve the performance evaluation protocol, such that it can be accepted by most people as a standard. We may also need more efforts to simulate more types of noise in real scanned images. Though we did not evaluate other aspects of the recognition performance, e.g., memory, time, and system robustness, we believe that these are also important factors that should be considered in future contests.

Acknowledgement

First of all, we thank all participants for their interest, support, and effort, which has made this contest meaningful. We also thank the workshop chairs, Dorothea Blostein and Yong-Bin Kwon for their support, Arnold Smeulders for his suggestion on stress testing, and Karl Tombre and Atul Chhabra for their discussion on the contest organization.

References

1. Kasturi R. and Tombre K. (eds.): Graphics Recognition: Methods and Applications, First International Workshop, University Park, PA, USA, August 1995, Selected papers published as Lecture Notes in Computer Science, volume 1072, Springer (1996)
2. Kong B., et al.: A Benchmark: Performance Evaluation of Dashed Line Detection Algorithms. In: Graphics Recognition: Methods and Applications, Lecture Notes in Computer Science, volume 1072, Springer (1996)
3. Dori D., Liu W., and Peleg M.: How to Win a Dashed Line Detection Contest. In: Graphics Recognition: Methods and Applications, Lecture Notes in Computer Science, volume 1072, Springer (1996).
4. Chhabra A. and Phillips I.: The Second International Graphics Recognition Contest - Raster to Vector Conversion: A Report. In: Graphics Recognition: Algorithms and Systems, Lecture Notes in Computer Science, volume 1389, Springer (1998).
5. Phillips I., Liang J., Chhabra A. and Haralick R.: A Performance Evaluation Protocol for Graphics Recognition Systems. In: Graphics Recognition: Algorithms and Systems, Lecture Notes in Computer Science, volume 1389, Springer (1998).
6. Chhabra A. and Phillips I.: A Benchmark for Graphics Recognition Systems. In: Proc. IEEE Workshop on Empirical Evaluation Methods in Computer Vision, Santa Barbara (1998).
7. Phillips I. and Chhabra A.: Empirical Performance Evaluation of Graphics Recognition Systems. IEEE Trans. on Pattern Analysis and Machine Intelligence 21(9) (1999) 849-870.
8. Liu W. and Dori D.: A Protocol for Performance Evaluation of Line Detection Algorithms. Machine Vision and Applications 9 (1997) 240-250.
9. http://graphics.basit.com/iapr-tc10/contests/contest97/pre-contest97.html
10. Liu W., Zhai J., Dori D., Tang L.: A System for Performance Evaluation of Arc Segmentation Algorithms. In: Proc. of third CVPR Workshop on Empirical Evaluation Methods in Computer Vision, Hawaii (2001) (http://www.cs.cityu.edu.hk/~liuwy/ ArcContest/NoiseModels.pdf).
11. http://www.cs.cityu.edu.hk/~liuwy/ArcContest/ArcSegContest.htm
12. Elliman D. TIF2VEC, An Algorithm for Arc Segmentation in Engineering Drawings. In: Post-Proc. of GREC2001, Lecture Notes in Computer Science (This volume), Springer (2002).
13. Hilaire X. RANVEC and the Arc Segmentation Contest. In: Post-Proc. of GREC2001, Lecture Notes in Computer Science (This volume), Springer (2002).

TIF2VEC, An Algorithm for Arc Segmentation in Engineering Drawings

Dave Elliman

University of Nottingham, UK
dge@cs.nott.ac.uk

Abstract. This paper describes a method for the recognition of arcs and circles in engineering drawings and other scanned images containing linework. The approach is based on vectorizing a binary image, smoothing the vectors to a sequence of small straight lines, and then attempting to fit arcs. The software was successful in winning first prize in the arc segmentation contest held at GREC 2001, and the results are presented in the context of this evaluation.

1 Introduction

Engineering drawings contain text, some special drawing symbols, outlines of views, shading, hidden detail, and dimensioning lines, some of which may have an arrowhead at one or both ends. Solid, dashed, and chained line styles will be found, and the thickness of lines will vary as a further indicator of type, with lines indicating faces and edges in a view ideally being twice as thick as dimensioning and leader lines. In practice this is not a reliable indicator of line type in scanned images of drawings, but it does provide useful evidence. Such edges are often straight lines, circular arcs and ellipses as these represent the edges of flat and cylindrical surfaces which are easily machined. However, parabolic curves are reasonably common, and any free form shape is possible.

Scanned images of drawings are usually held in the form of compressed bilevel images, for example using the TIFF Group IV compression standard which may be found in [1]. Such images are reasonably compact, and more convenient to store and retrieve than their paper equivalents. However, they are inconvenient to edit and unsuitable for further processing. There is a strong incentive to convert them into the more useful form of a CAD file. The ability to recognize arcs and lines in bilevel images of all kinds would useful for many purposes as there is a huge variety of drawings and diagrams in common use. The algorithm described in this paper is generic and may be used with any linework image.

The algorithm described converts a TIFF group IV image to a simple CAD format known as a *vec* file. It can also produce the AutoCad © *dxf* format, as well as a postscript output which was used to produce the figures in this paper. It was given the rather obvious name *tif2vec*.

The program described here was successful in winning the fourth in the series of graphics recognition contests organized by the International Association

D. Blostein and Y.-B. Kwon (Eds.): GREC 2002, LNCS 2390, pp. 350–358, 2002.
© Springer-Verlag Berlin Heidelberg 2002

for Pattern Recognition's technical committee on graphics recognition (IAPR TC10). The competition took place as part of Grec 2001 in Kingston, Canada, in September 2001. It is reported in [12].

2   The Algorithm Used

The starting point for the recognition process is a binary image, and we have stored these in TIFF group IV format. The final result is a vector description in terms of a sequence of straight line and arc segments which may be written to a file in AutoCad©*DXF* format.

The binary image is vectorized using the algorithm of Elliman [5]. This produces a representation that comprises chains of very small vectors, often a few pixels in length. Before attempting to fit lines and arcs to this data it is smoothed by a simple method that removes intermediate points that do not indicate a consistent curvature, but which merely straddle either side of a least-squares line. This results in rather longer vectors on average, and makes the rest of the processing more efficient.

The next step is an attempt to fit the largest possible arc to a curve, that is one that accommodates all n points in the sequence. First a check is made to see if the curve is adequately represented as a straight line. If this is not the case then a putative arc is fitted. This is formed by making the cord between the end-points of the curve and then finding the point d, most distant from this line. An arc is fitted through this point and the two end points. The maximum distance of the remaining points of the curve from this arc are then calculated as shown in Fig. 1.

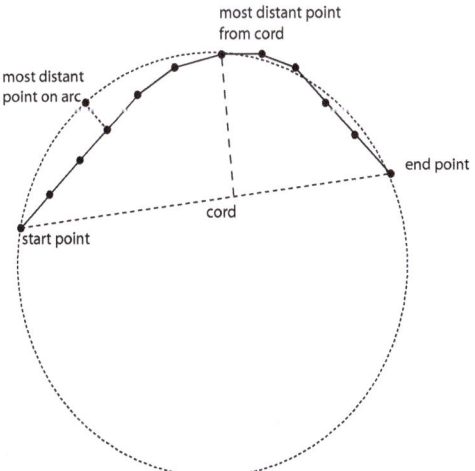

Fig. 1. The construction for putative arc fitting

If this is less than a threshold then the arc is accepted as an adequate representation of the curve. However, if this fails an attempt is made to fit the two possible arcs which contain $n - 1$ points. That is one that is found by discarding the first point, and another that is found by discarding the last. Three possible arcs may be tried on discarding two points from an end of the curve, or one point from each end. There are four possible arcs that may be tried by discarding three points, and five ways of discarding four and so on.

The algorithm terminates if there are only three points left in the chain as it is then guaranteed that an arc will be returned, but this may not be a plausible representation of the curve. The method has similarities to that of [8] but is believed to be more efficient. The worst case complexity of the algorithm occurs for the case when no arc is found. If there were n points in the curve then $n - 4$ passes of the procedure will have been attempted. On the first pass two putative arcs will have been fitted, with one extra arc fit being attempted on each successive pass. This is $O(n^2)$ worst case complexity, and with judicious reuse of distance and intersection calculations it seems a highly efficient procedure for finding the longest likely arc primitive within a curve. The procedure can then be applied recursively to the remaining segments of the polyline on either side of the arc.

A useful description of other methods of arc segmentation applied to engineering drawings can be found in [4]. Many authors have used the Hough transform to identify arcs in images of all types. An excellent reference on the approach is that of Illingworth and Kittler [7]. However this approach was not found to be very successful by the present author in that the end points and centre of the arc were often predicted with poor accuracy. This may have been as a result of finding arcs after extracting smoothed vectors. However the Hough transform performs very slowly if applied at the pixel rather than line segment level.

3   Evaluating Line and Arc Finding Algorithms

In order to make progress in improving algorithms for vectorization, and for spotting drawing entities such as lines, arcs and circles, it is necessary to have ground truthed images which may be used as a reference in giving an objective measure of the success of the process. It is a labour-intensive process to generate suitable data for this purpose, and a simpler alternative is to generate artificial images along with a ground truth data file. Synthesized images generate vectors first and then covert these to a clean bilevel image. This information can be in the form of a *vec* file of the same kind as is produced by the algorithm. A comparison of the two *vec* files then gives a measure of the accuracy of the process. This is a non-trivial operation and the method used will be described later.

A problem with automatically generated images is that they tend to be *too perfect*. That is they contain much less distortion and noise than would be the case with scanned images. To overcome this limitation it is possible to add various kinds of noise to images to make them more realistic. The ways in which this was done for the competition is described in [12].

The performance was evaluated on the accuracy of detection of solid arcs alone. In practice such arcs were often broken into a series of segments in the test images, but the intention was that a single solid arc should be found by the algorithm in these cases. The algorithms were executed as *black boxes* with fixed parameters and configuration. No human intervention was permitted in the on-site contest, as this also tests for robustness. A zero score was awarded for images that cause the algorithm to fail, or to loop forever. *tif2vec* processed all the images successfully taking between two and ten seconds on each using 700MHz Pentium III machine with 512 MB of RAM. For full details of the competition, again refer to [12].

4   Geometrical Constraints

The recognition process described so far has been a data-driven or bottom-up process with the implicit top-down model that the binary image comprises stroke-like features that are a sequence of straight lines and circular arcs. In real drawings the vast majority of curves have additional constraints on the end points of the constituent strokes. The first one is the obvious one of continuity:

$$E(c_i) = S(Next(c_i))$$

More interesting is the constraint of *smoothness*. This does not always apply. Many of the junctions between the arc and line primitives represent a *corner*. The angle between two line segments or between a line segment and an arc tangent or between two arc segments can easily be found. A *cornerness* measure η is defined as follows:

$$\eta = l_1 l_2 \theta$$

Where l_1 and l_2 are the lengths of the primitives if they are a line, or $0.7r$ where r is the radius of the arc, if they are arcs. This value arrived at in my experiments, but the results are not very sensitive to the value used. θ is the angle between the lines or tangents. If the value of η is below a threshold, then it is assumed that the intention of the draughtsman was to produce a smooth line. This insight allows an adjustment to be made to the start and end points of the primitives so as to make the curvature zero at this point. This improves the visual quality of the resulting vector image markedly. In fact the first commercial application of this technique has been in sign-making. An artist can draw the outline of the sign to his satisfaction, and this program will turn a scanned image of the artwork into a vector representation which may then be machined using an NC router. The enforcement of the smoothness constraints resulted in an acceptable quality for most of the cases scanned.

Enforcing the smoothness constraints required the identification of five distinct cases, and some complex coordinate geometry which is described, together with corresponding Java code in [6]. These are as follows:

- Find an arc that is tangential to a line and passes through two points (two solutions).
- Find an arc that is tangential to two other arcs and passes through a point (two solutions).
- Find an arc that is tangential to two lines and passes through a point (four solutions).
- Find an arc that is tangential to a line and to another arc and which passes though a point (two solutions).
- Find a line that is tangential to two arcs and passes through a point (four solutions).

In each case there are either two or four solutions to be considered. Imaginary solutions are discarded, and a simple test of proximity will identify which of the remaining solutions applies.

5   Further Constraints Based on a Drawing Model

The statistical properties of real drawings in comparison with the results of this arc and line finding make it tempting to apply rules to further *correct* the curves generated. These rules may be wrong from time to time, causing an increase in the error of a curve. They result in an improvement many times more often than they result in a mistake, however, and will thus greatly improve the recognition performance of a ground-truthed document on most plausible metrics. The rules that have been applied are:

- Near horizontal and near vertical lines are horizontal or vertical.
- Lines near to 45°, 30°, or 60° are set to those exact angles.
- Arcs that are close to a full semi-circle or quarter circle are adjsuted to that.
- Cords of arcs that are close to a multiple of 45° are at that multiple of 45°.
- lengths that are close to a whole multiple of (say) 5mm are a length that is the closest exact multiple of 5mm.

The algorithm for applying these rules needed careful thought. Each rule means that one or other or both end points need to be moved. Unfortunately there are an infinite number of possible changes that will satisfy an individual rule. What is needed is a change that simultaneously satisfies each rule that is to be applied. Again there may be anything from zero to infinity ways of doing this. If there are no ways in which all the rules can be satisfied then the rule that is least certain is dropped, and another attempt made. The best adjustment to be chosen from infinity is the one that makes the smallest changes to the position of the end points. The sum of absolute distances is used for this metric. In some cases the application of these rules results in a perfect view, although this has only been achieved for fairly simple cases so far. The approach used was to take an arbitrary start point and then form a description of the curve in terms of these corrected primitives. If the curve is closed, then the rules may be inconsistent if the final end-point is not the same as the start-point. The average movement of all vertices can then be calculated, and each point adjusted by this value to produce a minimal adjustment. This algorithm is often successful, and it is impressive when a perfectly correct view is produced from a poor quality scanned image. However there are cases where the rules do have a consistent solution which is not found by this method. Other methods, such as a gradient search of the state space are more powerful, but lose the simplicity and high efficiency of this simple approach. It is probably worth using this as a first pass in any case, and it would solve perhaps ninety percent of cases with minimal computation.

6   Evaluating Results against Ground Truth Data

Considerable thought has been given by various authors to the problem of evaluating the performance of recognition against ground-truthed data [10] [9] [11] [2] [3] and [13]. The method of evaluation used in the contest was based on [11] and summarised in [12]. This paper provides a useful historical background to the derivation of the method, and should be consulted for an in-depth understanding of the evaluation process. Refer to [12] for full details about the tests used in this competition.

The results obtained on the synthesized image with high frequency noise that was shown earlier are illustrated using the postscript output in Fig. 2. The lighter areas have been detected as straight line segments while darker ones represent segments of an arc.

The arcs and circles in the top row have been fragmented into many small pieces by the intensity of the noise added. These segments are too small to have significant curvature and are thus detected as many small lines. The circles are

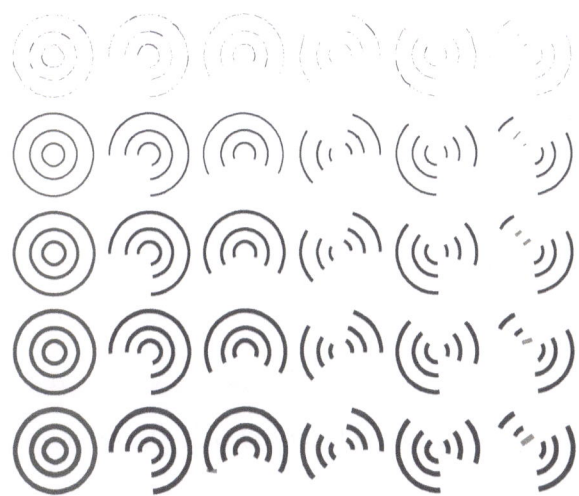

Fig. 2. The result of applying *tif2vec* to a synthesized image

detected accurately on the remaining rows. The leftmost column of this image comprises circles. The arcs are mostly detected correctly, but in some cases as two arcs rather than one, with some small stubby segments reasonably labelled as lines. The algorithm scored 0.853 on this image as presented in [12].

The results obtained on the third scanned image can be seen in the postscript output of Fig. 3. This image achieved a rather poor result of 0.371 and has a rather large number of arcs labelled as lines. No attempt has been made to segment text from the image, and this is the source of a number of false positives in the arc recognition. A continuous arc is broken at the point where shading touches the outline, and this fragments the arc in the same way that gaps in the boundary would do.

The *tif2vec* algorithm is efficient and works reliably on high quality scanned and synthesized images. It is robust to Gaussian noise up to and beyond the level seen on typical scanned images. Beyond this level the lines start to fragment, and holes result in a duplication of vectors, which results in false matches in the scoring system. This latter effect also happens with high levels of *pencil* noise, and the fragmentation effect also occurs as the level of high frequency noises increases.

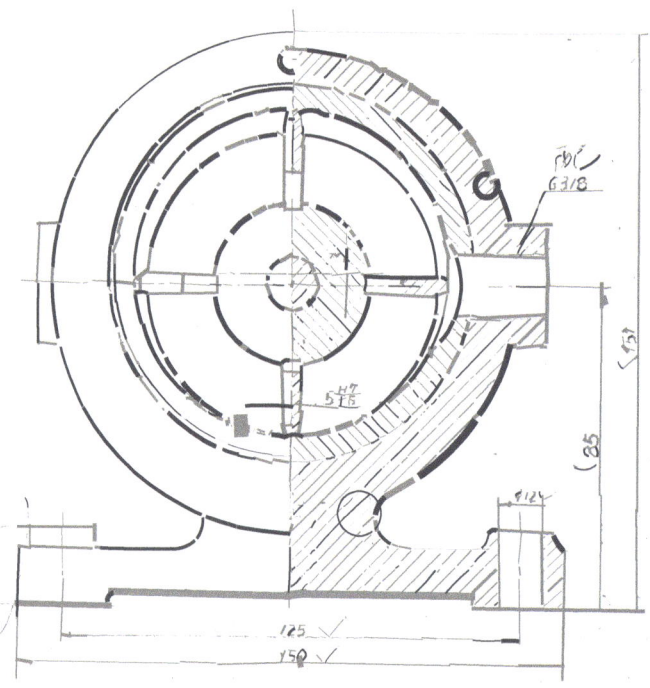

Fig. 3. The results of applying *tif2vec* to the third scanned image

A shortcoming of the current algorithm is that outlines can also become fragmented in the presence of shading. It is expected that this can be corrected by a post-processing stage.

Geometrical distortion results in several arcs being generated in place of one before the distortion. The arcs reflect the change in geometry, and this effect seems reasonable and appropriate.

7   Discussion and Conclusions

The scanned images and test images were all of a much lower quality than those on which the algorithm had been developed. All Lines contain sizeable holes, and are often broken. The edges of lines were extremely ragged and the images contained high levels of random noise. It seems that the scanned images were also subject to the addition of noise of various kinds. A Fujitsu Scan Partner 600c was used in development at a resolution of 12 pixels per mm. The test images represent a difficult test for arc finding algorithms, and one which is probably unrealistic in its demands given the current performance of low-cost scanners. However it was a fair competition as the same unseen images were used for all participants. The evaluation of the *tif2vec* algorithm has been a useful exercise. Although the algorithm was successful in the competition, the exercise showed up serious limitations which need to be addressed.

A higher score could have been achieved rather easily by smoothing the outlines more aggressively. one pixel high undulations are optionally smoothed at present, but it would not have been difficult to attempt to remedy the deleterious affect of high frequency noise. Another simple adjustment would have been to increase the size of holes that are patched and small blobs dark areas that are discarded. This is accomplished in a highly efficient manner by the vectorization algorithm, and could have been achieved by a single parameter change. However, it seemed unreasonable to tune this parameter for the competition rather than for typical drawings that have been encountered in practice.

The fragmentation of arcs was one of the largest factors in reducing the score achieved by the algorithm. Simply patching together the ends of small lines would have made a substantial difference to the score achieved. In fact this was not done due to an oversight. The facility to patch lines together if they are closer than a given threshold existed in the code, but the parameter had inadvertently been set to zero. Arcs of similar radius and centre that could form a contiguous arc are merged by the software. However this is done under rather strict conditions, and an improvement would have resulted from relaxing these constraints.

A higher score could also have been obtained by relaxing the accuracy of arc fitting, as this would reduce fragmentation especially in the presence of geometrical distortion. This was not done as it would reduce performance on real drawings, and that is the real goal of the work.

The results obtained by applying smoothing constraints and by applying a rule-based drawing correction process were considered more pleasing and acceptable by human evaluators of the program. However these methods reduced the score obtained in comparison with ground truth, as small reductions in geometrical accuracy resulted, and incurred a penalty. This raises a debatable issue in that a commercial product would undoubtable find more acceptance in the market with the correction algorithms included. Perhaps conformance to geometrical ground truth is not always the most significant criterion for system evaluation. It is certainly extremely useful as a benchmark for the comparison of algorithms.

References

1. Adobe Developers Association. Tiff revision 6.0, June 1992. 350
2. A. Chhabra and I. Phillips. A benchmark for graphics recognition systems. *Proceedings IEEE Workshop on Empirical Evaluation Methods in Computer Vision, Santa Barbara, CA*, 1998. 355
3. D. Dori and W. Liu. Sparse pixel vectorization, an algorithm and its performance evaluation. *IEEE Transactions on Pattern Analysis and Machine Intelligence*, 21(3):202..215, 1999. 355
4. Dov Dori. Vector-based arc segmentation in the machine drawing understanding system environment. *IEEE Transactions on Pattern Analysis and Machine Intelligence*, 17(11), 1995. 352
5. Dave Elliman. A really useful vectorisation algorithm. *GREC 1999, Jaipur, India*, September 1999. 351
6. Dave Elliman. The coordnate geometry of smooth sequences of lines and arcs. Technical Report 1002, University of Nottingham School of Computer Science, Jubilee Campus, Nottingham, NG8 1BB, UK, March 2001. 353
7. John Illingworth and John Kittler. The adaptive hough transform. *IEEE Transactions on Pattern Analysis and Machine Intelligence*, 9(5):690..697, 1987. 352
8. D.G. Lowe. Three dimensional object recognition from single two dimensional images. *Artificial Intelligence*, 31:355..397, 1987. 352
9. I. Phillips and A. Chhabra. Empirical performance evaluation of graphics recognition systems. *EEE Transactions on Pattern Analysis and Machine Intelligence*, 21(9):849..870, 1999. 355
10. I. Phillips, J. Liang, A. Chhabra, and R. Haralick. *Graphics Recognition: Algorithms and Systems*, volume 1389 of *Lecture Notes in Computer Science*, chapter A Performance Evaluation Protocol for Graphics Recognition Systems. Springer, 1998. 355
11. Liu Wenyin and Dov Dori. A protocol for performance evaluation of line detection algorithms. *Machine Vision and Applications*, 9(5/6):240..250, 1997. 355
12. Liu Wenyin, Jian Zhai, and Dov Dori. Extended summary of the arc segmentation contest. *This Publication*, 2002. 351, 353, 355, 356
13. Liu Wenyin, Jian Zhai, Dov Dori, and L. Tang. A system for performance evaluation of arc segmentation algorithms. *Proc. CVPR Workshop on Empirical Evaluation in Computer Vision, Hawai*, 2001. 355

RANVEC and the Arc Segmentation Contest

Xavier Hilaire[1,2]

[1] LORIA
B.P. 239, 54506 Vandœuvre-lès-Nancy, France
hilaire@loria.fr
[2] FS2i,
8 impasse de Toulouse, 78000 Versailles, France

Abstract. This paper briefly describes an experimental arc extraction algorithm that ran the Arc Segmentation Contest at GREC'2001. As the proposed method is based on the one detailed in [5], this paper only describes the improvments we brought to the original method. We first review some rules from the evaluation protocol that helped us to make major assumptions while designing the algorithm. We then explain the method, and discuss the results we obtained in various cases. Finally, we give some conclusions and introduce a possible extension to this method.

1 Introduction

We present an improved version of a general vectorization method, called RANVEC, that ran the Arc Segmentation Contest. Since RANVEC handles vectorization of both arcs and straight line segments, our motivation in running the contest was to evaluate the performance of the method for the specific case of arcs.

The performance evaluation protocol used during the Contest was that of Liu and Dori[4]. We make a few remarks about it.

First, the Contest only took *solid arcs* into account. Keeping the same notations as the authors, this means that $Q_{sh} = Q_{st} = 1$ for each detected vector. As a result, the expression of the vector detection quality indicator $Q_v(c)$ for a given vector c is reduced to

$$Q_v(c) = e^{\frac{d_1(c)+d_2(c)+2d_{overlap}(c)+|W(k)-W(g)|}{-5W(g)}} \tag{1}$$

An interesting point in equation 1 is that if we commit a *constant* error on the position and the thickness of a given vector, the penalty due to this error will rapidly decrease as the thickness of the corresponding *ground truth* vector increases. In other words, this means that *thick* vectors bring higher detection rates than thin ones.

A similar rule may also be observed in the overall Vector Detection Rate indicator, whose expression is

$$D_v = \frac{\sum_{g \in V_g} Q_v(g)l(g)}{\sum_{k \in V_d} l(k)} \tag{2}$$

D. Blostein and Y.-B. Kwon (Eds.): GREC 2002, LNCS 2390, pp. 359–364, 2002.
© Springer-Verlag Berlin Heidelberg 2002

In a similar way, we can notice from equation 2 that *long* vectors have greater impact than small ones in the performance evaluation. Taking this new remark into account, we may conclude that our detection method should *first worry about thick, long arcs or full circles*.

Next, whereas thick and long vectors present an obvious advantage in the overall score, the penalty associated with *fragmentation* is probably the worst we may encounter, as stated in the following expression (see [4] for detailed explanations): $Q_{fr}(g) = \frac{\sqrt{\sum_{k \in D(g)} l(k \cap g)^2}}{\sum_{k \in D(g)} l(k \cap g)}$. This equation merely states that if a single vector is detected as a set of n equal length vectors, then we have $Q_{fr}(g) = \frac{1}{\sqrt{n}}$. Already in the case of $n = 2$, we have $Q_{fr} \approx 0.7$, and about 30% of the quality indicator is definitely lost. This additional remark leads us to conclude that we *should also pay attention to fragmentation*.

Finally, a last observation may be made on the way vectors are represented. When using a medial axis to represent vectors, the problem that arises is how to handle vectors of odd and even thicknesses in the same ground truth file. The proposed solution, for the purpose of the Contest, was to tolerate a one unit error on thicknesses, therefore our method also benefits from a small error margin.

2   The Algorithm

Taking the above remarks into account, our idea was to propose an algorithm that could find long, thick arcs with good accuracy, and without fragmentation. Since the input images were supposed to be noisy, we used a preprocessing stage to eliminate noise and restore the shapes at the same time. The complete algorithm consists of three stages: (1)Filter the input image; (2) Skeletonize the filtered image, then segment the skeleton chains and instantiate each vector using RANVEC [5]; (3) Recover fragmented vectors.

It mainly bets on the fact that despite the loss of small or thin arcs, the filtering stage is very helpfull to recover thick and long shapes, which are the most important, as discussed in section 1. We now describe each of these stages in detail.

2.1   Filtering

The preprocessing stage aims to correct – or at least reduce – noise introduced in the input images for the specific type of pixel noise, and to reconnect components disconnected by hard pencil noise. Clearly, the goal is to make images suitable for segmentation. Indeed, adding pixel type noise in an image is roughly equivalent to add random white or black pixels. Thus, we may get groups of black pixels in the background as well as groups of white pixels in the foreground.

Groups of black pixels in the background are not awkward at all as long as they do not overlap the foreground, nor are groups of white ones with respect to the background. As a result, the filter we designed simply consists of a trivial hole removal technique : having the input image labelized, we eliminate black, 4-connected components whose area is below a predetermined threshold, and process the complementary image in the same manner to fill white holes in the foreground. In order to handle the specific hard pencil noise as well, we also apply a morphological closure immediately after filling the holes.

Pixels defining the frontier of a given source shape, however, are not always properly restored by the previous filters. Despite the disturbances frontiers are subject to, the point is we may expect the features of any detected arc to be very close to the ground truth, except its thickness. This is due, in part, to the fact that only two kinds of noise may have survived after the filtering stage, as suggested in figure 1(b):

- localized, well-marked noise, which may be a survival of hard pencil noise after filtering. Skeletonization of such damaged parts results in outlier points that are simply detected and thrown away during segmentation. Thus, results after fitting only suffer from the lack of a small number of points.
- scattered, here and there noise, which consists of small, uncoordinated displacements of pixels forming a frontier, for which the computed equation of the circle after fitting is very close from the ground truth one. This is shown in figure 1(a), where we consider a full circle with thickness $t = 2n+1, n \geq 1$ and radius R. Let p denote the probability that a point of the image is corrupted. If p is constant (e.g, does not depend of x and y image coordinates), then estimations of the amount of corrupted points belonging to the external and internal frontiers are given by $2\pi(R+n)p$ and $2\pi(R-n)p$, respectively. Obviously, when $R >> n$, the quotient $\frac{R-n}{R+n}$ tends to 1.

To evaluate the thickness t of an arc, we simply use the proportionality relation there exists between the number of pixels that belong to this arc and its area. Keeping the same notations as in figure 1(a), and calling θ the opening angle of the arc, we have $\theta\left((R+n)^2 - (R-n)^2\right) \propto N$ where N is the number of counted pixels. As a result, we simply get

$$t = 2n \propto \frac{N}{R\theta} \qquad (3)$$

An important consequence of equation 3 is that despite its ability to estimate the position and the radius of a given arc with good accuracy, our method may estimate thicknesses with poor accuracy if the number of pixels belonging to a shape is slightly modified by noise, which is the case for gaussian noise, in particular.

2.2   Skeletonizing, Segmenting, and Instantiating

After the input image has been cleaned, we skeletonize it using Saniti di Baja's skeletonization method [1]. The resulting skeleton is then chained using the algorithm described in [2], which provides a crude segmentation of the skeleton.

Accurate vector segmentation and instantiation is then performed by RAN-VEC [5]. For each input skeleton chain, the method finds reliable subsets of

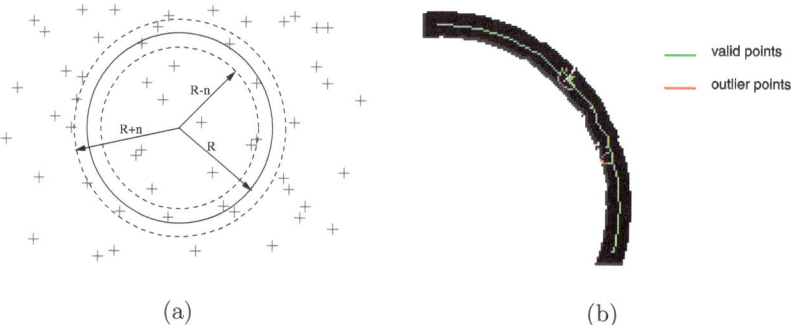

                    (a)   (b)

Fig. 1. Characterization (a) and effects (b) of scattered noise

points forming a primitive (e.g, either an arc or a straight segment), groups the
primitives forming a junction, and instantiates each group using the reweighted
least squares technique, with inlier points only.

2.3   Recovering Fragmented Vectors

This last post-processing stage attempts to recover fragmented vectors. A vector
may be fragmented when it overlaps at least one other vector. A typical example
is shown in figure 2: in this example, the ground truth states there are only two
circles in case a, and one segment and one arc in case b, whereas the regular
output of RANVEC claims there are four arcs, and three segments and two
arcs, for the a and b cases, respectively.

   To solve this problem, we use a very simple, incremental approach which
operates as follows. Consider a set P of vectors forming a single junction, for
example $P = \{s_1, s_2, a_1, a_2\}$ in the case b of figure 2. It is clear that a_1 and a_2, for
example, form a single circle, as well as s_1 and s_2 belong to the same segment.
Since each vector $v \in P$ has been instantiated using least squares, the variance
$Var(v)$ of the distance between the set of skeleton points defining v and v itself,
is known. Given two vectors u and v of P, and denoting $u + v$ the vector defined

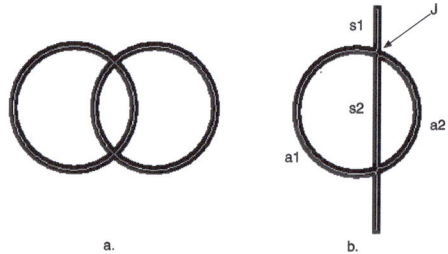

Fig. 2. Some examples of fragmented vectors

over the union of points forming u and v, the unifiability test of u and v we suggest is

$$u \text{ and } v \text{ are unifiable} \iff Var(u+v) \leq \max(Var(u), Var(v)) \qquad (4)$$

This allows to design a very simple procedure for recovering fragmented vectors:

1. For each junction, build P and its associated matrix M formed by the values of $Var(u+v) - (Var(u) + Var(v))$ for each pair of vectors $(u, v) \in P \times P$, such as $u \neq v$
2. Search for the lowest value of all matrices M and peak the corresponding pair (u_0, v_0) of vectors.
3. If u_0 and v_0 do *not* pass the above test, stop the procedure.
4. Otherwise, unify u_0 and v_0, update the corresponding matrix and junction, and jump to step 2.

3   Results

Figure 3 reports the results we obtained on the training set of images provided by the contest organizers [3]. As we could expect from equation 3, performance on

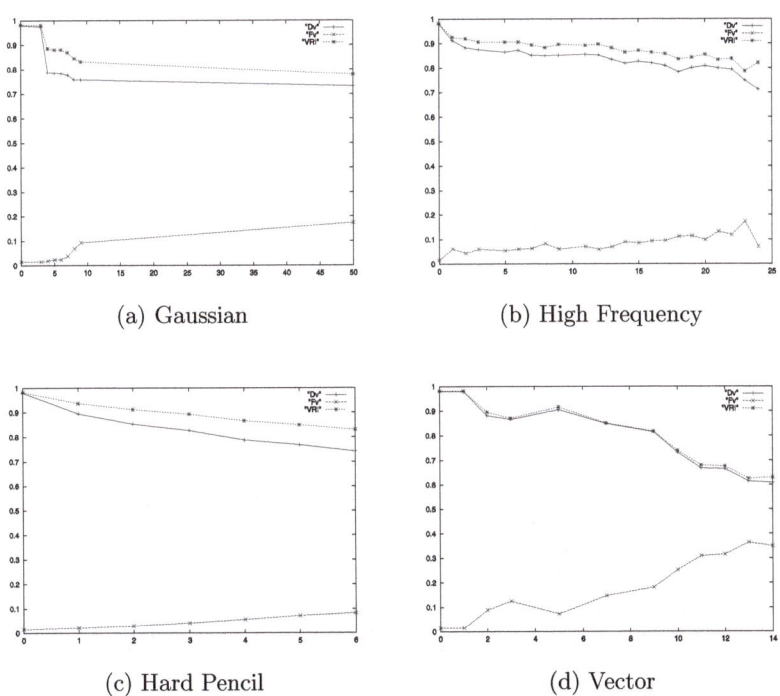

(a) Gaussian

(b) High Frequency

(c) Hard Pencil

(d) Vector

Fig. 3. Performance evaluation on various types of noise. X axes: levels of noise. Y axes: Dv, Fv, and VRI indices

gaussian noise type rapidly decreases, because this type of noise adds many extra pixels to the shapes and severely disturbs the estimation of their thicknesses, as discussed in section 2.1. On the contrary, high frequency and hard pencil noise types do not suffer from this drawback, and, accordingly, the behaviour of the method seems to be quite stable for these two classes of noise.

The worst results are obtained for vector type noise. The main problem in this case is that estimated positions of arcs are far away from those of the ground truth, because local displacements of medial axis points suffice to completely bias the solution estimated by the reweighted least squares technique. Furthermore, test 4 used in the recovering procedure of fragmented vectors, appeared not to operate over a certain level of noise in the specific case of pixel type noise. Of course, this increases overfragmentation.

In all cases, one may observe the presence of a "floor" inside which Dv and VRI decrease faster than anywhere else: in fact, we probably simply observe the impact of the "one unit tolerance" introduced in these indicators, as discussed at the end of section 1.

4   Concluding Remarks

We have presented an adapted version of a general purpose vectorization method, called RANVEC, that ran the Arc Segmentation Contest. The method exhibited reasonable results, and seemed to have quite a good behaviour for very noisy images.

Efforts should now be made to reduce fragmentation in the case of vector type noise. Attempts to unify fragmented vectors with random sampling and similar techniques are also in progress. Unifying fragmented vectors is not only useful for reducing oversegmentation, but also to handle the case of *dashed* arcs detection, which constitutes an interesting extension to the method.

References

1. G. Sanniti di Baja. Well-Shaped, Stable, and Reversible Skeletons from the (3,4)-Distance Transform. *Journal of Visual Communication and Image Representation*, 5(1):107–115, 1994.   361
2. K. Tombre, Ch. Ah-Soon, Ph. Dosch, G. Masini, S. Tabbone. Stable and Robust Vectorization: How to Make the Right Choices. In A.K. Chhabra and D. Dori, editors, *Graphics Recognition – Recent Advances*, volume 1941 of *Lecture Notes in Computer Science*, pages 3–18. Springer Verlag, 2000.   361
3. http://research.microsoft.com/users/wyliu/arcsegcontest.htm.   363
4. Liu Wen-Yin and Dov Dori. A Protocol for Performance Evaluation of Line Detection Algorithms. *Machine Vision and Applications, Special Issue on Performance Characteristics of Vision Algorithms*, 9(5/6):240–250, 1997.   359, 360
5. Xavier HILAIRE and Karl TOMBRE. Inproving the Accuracy of Skeleton-Based Vectorization. In *Proceedings of the Fourth IAPR International Workshop on Graphics Recognition*, pages 381–394, 2001.   359, 360, 361

Summary of Panel Discussions at GREC'2001

Atul / hhabra[1] and Karl Tombre[2]

[1] Verizon Communications
500 Westchester Avenue, White Plains, NY 10604, USA
[2] LORIA–INPL
B.P. 239, 54506 Vandœuvre-lès-Nancy CEDEX, France

Traditionally, the HRJ / workshops give the opportunity for panel discussions on the various topics addressed by the participants during the presentations. This report tries to summarize the main points discussed during these panel discussions.

't is obvious that our area has moved a little bit away from pure reconstruction and recognition problems, towards issues such as *indexing, retrieval and navigation*. The graphics recognition methods may be the same or at least similar, but the emphasis is not so much anymore on being able to recognize *everything*, but on the ability to find useful features for searching and navigating graphics-rich documents. We have also seen an evolution in the place given to the user in the graphics recognition process: a decade ago, the dream was to build fully automatic systems, nearly without user interaction. Then, it became obvious that the user had to be placed "in the loop". 1 ow, the trend seems to be to put the user not only in the loop, but at the center. This is especially true for interactive symbol or math recognition tools, but also for more general graphic recognition tasks, such as that of tracking lines.

This being said, the traditional areas of graphics recognition are still there, with progress reported at the workshop. 'n the case of *symbol recognition*, some people wondered if progress could be measured in years, or if it made more sense to measure it in decades! 'ndeed, whereas there are well-known methods for the recognition of isolated symbols, a lot remains to be done for tackling the general problem of localizing, extracting and recognizing the symbols in their context.

The basic problem of *vectorization* is also still there... Many approaches use skeletons, but it was mentioned at the workshop that a skeleton is not wrong, it just gives something else than what we usually expect from a vectorization. Actually, it still seems hard to give a clear definition of what vectorization ought to be. What do we expect from it? The technique you use must be driven by what you expect, as there is no such thing as *the* correct vectorization. Also, as previously mentioned, it may make more sense to track lines semi-automatically, instead of automating everything.

The question came up several times of the *pertinence* of some applications. One participant even publicly asked whether map conversion should be done or considered as sinful—shouldn't we aim at producing wonderful and useful maps, rather than taking them apart? Similar questions were raised for the analysis of music or of business charts.

Some domains should be better taken into consideration; one of them is the handling of the specific issues related to *color segmentation*, although we usually

D. Blostein and Y.-B. Kwon (Eds.): GREC 2002, LNCS 2390, pp. 365–366, 2002.
© Springer-Verlag Berlin Heidelberg 2002

deal with simple color problems. There was also some discussion on the usefulness of general *perceptual organization* tools in graphics recognition problems. Some participants took a pragmatic view, using the perceptual groups they needed for their specific applications, whereas other participants thought that there might be room for having a common base of perceptual organization.

The place of *knowledge* in the graphics recognition process was also discussed. There are many approaches, but all of them do not have the same expressiveness. We should probably aim at better ways to take into account the domain knowledge in all the areas. As an example, it was mentioned that few people working on map recognition have actually worked with cartographers.

Questions about *evaluation and validation* were addressed several times. A debate took place about ground truth, which was deemed to be in the eye of the beholder. We have organized several contests, and continue organizing them, but we must never forget that contests are not the only useful way for making progress. Actually, you can win the contest without solving the real problem. Still, it is felt that the ground-truthed data produced for the contests are very useful, and should be made available to other researchers, together with the performance evaluation tools, so that they can be used for other purposes than contests, especially for assessing and comparing the qualities of various graphics recognition methods and algorithms. T/10 is considering making these data and tools available on its web site. Another interesting question was: Should we analyze a process to evaluate it, or should we rather be able to analyze the data, to find the right process? Finally, in some areas such as math or music notation, there is a general lack of reference databases.

A pervasive topic in the various panel discussions was that of software. ̈t was felt that a lot could be gained from defining the right set of AP ̈s, as has been done in the image processing community, and in making software modules available to other researchers. Also, when we design larger systems, systems engineering issues become important: how should we assemble complete systems using such modules...

At the end of the workshop, Heorge ı agy was asked what the major advances had been over the last 20 years. Z e cited the following topics:

- The definition of target formats
- Better tools for O/R, domain-specific vectorization and thresholding, connected component analysis...
- The emphasis on interaction
- Progress in the technology we use: storage, digitizers, displays...
- The ability to design software architectures which are not pipeline processes, but which include feedback mechanisms.

Author Index